Library of
Davidson College

THE
BRITISH FISHERY
AT
NEWFOUNDLAND
1634–1763

BY
RALPH GREENLEE LOUNSBURY

ARCHON BOOKS
1969

COPYRIGHT 1934 BY YALE UNIVERSITY PRESS
©1962 BY MRS. LYNDON R. CONNETT
ALL RIGHTS RESERVED

REPRINTED 1969 WITH PERMISSION
IN AN UNALTERED AND UNABRIDGED EDITION

SBN: 208 00795 4
LIBRARY OF CONGRESS CATALOG CARD NUMBER: 69-19217
PRINTED IN THE UNITED STATES OF AMERICA

TO
MARGARET, CHARLES, AND SALLY

PREFACE

SOME years ago the author's interest was aroused in the early history of northeastern British America. Research in this field led eventually to a study of British maritime enterprise and settlement in Newfoundland. The fruits of this work are presented in the following chapters, which trace the development of the cod-fishery from the earliest times to the end of the Seven Years' War and lay particular emphasis upon the period following 1660. The treatise was originally prepared as a doctoral dissertation, which was submitted at Yale University in 1928. In the intervening years the work has been thoroughly revised, some deletions made, and much new material added, and the entire study completely rewritten. It is hoped that the volume will prove a fitting contribution to the early history of North America and will throw new light upon a more or less obscure and little understood phase of British overseas enterprise.

The author is indebted to a great many people for assistance in gathering the material and in the preparation of the volume for publication. In the first place he must acknowledge his deep appreciation for the kindly advice, scholarly criticism, and friendly encouragement given him by Professor Charles M. Andrews of Yale University under whom the study was begun and brought to a conclusion. Thanks are due also to Professor Leonard W. Labaree, present editor of the Yale Historical Series, for his advice and comment. Valuable suggestions relative to the social development of Newfoundland were made by Professor Ralph H. Gabriel, also of Yale; while helpful criticism of the early chapters was given by Professor Theodore F. Jones of New York University.

The author must acknowledge the kindness shown him when gathering material in the various public and private depositories. Especially are thanks due to Dr. Arthur G. Doughty, Director of the Public Archives of Canada and the members of his staff at Ottawa; the officers of the Public Record Office in London; the staff of the British Museum; Dr. Max Farrand and the staff of the Henry E. Huntington Library and Art Gallery, San Marino, California; and the Office of the Colonial Secretary at St. John's Newfoundland, for access to original documents and transcripts. Furthermore, he desires to express appreciation for the many courtesies shown him by the staffs of the Yale University Library and the General Library of New York University. The conscientious and painstaking work of Miss Dorothy Shilton and Mr. Richard Holworthy of London in preparing transcripts of manuscripts in the Public Record Office deserves particular notice. The kindly patience of the author's wife, who has been obliged to listen to dissertations on dried codfish for a large part of her married life, is also acknowledged with deep gratitude. Her quiet but steady encouragement has been most helpful in bringing this work to a conclusion.

R. G. L.

New York University,
  University Heights,
    New York City,
      November 26, 1933.

CONTENTS

	Preface	vii
	Introduction: The Physical Environment	1
I.	The Beginnings of the Fishery	19
II.	The First Western Charter	55
III.	The Fishery under Cromwell and Charles II	92
IV.	The Problem of Regulation, 1660-1676	126
V.	The Collapse of Charter Regulation, 1676-1699	149
VI.	The Menace of France and New England, 1660-1699	182
VII.	Law and War, 1699-1713	204
VIII.	Peace and Disorder, 1713-1729	245
IX.	Civil Government, 1729-1763	273
X.	The End of an Era: Recapitulation	310
	Bibliographical Note	337
	Index	353

MAPS

The Region of the Banks *opposite page* 8
Newfoundland *following index*

The author wishes to acknowledge the courteous assistance of Mr. Julian W. Rothery of the International Paper Company, and Dr. John K. Wright, Librarian of the American Geographical Society, New York, who furnished valuable suggestions which were followed in the preparation of these maps.

BRITISH FISHERY AT NEWFOUNDLAND

INTRODUCTION

THE PHYSICAL ENVIRONMENT

THE natural environment of Newfoundland has affected its history profoundly. Some of the physical influences were comprehended in the earliest times, but others were not understood until recently. Although the island is closely adjacent to the North American mainland and forms part of the great North Atlantic seaboard region, it possesses sufficient local diversities to produce marked contrasts with neighboring Nova Scotia and New England, lands equally famous with Newfoundland as the centers of great fishing industries. Nowhere in northeastern America have local physical peculiarities played such an unrelenting part in history as has been the case at Newfoundland.

The great island lies at the mouth of the Gulf of St. Lawrence. At its northwest corner it is separated from the mainland by the Strait of Belle Isle, which is only nine miles wide at its narrowest point. The southwest corner is not immediately adjacent to the continent, but Cape Ray is only sixty miles across Cabot Strait from Cape North, the northernmost promontory of Cape Breton Island. It is by way of Cape Breton Island that modern Newfoundlanders are in closest contact with the more populous parts of North America.[1] But in spite of the nearness of its western extremities to other parts of the continent, the bulk of the island projects farther out into the Atlantic than any other portion

[1] The sailing distance between North Sydney, Nova Scotia, and Port aux Basques, Newfoundland, is only 110 miles.

2   BRITISH FISHERY AT NEWFOUNDLAND

of habitable America, and its eastern coast is consequently closer to Europe than the more developed maritime regions of Canada and the United States.[2] Even its nearness to the transatlantic shipping lanes has not brought Newfoundland into very close contact with the nearby continental centers. Isolation from its North American neighbors was even greater in the early days than at present, but its proximity to the Old World has been an important determinant throughout the island's history.

Although Newfoundland has roughly the form of an equilateral triangle, its length from Cape Ray to Cape Norman being 317 miles and its breadth from Cape Spear to Cape Anguille 316 miles, it has probably the greatest coastline in proportion to its size of any country in the world.[3] The general outline is very irregular, the shore being diversified by many bays and headlands, with the result that the coastline is tripled if not quadrupled. There are numerous deep and thoroughly landlocked harbors, many of which extend long distances inland. These

[2] St. John's is 1,672 nautical miles from Fastnet, Ireland, and 1,808 miles from Bishop's Rock, England.

[3] The account of the physiography of Newfoundland which follows is based largely upon William H. Twenhofel, "The Physiography of Newfoundland," *American Journal of Science*, Fourth Series, XXXIII, 1-24. Other authorities consulted were: Nelson C. Dale, "Pre-Cambrian and Paleozoic Geology of Fortune Bay, Newfoundland," *Bulletin* of the Geological Society of America, XXXVIII (1927), 411-430; Alfred K. Snelgrove, "The Geology of the Central Mineral Belt of Newfoundland: A Collation and Contribution," reprinted from the *Transactions* of the Canadian Institute of Mining and Metallurgy, 1928; "Geology and Ore Deposits of the Betts Cove-Tilt Cove Area, Notre Dame Bay, Newfoundland," *Princeton University Contribution to the Geology of Newfoundland*, No. 9, reprinted from the *Bulletin* of the Canadian Institute of Mining and Metallurgy, April, 1931; Albert O. Hayes, "Structural Geology of the Conception Bay Region and of the Wabana Iron Ore Deposits of Newfoundland," *Economic Geology*, XXVI (1931), 1-19; G. H. Ashley, "The Age of the Appalachian Peneplains" (abstract), *Bulletin* of the Geological Society of America, XLI (1930), 101.

INTRODUCTION—ENVIRONMENT

bays and their projecting headlands are elongations of the structural features of the interior and are most prominent on the northeastern, eastern, and southern coasts. Among the most notable indentations of the coast are Notre Dame, Bonavista, Trinity, Conception, Trepassey, St. Mary's, Placentia, and Fortune bays, most of which have been important fishing resorts since the early sixteenth century.

The major topographic feature is the marked parallelism of the peninsulas, bays, lakes, rivers, ridges, and faults, all of which trend generally from southwest to northeast, and this same tendency is reflected in the adjacent submarine topography. Even a cursory examination of a physical map will reveal this striking characteristic. The backbone of the island is the Long Range, the highest elevation, which extends from Cabot Strait to the Strait of Belle Isle, following close to the western shore. Geologically Newfoundland is very old. The most ancient rocks which date back to the Laurentian period or earlier, are igneous or metamorphic. These have since been overlain with sedimentary rocks, through which igneous rocks have intruded as a result of volcanic action in more recent geologic times. Generally, the rivers have cut through the softer sediments, with the result that the more fertile land is found in their valleys, while the highlands are hard and barren.

The average elevation of the Long Range is about 2,000 feet. From its middle an axis of high land runs eastward, following the south shore closely, and extending into the Avalon Peninsula. The land decreases in altitude northeastward and southwestward from the axis. The Long Range rises like a wall along the west coast, being broken only in a few places such as at Bay St. George and Bay of Islands where the St. George's and Humber rivers have cut their way through from the interior plateau. East of this flat-topped range there is a series of parallel valleys which are separated from one another by flat-topped ridges, upon which stand conical monadnocks or "tolts." Some of the tolts are over 2,000 feet high while others exceed 1,500 feet, and they comprise the most prominent topographic features of the central plateau, the average elevation of which is only about 1,000 feet.

There are several large river systems, of which the Exploits, Gander, and Terra Nova are the principal ones flowing to the northeast, while others such as the shorter La Poile and Bay d'Espoir drain to the southwest. Those flowing to the northwest are generally longer and their descent to the sea is less abrupt than those draining to the southwest. Although the upper reaches of the Humber and St. George's conform to the general tendency to parallelism, both have broken through the Long Range and empty into the Gulf of St. Lawrence. Near the coast all the rivers are youthful and falls and rapids are frequent, but in the interior they flow in broad valleys, consisting mainly of chains of lakes separated by stretches of rapids. Although many of them flow into the great bays, penetration of the interior has been hindered by the proximity of the fall line to the coast and by the fact that the longer rivers discharge into those bays which are most likely to be ice-bound for a considerable part of the year. None of the important rivers have their outlets in the great fishing harbors originally frequented by the English, and the island was not traversed from east to west until the early nineteenth century.

Newfoundland has sunk below its former level and again been relatively uplifted in recent geologic times. The process of rising still continues as fishing stages on the west coast have to be lengthened repeatedly and new passages have to be found among the shoals as the older

INTRODUCTION—ENVIRONMENT 5

ones become too shallow for navigation. In spite of the recent uplift, there are indications that the island is more submerged than formerly as is shown by the submerged foreland on the east and west coasts. Glaciers covered Newfoundland on at least two occasions, and erratics, glacial drift, and finer morainal material are found everywhere except where they have been swept into the sea. Altogether, the present-day physiography indicates that the island was formerly an almost perfect peneplain, subsequently dissected by subaërial erosion, and then tilted from west to east. Geologically, Newfoundland occupies an intermediate position between the two great physiographic regions of eastern North America, the Laurentian and the Appalachian, possessing characteristics of both.

The plant and animal life of Newfoundland also occupies an intermediate position. Some species of its flora have been derived from the Arctic-Hudsonian region, other are identical with plants found in seaboard New England, Long Island, and New Jersey, from which they probably entered when the coastal plain extended farther northward and eastward than at present. Other types of plants are common to the Atlantic and Mediterranean shores of Europe and consequently must have crossed the ocean. Strangely enough, there are few Canadian species.[4] Black spruce, fir, and larch are the principal timbers. There is little wood near the coast, partly owing to the cold winds and partly to the destructiveness of man. In many exposed places the trees are so stunted as to form only miniature forests, while elsewhere forest fires have wrought havoc with some of the best and most accessible woodlands. Except in formerly timbered regions the land is ill suited to agriculture. The best arable is not situated

[4] M. L. Fernald, "The Contrast in Floras of Eastern and Western Newfoundland," *American Journal of Botany*, V (1918), 237-247, *passim*.

near the places of original settlement, and until the nineteenth century little attention was paid to agriculture. Pulpwood is the most important forest product, but the many berrybushes which cover the ill-drained swamplands of the interior are economic assets of some value.

The fauna also differs markedly from that of neighboring parts of Canada and is more in agreement with that of the Arctic and Hudsonian areas than with others. Many of the common Canadian mammals and resident birds are absent, although woodland caribou, bear, fox, and beaver are common. Hunting and trapping are restricted to a limited number of species, and the fur trade never attained the relative importance that it enjoyed in nearby parts of Canada and the United States.[5]

The original inhabitants were the Beothuks or "Red Indians." Little is known concerning them because no satisfactory records were made of their customs or language before they became extinct early in the nineteenth century. They may have occupied all of Newfoundland at one time, but just before their disappearance they were confined to the region of the Exploits River and Red Indian Lake.[6] Always a shy race, they avoided contact with the whites, and mention of them by early writers is infrequent. The present Indians are Micmac-Montagnais, the mixed descendants of immigrants from Cape Breton and Labrador. They live on the southern and western coasts and hunt in the interior. They appear to have been on friendly terms with the Beothuks.[7] The aborigines were extremely primitive because of the paucity of natural advantages afforded by the island.

[5] Frank G. Speck, *Beothuk and Micmac* in *Indian Notes and Monographs: A Series of Publications Relating to the American Aborigines*, edited by F. W. Hodge, Museum of the American Indian-Heye Foundation (New York, 1922), p. 131. Cited hereafter as Speck, *Beothuk and Micmac*.

[6] *Ibid.*, pp. 11-12, 14-15.

[7] *Ibid.*, pp. 25-27.

INTRODUCTION—ENVIRONMENT

In spite of its barrenness, Newfoundland is very favorably situated for its inhabitants to gain a livelihood from the sea. The surrounding undersea topography, the nature of the ocean currents, and the character of the climate all contribute to the growth of an abundant marine life, while the shores offer unsurpassed facilities for the exploitation of the valuable food fishes which dwell in the nearby waters. In spite of the dangers from wind, fog, and ice, the presence of this vast marine storehouse determined from the first that the destinies of Europeans at Newfoundland should be linked with the sea.

Newfoundland is the highest of a great system of plateaus most of which are submerged but a few fathoms under the sea. This group of submarine tablelands, called the Banks, forms a series of archipelagoes extending from Cape Cod northward to Cape Chidley, and eastward to Flemish Cap, 200 miles east of Cape Race. They lie upon the submerged coastal plain or continental shelf, and the higher areas are covered by scarcely more than 300 feet of water and sometimes by as little as fifty feet. The shallowness of the water permits the sunlight to penetrate to the higher points and thus encourages the growth of marine life, both in the water and on the bottom. The presence of tremendous numbers of small organisms which serve as food for many varieties of fish, makes the region one of the world's leading fishing grounds.[8]

[8] The description of the submarine topography of the region of the banks has been drawn largely from the following: Robert Perret, *La Géographie de Terre-Neuve* (Paris, 1913), pp. 78-83, *passim*; J. W. Spencer, "Submarine Valleys off the American Coast and in the North Atlantic," *Bulletin* of the Geological Society of America, XIV (1913), 207-226, *passim*; Hydrographic Office, United States Navy, *Nova Scotia Pilot: Bay of Fundy, Southern Coast of Nova Scotia and Cape Breton Island* (7th ed., Washington, 1930), pp. 239, 339-340, 344-347; *Sailing Directions for Newfoundland Including the Coast of Labrador from Cape St. Lewis to Long Point* (5th ed., Washington, 1931), pp. 316-321, *passim*. Cited hereafter as Perret, *Terre-Neuve*; *Nova Scotia Pilot*, and *Sailing Directions for Newfoundland*, respectively.

The high plateaus of the continental shelf fall into four main groups: first, the New England Banks, which lie east of Boston and in the Gulf of Maine and the Bay of Fundy; second, the Nova Scotia Banks, which lie off the southeastern coast of that province between Cape Sable and Cabot Strait; third, the Newfoundland Banks, which lie south and east of that island, and include beside the important Grand Bank, Green, St. Pierre, and Burgeo banks, Flemish Cap, and the unnamed submerged coastal plain lying off the eastern and northeastern coasts of Newfoundland between Cape Race and the Strait of Belle Isle; and fourth, the Labrador Banks, which lie parallel to the coast between the Strait of Belle Isle and Cape Chidley. These groups have many characteristics in common, but there are a number of important local differences which give them individuality.

The Newfoundland Banks are separated from those of Nova Scotia by a great submarine channel, the Laurentian Valley, which is from fifty to seventy miles wide. It is broader near the edge of the continental shelf and attains a depth of 15,000 feet. The banks which lie north and east of the great channel and south and southeast of Newfoundland are Burgeo, St. Pierre, Green, the Grand Bank proper, and Flemish Cap. Green and St. Pierre, which are separated from each other and from the main bank by fairly deep channels are usually considered as forming part of the Grand Bank because of their similar characteristics and the essential unity of the region. Flemish Cap is separated from the rest of the group by an interval of very deep water which normally serves as a passage for the Arctic Current.

The Grand Bank, including those mentioned above, is the most extensive of all the submarine archipelagoes. It lies between 48° 35′ and 42° 54′ north latitude and 47° 35′ and 57° 20′ west longitude. Its shape is approximately

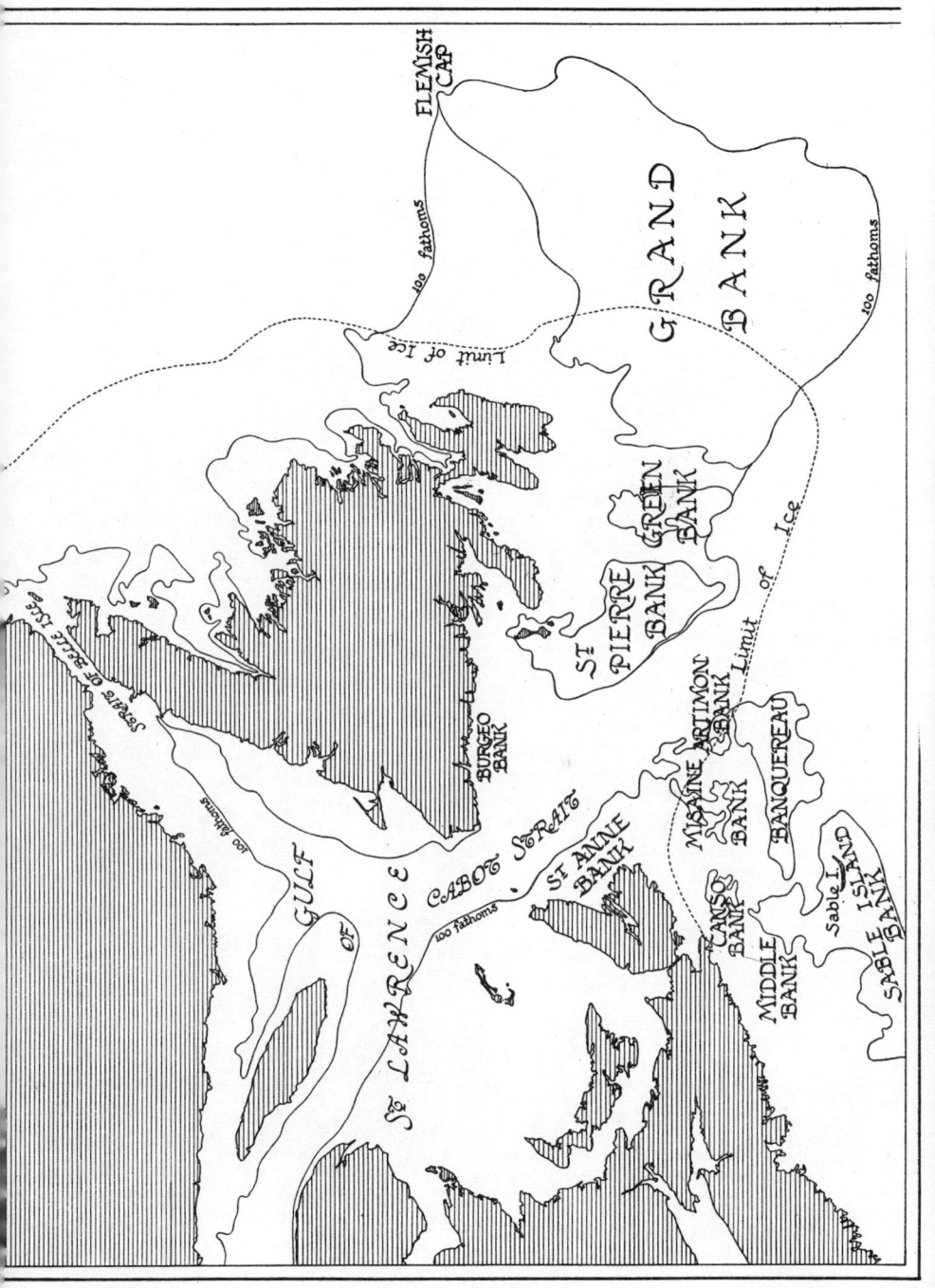

INTRODUCTION—ENVIRONMENT

that of an equilateral triangle with its base parallel to the southern coast of Newfoundland and its apex at 42° 54′ north latitude and 50° west longitude. East of 52° west longitude the general depth ranges from thirty to forty-five fathoms, but west of that line it is shallower, ranging from twenty to fifty fathoms. The undersea topography of the bank is similar to that of the island. It is an extensive plateau elevated from 150 to 200 feet above a lower plain, which is slightly tilted from east to west, its southern edge being an abrupt declivity. The surface of the upper plateau is gently undulating in some places and roughly irregular in others. There are many deep steep-sided holes produced by the action of eddying currents upon the materials deposited upon the bottom, the most notable being the Whale Hole which is about 400 feet deep. Besides Flemish Cap, there are two well known shoal areas on the Grand Bank: the Virgin Rocks and Banks and the Eastern Shoals, which are the summits of rocky ridges lying only a few feet below the surface. The sea breaks upon them in bad weather and is even confused in their vicinity when strong breezes blow. These shoals are similar to the ridges and tolts of the island. In spite of the dangers to navigation, the Virgin Rocks are the site of an important hook-and-line fishery. The Eastern Shoals have more water over them but are considered more perilous because of the unevenness of the ground for some distance around them. Likewise, the submerged coastal plain off the eastern coasts of Newfoundland and Labrador between Cape Race and Cape Chidley is a region of alternating ridges and gullies. Indeed, the entire bank area has topographic features which resemble those of the neighboring island. The tendency to parallelism is also reflected in the undersea topography, the submarine ridges and channels in many cases being extensions of the ridges and bays of the island.

The peculiarities of the surrounding waters have been significant factors in the development of the great fishery in this region. Currents, tides, ice, salinity, and temperature have all affected the marine life of the area and many of these phenomena have also influenced navigation and other human activities.

There are two great currents, the Gulf Stream and the Arctic Current, which intermingle on the banks and finally pass each other. The warm current passes south of the Grand Bank during the winter, but during the summer it extends over the southern part of the submarine plateau. The motion of the currents is oscillatory, so that streaks of warm water alternate with cold. The Gulf Stream modifies the temperature of both sea and air, but its influence near Newfoundland is not as pronounced as that of the cold northern current. The Arctic or Labrador Current, after passing through Davis Strait, proceeds down the coast of Labrador and eastern Newfoundland. Near the land it is influenced greatly by the wind, which affects both its direction and velocity.[9] A western branch sometimes enters the Gulf of St. Lawrence through the Strait of Belle Isle, and rejoins the main current off the coast of Nova Scotia after passing through Cabot Strait. The main current usually turns west-southwest at Cape Race, but sometimes it follows a northwesterly course as far as Cape Ray, and may even enter the Gulf of St. Lawrence. Many wrecks have occurred in the fog on the southern coast of Newfoundland owing either to the indraft of the current into the great bays or to its setting into Cabot Strait. The usual course is in a southwesterly direction across the Newfoundland Banks and thence down the coast of Nova Scotia. The

---

[9] The paragraphs devoted to the hydrography of the Newfoundland seas are based largely upon the following: *Sailing Directions for Newfoundland*, pp. 30-45, *passim*, and *Nova Scotia Pilot*, pp. 40-41.

INTRODUCTION—ENVIRONMENT

variability of the Labrador Current makes navigation dangerous in thick weather. The currents in the Strait of Belle Isle, Cabot Strait, and the open area of the Gulf of St. Lawrence are also extremely variable, being influenced greatly by wind and tide.

The temperature of the waters surrounding Newfoundland is generally quite low. Off the southeastern coast between St. John's and Placentia Bay the water is practically at the freezing point at a depth of thirty fathoms even in summer, although the surface water of the Arctic Current will become as warm as that which surrounds it. When strong offshore winds prevail the warm surface water is driven away from the coast and the cold water rises to the surface. The same contrast between the temperature of the surface and the lower layers of water prevails in Cabot Strait, the Strait of Belle Isle, and the Gulf of St. Lawrence.

Ice not only affects the climate and obstructs navigation, but also influences the temperature and salinity of the water, and thus is a factor of importance in the lives of the marine organisms. Every sheltered harbor is likely to be frozen over during the winter, but no definite time limit may be set for the appearance or disappearance of harbor ice because the severity or mildness of the winter is the determining factor. Field ice and icebergs are of greater importance, for they cover the sea for a considerable part of the year. During the spring and fall navigation is difficult and in winter impossible in the Gulf of St. Lawrence and the Strait of Belle Isle. Although Cabot Strait is never completely frozen over it is often jammed with ice from the gulf to such an extent that in early spring a blockade extends from St. Paul Island to Cape Ray. This blockade, known as "the Bridge," sometimes lasts for three weeks, holds up vessels anxious to enter the gulf, and causes wrecks on the Newfoundland coast.

Ice is a constant menace to navigation on the eastern coast of Newfoundland and in the transatlantic shipping lanes. Icebergs or extensive fields of solid, compact ice are carried into this region by the Arctic Current. North of 43° north latitude bergs are seen during most of the year, except in August. Field ice reaches a southerly limit at about 42° north latitude, 45° west longitude, but it is impossible to determine its limits with accuracy because its position and quantity vary from year to year. Field ice is formed in the Arctic seas and carried southward by the current. The harbors of Labrador and northeastern Newfoundland are blocked in the fall, and the field usually reaches Cape Race by the end of January. Easterly winds sometimes cause the ice to block the eastern harbors until June or even July, although most of them are usually free by the end of May. The southern ports are freer from field ice than those elsewhere. Sometimes the pack may obstruct the east coast completely. The first field ice to appear is "white" or "northern slob" formed off Labrador in the early winter, but it is followed almost immediately by "sheet" ice which is more dense and solid though not as thick. The fields, which are often from five to fifteen or more miles wide in January, may increase to between 80 and 130 miles wide in the following month. There is no uniform breadth and the pack is often intersected by leads and open areas. The fields drift offshore for some time before closing in on the coast. The seasonal character of field ice and its susceptibility to the influence of winds and currents makes it a serious, though temporary menace to navigation.

Icebergs, which follow the field ice, are detached portions of glacial ice carried away from the polar regions by the currents. Although they melt rapidly on their

INTRODUCTION—ENVIRONMENT

southward journey they are still sufficiently large in the vicinity of Newfoundland to endanger shipping. Their disintegration is rapid owing to internal expansion. Their frequency varies from year to year, but they are apt to be most numerous in June. The bergs usually move across the banks in a southerly or southwesterly direction but their progress is often arrested by the eddies at Flemish Cap and the Virgin Rocks. Were it not for these two obstacles, the difficulties of navigation in late spring would be increased immeasurably. Bergs are less susceptible to winds than to currents. Sailors can locate field ice on clear days and moonlight nights because of the diffused light or "ice blink" which appears over the pack. Its proximity may also be determined by sounds resembling those of distant gunfire or breakers, and by the presence of herds of seal far from land, but there are no warnings of the presence of bergs. Danger from them is increased by fog, so that ships often approach fatally close before apprehending their presence. Altogether, ice in the form of fields or bergs hampers communications and imperils the lives of fishermen and sailors at Newfoundland.

Innumerable minute varieties of marine animal and vegetable life exist in the waters over the banks and on the sea-bottom. Their existence makes this region one of the world's great fishing grounds because they serve as food for the larger fish and even the cetaceans. "Plankton," composed of floating organisms unprovided with means of locomotion, forms the great organic mass. The plankton is borne from place to place by the currents, and thus influences the movements of the species which feed upon it. During the act of spawning the hunger of both male and female fish is increased and, therefore, at that period they turn toward the region which is most abundant in plankton. However, the fish approach their

14 BRITISH FISHERY AT NEWFOUNDLAND

food by swimming against the current so that they have only to open their mouths in order to be fed.[10]

Plankton is the only food eaten by many of the smaller species of fish such as herring and caplin, as well as by the fry of larger varieties such as the cod. It also forms an important part of the diet of the adult cod and the marine mammals. The smaller fish, which are absolutely dependent upon it, serve in turn as the prey of the larger ones. Three kinds of plankton are found on the banks: first, that which is rich in animal life but poor in vegetal organisms, brought northward by the Gulf Stream; second, that which is almost exclusively vegetal, brought south by the Arctic Current; and third, a mixed type found below Newfoundland where the warm and cold currents mingle, which is equally rich in both kinds of life. The distribution of the various kinds of plankton corresponds directly with the temperature and salinity of the water. Conditions which mean life to one species may mean death to another. Variations in these two respects cause many organisms to die and fall to the sea-bottom where they serve as food for the more sedentary fish and the marine life attached to the bottom. The plankton is divided into swarms of identical species. The position of these groups depends upon equilibrium which exists between their specific gravity and that of the water. Although the penetrating sunlight encourages their growth, the melting ice continually modifies the temperature and salinity of the water, changing its specific gravity and thus upsetting the balance necessary to the maintenance of life. Their equilibrium is also affected by the degree of the intensity of light, and the changes in the position of the plankton are, therefore, periodic and seasonal, dependent upon climatic and hydrographic conditions.

[10] Information regarding marine biology is based almost entirely upon Perret, *Terre-Neuve*, pp. 169-179, *passim*.

INTRODUCTION—ENVIRONMENT

Variations in the physical conditions occasion vertical as well as horizontal displacement of the plankton, and, consequently of the fish which depend upon it. Thus, the habits of the fish are determined by the conditions affecting their food supply.

Most of the minute organisms are localized, but owing to the constant changes mentioned above, the various types of plankton move from place to place. The continual movement of its food supply determines the route followed by the larger species such as the cod. The path followed by these fish from their spawning place to their winter home cannot be traced on the map because it is neither direct nor constant. Moreover, the cod do not follow identical customs in the various regions they frequent, because the waters which they find habitable are not found everywhere in the same horizontal plane, but vary in position according to the latitude, the nature of the currents, and the character of the local climate. One reason why Newfoundland's great bays are such excellent fishing grounds is because the surface water from the land runs off rapidly in the spring and the salt water can penetrate the deep inlets without appreciable loss of salinity. The bays are, therefore, frequented by schools of cod in search of food and a place to spawn, thus bringing the fishery close inshore. At Newfoundland the cod spawns between May and September, the interval between the disappearance and reappearance of the ice. This is the fishing season, for during these months the cod frequents the many inlets and occupies the upper layers of water on the banks. The fish migrates simultaneously both in horizontal and vertical directions, the coastal variety moving between the inshore zone and the bottoms of the banks and wintering in the harbors which are relatively free from ice; while the oceanic type appears on the surface or descends into the depths according to season. The northern or Labrador cod spends the winter in the deep waters of the open sea, ascends the Arctic Current in the spring, and permits itself to be carried back by the current when bad weather impends. In each region the cod reacts in accordance with the physical phenomena which affect its food.

Temperature and saltness have a direct bearing on the reproductivity of the fish. The roe, after spawning, temporarily becomes part of the plankton, and sometimes serves as food for the adults. The buoyancy of the roe determines the location of the spawning places. The Gulf of St. Lawrence and Cabot Strait are poor fishing grounds because of the large admixture of river water found there. For similar reasons Conception Bay has always been poor compared with Trinity and Placentia bays and the banks where conditions are more favorable. The habitat of the cod varies with its age. The eggs maintain only a floating existence; the young fry, which is absolutely dependent upon plankton for its food is considerably restricted in its movements; but the adult is a pelagic fish which preys upon other species and consequently has a more extended range. The species of fish which serve as food for the cod are weaker and more susceptible to physical changes than their enemy. Herring is the hardiest, but caplin cannot bear a temperature lower than 35° even for a short time. Other species to which the cod is partial are more robust than the caplin and some are quite as voracious as the cod. The continual chase of one species after another causes a constant displacement of marine life. Sometimes after spawning, the famished cod hurls itself upon schools of caplin or other small fish with such force that "banks" are formed, rising several feet above the surface and extending for long distances. The distribution of the cod can be determined by noting the presence or absence of the objects upon which it feeds.

INTRODUCTION—ENVIRONMENT

Individuals sometimes travel long distances in search of food, but the cod is not usually considered a migratory fish, the majority living in comparatively restricted areas.

The climate of Newfoundland and the banks is extremely variable. The island lies midway between the continental and oceanic climatic areas, with the result that constant struggle takes place between the meteorologic forces originating in the two regions. The climate of the west coast is predominantly continental, but that of the east coast possesses characteristics which are definitely oceanic. Consequently, the climate of St. John's is more moderate than that of the Laurentian shore or the interior, where the extremes are typically continental. Because the island lies southwest of the permanent Icelandic low pressure area and because it is also immediately in the path of temporary low pressure centers moving seaward from the interior of North America, its weather is subject to frequent and violent changes. The summer is less stormy than the winter, but during the period from November to April the island and surrounding seas are disturbed by a series of cyclonic storms, called either "southers" or "northers," which bring rain, snow, and sleet, and which are often so violent that they cause the loss of ships and men. Usually stormy weather ensues when easterly and southeasterly gales prevail, but westerly winds bring clear skies. The terrific storms are most dangerous because of the presence of fog, icebergs, field ice, and a rock-bound coast, and add greatly to the perils of fishing and navigation.[11]

Fog is a terrible menace to navigation on the banks and on the coast of the island, and even interferes with

[11] Data relating to the climate are derived largely from the following: *Nova Scotia Pilot*, pp. 487-494, *passim; Sailing Directions for Newfoundland*, pp. 27-30, 664-669, *passim;* and Perret, *Terre-Neuve*, pp. 116-127, *passim.*

the proper curing of codfish. It is produced by the contact of the warm, moist air over the Gulf Stream with the cold air over the Arctic Current. The chilling process sometimes produces rain or snow, but more often fog, which may be anything from a wet drizzle to a dry mist. Fog is prevalent during most of the year, but persists through June and July on the eastern and southern coasts at the height of the fishing season. It withdraws to the eastern edge of the Grand Bank during the fall. It is usually borne by easterly winds and dissipated by those from the opposite direction. Generally, fog is less heavy north of St. John's than south of that port, and is seldom encountered on the west coast, although it is a serious menace in the Strait of Belle Isle. Usually there is a clear belt between the fog bank and the shore and fog seldom extends far inland. It occurs in all sorts of weather, but is most quickly dissipated by rain. Though increasing the maritime hazards it has a moderating effect upon the climate, and has made Newfoundland more suitable for human settlement and activity than the continental regions in the same latitude.

Nature has distributed her riches very unevenly at Newfoundland, with the result that man has turned his eyes toward the sea and concentrated upon exploiting the wealth contained there. The barrenness of the interior, the absence of navigable rivers flowing into the popular fishing harbors, and the remoteness of these harbors from the more fertile and forested areas, all have tended to hinder men from venturing far from the coast. In spite of the perils of the sea the inhabitants have found it more advantageous to pursue maritime activities than to pioneer in the inhospitable and inaccessible interior. In the following chapters we shall follow the development of the codfishery, the earliest, and still the most important of Newfoundland's industries.

CHAPTER I

THE BEGINNINGS OF THE FISHERY

THE exploitation of the rich harvest of the sea in the vicinity of Newfoundland began immediately after the return of the earliest voyages of discovery to the northeastern coasts of North America. Although John and Sebastian Cabot did not make a landfall at Newfoundland they sailed very near it, and reported on their return to England in 1497, that the region abounded in large codfish, salmon, sole, and seal.[1] Sebastian Cabot called the country which he discovered "Baccalaos," the native name of the cod. He alleged that these fish were so plentiful that they sometimes stopped the progress of his ships.[2] The cod were so numerous that his sailors caught them by the simple expedient of lowering baskets overside. A member of Cabot's expedition said that the supply was sufficiently great to make England independent of Iceland for her importations of stockfish.[3] There are legends that European fishermen had

---

[1] "An extract taken out of the map of Sebastian Cabot, cut by Clement Adams, concerning his discovery of the West Indies," in Richard Hakluyt, *The Principal Navigations Voyages Traffiques and Discoveries of the English Nation Made by Sea or Over-land to the Remote and Farthest Distant Quarters of the Earth at any time within the compass of these 1600 Yeeres* (Maclehose Edition, Extra Series [for the Hakluyt Society], 12 vols., Glasgow, 1903-1905), VII, 145-146. Cited hereafter as Hakluyt, Extra Series.

[2] "Another testimonie of the voyage of Sebastian Cabot to the West and Northwest, taken out of the sixt[h] Chapter of the third Decade of Peter Martyr of Angleria," Hakluyt, Extra Series, VII, 150-152.

[3] Second dispatch of Raimondo di Soncino to the Duke of Milan, Dec. 18, 1497, Henry P. Biggar, *Precursors of Cartier* (Canadian Archives, no. 5), p. 20. Cited hereafter as Biggar, *Precursors*.

visited Newfoundland previous to the voyage of the Cabots, but none of these are supported by documentary evidence. The discovery of the island of Newfoundland cannot be credited to Cabot, and there is nothing to indicate the presence of any Europeans on the island previous to 1501.[4] The Portuguese, Gaspar Cortereal, deserves credit for discovering the island in the first year of the sixteenth century.[5]

The Portuguese became interested immediately in the newly discovered region. They fitted out expeditions at Lisbon and the Azores for the purpose of fishing and trading with the natives of northeastern America, while their whaling expeditions visited the coast of Labrador and penetrated into Hudson Strait.[6] By 1506 Portuguese interest in the fishery was so keen that the crown found it profitable to place a tax on codfish imported from Newfoundland.[7] During these early years the island and the lands adjoining were known as Terra de Cortereal de Portugal, Terra de Pescaria, and sometimes Baccalaos. The last two names indicate the object of nearly all the voyages made there, and the discovery of a large part of northeastern North America may be attributed to these early fishing expeditions.[8] Portuguese interest was so aroused that attempts were made to establish settlements on the island in the early sixteenth century.[9] None of these were permanent, and all traces of them are lost, but Portuguese influences remain in the nomenclature of south-

[4] Biggar, *Precursors*, pp. xiii-xiv.

[5] Henry Harrisse, *Découverte et Évolution Cartographique de Terre-Neuve et des Pays Circonvoisins, 1497-1501-1769* (Paris, 1901), p. xxi. Cited hereafter as Harrisse, *Terre-Neuve*.

[6] *Ibid.*, p. xxiii.

[7] Copy of a letter from . . . the king (of Portugal) in regard to the tithe on codfish, Oct. 14, 1506, Biggar, *Precursors*, pp. 96-98; Harrisse, *Discovery of North America* (London and Paris, 1892), p. 174. Cited hereafter as Harrisse, *Discovery*.

[8] *Ibid.*, p. 180.   [9] Harrisse, *Terre-Neuve*, p. xxiv.

THE BEGINNINGS OF THE FISHERY

eastern Newfoundland, particularly in the zone lying between Cape Race and Cape Bonavista.[10]

Although the Portuguese are generally regarded as having been the first Europeans to visit Newfoundland, the Normans and Bretons were there almost as early.[11] Indeed, large numbers of Frenchmen were fishing at Newfoundland in the early sixteenth century.[12] Both French and Spanish Basques fished for cod and whaled there at a very early date.[13] Beginning their fishery on the Grand Bank and in the bays and inlets of Newfoundland, the French rapidly extended their operations to the southwestern banks, the coasts of Acadia, and the Gulf of St. Lawrence.

English fishermen did not take much interest in the region until about the middle of the sixteenth century. In spite of alleged priority of discovery and of early visits to Newfoundland in the interests of trade or discovery, Englishmen remained aloof.[14] In the early part of the century there were several joint Anglo-Portuguese expeditions, and a few undertaken by Englishmen alone, but there was no concerted effort on the part of the fishing interests to exploit Newfoundland until about 1550.[15] During the first half-century the English government was unconcerned about its title to the island, and made no

[10] Harrisse, *Discovery*, pp. 128, 693-694.

[11] Harrisse, *Terre-Neuve*, pp. xix, xxxiii.

[12] French preponderance is indicated by the application of the fishermen of Guernsey to the Bretons, asking for permission to fish at Newfoundland, Harrisse, *ibid.*, p. xxxiv.

[13] *Ibid.*, pp. lviii, lix; Harold A. Innis, "The Rise and Fall of the Spanish Fishery at Newfoundland," *Transactions* of the Royal Society of Canada, XXV, section II (1931), 51-70. See especially, pp. 51-54. Cited hereafter as Innis, "Spanish Fishery."

[14] Harrisse, *Terre-Neuve*, pp. iv-viii.

[15] Hakluyt, Extra Series, II, 166; VIII, 3-7; Harrisse, *Discovery*, pp. 687-688; Biggar, *Precursors*, pp. 165 ff.; Biggar, *Voyages of Jacques Cartier* (Canadian Archives, no. 11), pp. 273-277. Cited hereafter as Biggar, *Cartier*.

attempt to assert its sovereignty.[16] The new fishery is mentioned in some of the legislation of the reigns of Henry VIII and Edward VI, but Newfoundland was considered of no more importance than Iceland, Shetland, and other distant fisheries in which Englishmen were interested.[17] The failure of their Iceland fishery, in which they employed a large number of ships annually, may have caused them to turn eventually to the transatlantic fishing grounds.[18] By 1548 the trade from England to Newfoundland was common and frequent, although fishing was not yet carried on very extensively.[19] However, by the beginning of the reign of Elizabeth, English fishermen were well established there.

England became more alive to the possibilities offered by Newfoundland in the last quarter of the sixteenth century. By this time people had begun to consider the suitability of the island and other parts of America for colonization. Many persons of prominence among the lesser feudal nobility and merchant classes became interested in schemes of trade and settlement in the new world. As early as 1566 Sir Humphrey Gilbert proposed a voyage to discover the northwest passage, and suggested the establishment of an English plantation in the lands which he expected to discover.[20] Twelve years later, in 1578, Anthony Parkhurst, who had visited Newfoundland,

[16] Harrisse, *Terre-Neuve*, p. viii.

[17] "The bill concernyng bying of fishe upon the See," 1541-1542, 33 Henry VIII, c. 2, *Statutes of the Realm* (ed. 1819), III, 827; "An Acte againste the exaction or other thinge by any officer to traffike into Iseland," 2 & 3 Edward VI, c. 6, *ibid.*, IV, pt. I, 44-45. Cf. Leo F. Stock, *Proceedings and Debates of the British Parliaments relating to North America*, I, 3-4. Cited hereafter as Stock, *Proceedings and Debates*.

[18] Harrisse, *Terre-Neuve*, p. viii.   [19] Hakluyt, Extra Series, VIII, 9.

[20] Petition of Gilbert, 1566, Carlos Slafter, *Sir Humfrey Gylberte, and his Enterprise of Colonization in America* (Prince Society, Boston, 1903), pp. 183-184; Gilbert to Elizabeth, Feb. 1567, *ibid.*, pp. 184-186. Cited hereafter as Slafter, *Gylberte*.

THE BEGINNINGS OF THE FISHERY 23

wrote enthusiastically to the elder Hakluyt of its possibilities. He pointed out its desirability as a resort for fishermen, remarked upon the excellence of the climate, the fertility of the soil, and the abundance of natural resources. He contributed some interesting figures concerning the relative size of the fishing fleets of the various nations and compared their equipment and processes of preparing the fish for market. According to Parkhurst, English shipping had increased from thirty to fifty sail in four years. There were a hundred Spaniards there engaged in making "wet" fish, which they pickled in brine and then dried on returning to Spain; besides twenty or thirty whalers from Biscayan ports. The Spaniards and Basques were better equipped than the English. Their tonnage he estimated at between 5,000 and 6,000. There were only about fifty Portuguese vessels, of 3,000 tons. France had the largest representation, 150 sail, but these vessels were small, and their tonnage did not exceed 7,000. Like the Spaniards, the Portuguese, and French also engaged in making wet fish. Parkhurst recommended that England establish a colony and build fortifications on the island in order to suppress these competitors.[21]

It is to Parkhurst that England is indebted for the first definite scheme of political control for Newfoundland, and for the earliest plan of economic organization for the fishery. In a manuscript, probably written in 1578 in support of Gilbert's plan for a settlement, Parkhurst enlarged upon his earlier suggestions contained in his letter to Hakluyt. The plan is significant because it shows the beginnings of the later mercantilist conception of the plantation as a supplement to the productive area of the mother country; introduced the idea that colonization in Newfoundland would improve and expand the English

[21] Anthony Parkhurst to Richard Hakluyt of the Middle Temple, Bristol, Nov. 13, 1578, Hakluyt, Extra Series, VIII, 9-16.

fishery there; and emphasized the development of a national mercantile marine. Parkhurst suggested that if Newfoundland were settled by Englishmen the inhabitants would be able to make larger quantities of fish, and do so more carefully, than did the fishing ships. Because of their fixed residence the colonists would enjoy a fishing season which would be longer than the short period of two months when the fishing ships were there. Moreover, the settlers would waste less time in quarrelling over the occupation of fishing rooms and in building and repairing boats, stages, and other necessary equipment. The colonists would be able to make enough salt to supply their own needs and to furnish it to the seasonal fishermen from England, thus relieving them of the necessity of purchasing it abroad. Besides participating in the fishery, the colonists would be able to develop the other resources of the island, engaging in mining, agriculture, stock-raising, hunting, and trapping.

Parkhurst felt that such an arrangement would prove of advantage to English shipping because many of the fishing ships could be converted into freighters to carry the products of the plantation and its fishery to foreign markets. The consequent increase in the efficiency of the fishery and the development of an active carrying trade to and from Newfoundland would improve the economic position of England. Seamen employed in this branch of the merchant marine would receive better wages than for voyages from England to the Continent, and they would be inclined to spend their earnings at home rather than in foreign ports. Thus the local tradesmen and merchants of the home ports would be benefited. A voyage paying high wages would attract a large number of men, reduce unemployment in the agricultural districts, and relieve the landed gentry and those responsible for the care of the poor. Most important of all, Parkhurst believed his

THE BEGINNINGS OF THE FISHERY

plan would increase the exports to foreign countries and tend to swell the importation of bullion. According to the scheme the West Country fishing ships would not be eliminated, but the plantation would assist them, and thus increase the productivity of the fishery many times.[22] With certain modifications, this plan for the colonization and economic development of Newfoundland formed the basis for the later colonization projects undertaken in the early seventeenth century. The story of the various attempts to apply these theories and the resistance which they encountered forms a large part of the history of Newfoundland in the two centuries that followed.

The first attempt to establish a colony in conformity with the ideas of Anthony Parkhurst was made by Sir Humphrey Gilbert and his associates. His expedition was planned partly as a voyage of exploration and discovery, and partly as an effort to colonize Newfoundland and to reassert England's claims to sovereignty over the island.[23] Although Gilbert and his friends applied for permission from the crown to undertake such a voyage in 1574, they did not become active until some time later.[24] In 1578 he finally obtained letters patent which granted him proprietary rights over the lands which he should discover or secure for the crown in America. The purpose of the expedition was set forth in the charter and indicates that colonization was the principal objective.[25] The

[22] Parkhurst's account of advantages arising from encouraging traffic at Newfoundland [1578], British Museum, Lansdowne MSS., 100, no. 10, fols. 95-97.

[23] Letters Patent granted by the Queen to Sir Humphrey Gilbert, Knight, "for the inhabiting and planting of our people in America," June 11, 1578, Hakluyt, Extra Series, VIII, 17-23.

[24] "Petition of divers gentlemen," Mar. 22, 1574, Slafter, *Gylberte*, pp. 221-222. See also the Petition of the West Parts to the Lord High Admiral, Lincoln, same date, *ibid.*, pp. 230-237; Gilbert's "A Discourse how her Majesty may annoy the King of Spayne," Nov. 6, 1577, *ibid.*, pp. 237-244.

[25] Gilbert's patent, June 11, 1578, Hakluyt, Extra Series, VIII, 17-23.

failure of the expedition, owing to the desertion of Knoles, resulted in the postponement of the attempt to establish an English settlement in the New World.[26] Later, in 1582, Gilbert entered into an agreement with some merchant adventurers of Southampton, and in the following year set out on his ill-fated voyage.[27]

In spite of the optimism of the participants, the dream of founding an English colony in Newfoundland was short-lived. On his arrival at St. John's, Gilbert asserted the right of the English crown to the island; made laws to govern the people, and attempted to found a colony by leaving a few of his men behind as a nucleus. He confirmed some of the masters of fishing ships in the tenure of their stages and drying places, and levied on the shipping in the harbor for provisions to carry his expedition farther. Unfortunately, his arrogance and partiality, as well as the exactions which he imposed upon them, aroused the antagonism of many of the fishermen at St. John's. As lord proprietor he confirmed the rights of certain shipmasters to their fishing rooms, thus upsetting custom which granted to the first arrival of the season choice of the stages and boats' rooms in any harbor, and permitted the later arrivals to make their choice in turn. The shipmasters whom he favored were probably the employees or associates of the Southampton merchants who were backing him. Soon after his departure on the disastrous voyage to the southwest his colonists became discouraged and returned home.[28] Unquestionably the enterprise was based on Parkhurst's plan, but was intended to

[26] Documents relating to the expedition of 1578, Slafter, *Gylberte*, pp. 245-257.

[27] Articles of agreement between Gilbert and those who adventured with him, Nov. 1, 1582, Slafter, *Gylberte*, pp. 278-295.

[28] Edward Hale's report of Gilbert's voyage of 1583, Hakluyt, Extra Series, VIII, 35-55; Sir George Peckham's report of Gilbert's discoveries, *ibid.*, pp. 89-131.

THE BEGINNINGS OF THE FISHERY 27

give Gilbert's associates a favored position in the fishery and carrying trade. Such favoritism could not be approved by the other English fishermen. Hence instead of obtaining their coöperation Gilbert created among them a stubborn opposition to the colonization of the island and the establishment of local government there. The fishermen did not forget their experiences with the first proprietor, and a traditional hostility toward settlement was established which was to influence the course of Newfoundland's history in later years. Thus they were on the alert to protect their interests when a more serious and better organized attempt was made to colonize the island nearly thirty years later.

Even though Gilbert's attempt failed through both his arrogance and his subsequent loss at sea, the possibilities of establishing an English settlement in Newfoundland continued to find advocates. Sir George Peckham, one of the principal backers of the expedition of 1583, was enthusiastic. He wrote a glowing account of the suitability of the island for colonization, and pointed out the benefits which would accrue to trade and navigation were a definite project undertaken.[29] Richard Clarke, master of one of Gilbert's ships, reported the southern coast as attractive and suitable for a plantation.[30] Steven Parmenius of Buda, also a member of the expedition, was the only adverse critic, doubting the favorableness of the climate for colonization.[31] In spite of the continuance of propaganda favoring a settlement for some time after Gilbert's failure, no further attempts were made in that direction until the second decade of the seventeenth century. About 1600, there was a temporary revival of in-

[29] Peckham's report, Hakluyt, Extra Series, VIII, 89-131; relation of Richard Clarke who accompanied Gilbert, *ibid.*, pp. 85-88.

[30] Clarke's relation, Hakluyt, Extra Series, VIII, 85-88.

[31] Steven Parmenius of Buda to Richard Hakluyt, "St. John's Port," [Newfoundland], Aug. 6, 1583, Hakluyt, Extra Series, VIII, 81-84.

terest in the colonization of America for the purpose of improving England's trade, but it was not for several years afterward that any practicable schemes were offered.[32] During the years which intervened between the voyage of Gilbert in 1583 and the establishment of John Guy's settlement in 1610, England was too preoccupied with the struggle with Spain, and her explorers and mariners too greatly attracted toward piratical raids on the Spanish dominions to devote any serious attention to Newfoundland or other parts of the neighboring mainland of northeastern America.

Meanwhile the English continued to fish at Newfoundland and to drive out some of their competitors by warfare and piracy. They made themselves secure as a result of the war with Spain. Gilbert's abortive expedition of 1578, although ostensibly a voyage of discovery, was undertaken with the object of attacking and destroying Spanish shipping at Newfoundland and in the West Indies.[33] Although this expedition proved a fiasco, Walsingham revived the project in 1585. He suggested that the seizure of the enemy's shipping at Newfoundland would injure the Spanish navy, and would be advantageous to England as an act of retaliation against the seizure of English ships in Spanish ports.[34] Bernard Drake was commissioned to proceed to Newfoundland with the twofold purpose of warning the English shipping there not to sail for Spanish ports, and of seizing all ships there belonging either to the king of Spain or his subjects.[35]

[32] "Yt beinge a verrey noble action to inlarge a dominion, &c.," [about 1600], Colonial Office 1: 1, no. 9.

[33] Gilbert's "A discourse how her Majesty may annoy the king of Spayne," Nov. 6, 1577, Slafter, *Gylberte*, pp. 237-244; *Calendar of State Papers, Domestic Series*, 1574-1580, p. 565. Cited hereafter as *Cal. State Paps., Dom.*

[34] A plot . . . for annoying the king of Spain, Mar.? 1585, *ibid.*, 1581-1590, p. 234.

[35] Commission to Bernard Drake to proceed to Newfoundland, June 20,

THE BEGINNINGS OF THE FISHERY 29

Drake, accompanied by Sir Walter Raleigh, attacked the Spanish shipping in the summer of 1585. Raleigh captured a number of vessels and took 600 prisoners back to England.[36] A similar raid was planned by Sir John Hawkins and Sir Francis Drake in 1587.[37] Raleigh and Drake wreaked such havoc upon the Spanish shipping in 1585 that the enemy hesitated to send any vessels to Newfoundland in the following year.[38] However, Spain was unwilling to suffer the loss of its position in the fishery without striking back. For several years the English fishing fleet returning from Newfoundland had to run the gantlet of Spanish ships of war which lay off the mouth of the Channel, seeking to intercept it. In the summer of 1591 thirty or forty of the enemy's ships were reported as cruising between the Scilly Islands and Ushant, and in the following year twenty ships were occupying the same station.[39] In 1594 Spain was so active that Sir Walter Raleigh expected that the entire Newfoundland fishing fleet of over one hundred sail would be captured unless steps were taken to drive the Spaniards from the coast. He felt that "if those should be lost it would be the greatest blow ever given to England."[40] During the last years

---

1585; proclamation by the Queen to the inhabitants of Newfoundland engaged in the fisheries there, same date, *Cal. State Paps., Dom.*, 1581-1590, pp. 28, 246.

[36] Privy Council to Sir John Gilbert, Oct. 10, 1585, *ibid.*, 1581-1590, p. 302.

[37] Translation of a statement furnished to the Queen of England by Francis Drake and John Hawkins relative to undertaking a voyage to ruin the Spaniards entirely, 1566, *Calendar of State Papers, Spanish Series*, 1587-1603, pp. 20-21. Cited hereafter as *Cal. State Paps., Span.*

[38] Advertisements out of Spain, Jan. 16, 1586, *Cal. State Paps., Dom.*, 1581-1590, p. 30.

[39] Two letters from the Privy Council to Sir Henry Palmer, Admiral at Plymouth, July 21, 1591, *Acts of the Privy Council, Domestic Series*, 1591, p. 303. Cited hereafter as *Acts, Priv. Coun., Dom.* Sir Walter Raleigh to the Lord [High] Admiral [Lord Howard of Effingham] Aug.? 1592, *Cal. State Paps., Dom.*, 1591-1594, p. 265.

[40] Raleigh to Sir Robert Cecil, July 20, 1594, Historical MSS. Commission,

of the sixteenth century and the early ones of the new epoch the Spaniards continued to threaten the English fishing fleet, but their efforts were sporadic because of the necessity of using their warships in other naval enterprises.[41]

The Spanish fishery at Newfoundland had been established at about the same time as that of the English, but it never received much encouragement from the government, and suffered from the restrictions which the crown imposed upon it from time to time. The Spanish fishermen were also the victims of the almost continual warfare with France and England. Besides the attacks of the enemy, they were never certain whether they would receive permission from the government to go to Newfoundland. Sometimes they were able to set forth, but often they were held back in order that the press gangs might do their work. On one occasion in 1602 the Spanish Newfoundland fleet was held in port for the purpose of transporting men for Philip's proposed invasion of Ireland.[42] The effect of the attacks, retaliations, and restrictions was to weaken Spain's fishery and to strengthen those of England and France. The Portuguese, then under the rule of Philip, lost their fishery to Spain's enemies just as they lost their colonies during this period. Although Spanish and Portuguese ships did not disappear altogether from Newfoundland waters, after 1610 England and France divided the control of the fishery

Calendar of the Manuscripts of the Marquis of Salisbury, IV, 566. Cited hereafter as *Salisbury MSS*.

[41] Sir John Gilbert to Cecil, July 15, Sept. 21, 1601, *Cal. State Paps., Dom.*, 1601-1603, pp. 68, 100.

[42] The confession of John Hill of Stonehouse, Mar. 19, 1596, *Salisbury MSS.*, VII, 123; William Stallenge to Cecil, Mar. 22, 1599, *ibid.*, IX, 111; A. White (?) to Robert Meagh, Lisbon, Apr. 17, 1602, *Cal. State Paps., Dom.*, 1601-1603, p. 176; ——— Pilgrim to Lord [High] Admiral Nottingham, St. Jean de Luz, Apr. 19, 1602, *ibid.*, p. 178.

THE BEGINNINGS OF THE FISHERY 31

between them. The English succeeded the Portuguese in the occupancy of the southeastern coast between Cape Race and Cape Bonavista, while the French established themselves on the southern, western, and northern shores of the island.[43] A few Spanish Basques continued to frequent the region in search of whales for many years to come, but from 1610 onward Spanish and Portuguese participation in the codfishery was infrequent. By the second decade of the seventeenth century England and France had discovered that their largest foreign markets for cod were to be found in Spain and Portugal.[44] The elimination of these countries from the Newfoundland fishery not only opened a valuable market for the fish caught by English and French fishermen, but also contributed directly to growing commercial competition between England and France.

During the first century of the Newfoundland fishery the English carried on warfare and piracy against the French fishermen. The first mention of the capture of a French vessel returning from Newfoundland occurs in 1524.[45] The religious wars in France gave Englishmen an opportunity to prey upon the commerce of their ancient rivals and to profit from their distress. In 1544, when war was contemplated with France, the Privy Council considered, but eventually abandoned, a proposal to attack the French ships returning from the Newfoundland

[43] The French and English fished together at St. John's as late as 1596. Report of Richard Clarke and others, 1596, *Calendar of State Papers, Colonial Series*, 1574-1660, p. 4. Cited hereafter as *Cal. State Paps., Col.* Privy Council to the Lord Deputy of Ireland, July 12, 1613; order in council, May 10, 1614, *Acts of the Privy Council, Colonial Series*, 1613-1680, pp. 3-4, 8. Cited hereafter as *Acts, Priv. Coun., Col.*

[44] Stock, *Proceedings and Debates*, I, 11; Innis, "Spanish Fishery," *passim*.

[45] Notice of the capture of a French vessel, Jan., 1524, Biggar, *Precursors*, pp. 163-164.

Banks.⁴⁶ The West Country ports often seized French vessels returning from the banks, brought them into port and sold their cargoes for what they would bring.⁴⁷ Seizures were made on the pretense that the vessels were owned by French merchants partial to the Catholic League.⁴⁸ But such claims did not prevent the West Country merchants from selling supplies, including Newfoundland fish, to the enemies of Henry of Navarre.⁴⁹ Isolated and sporadic though these attacks may have been they are indicative of the national rivalry of England and France, which was destined to increase in the North Atlantic fisheries as well as in other branches of commerce and colonization during the seventeenth and eighteenth centuries. The West Country fishing interests had heeded the advice of Anthony Parkhurst and were seeking to extend their fishery at the expense of other nations.⁵⁰

Dutch competition in the fisheries surrounding the British Isles was an incentive to the development of an English fishery in North American waters. During the

⁴⁶ The council with the king to the council with the queen, Sept. 26, 1544, *Letters and Papers, Foreign and Domestic, Henry VIII*, 1544, Part II, 159. Cited hereafter as *Lets. and Paps., Henry VIII*.

⁴⁷ Complaints by Frenchmen, Dec. 19, 1542, *Lets. and Paps., Henry VIII*, 1542, §§738, 1220. Petition of Roger Poitow to the council, Aug.?, 1577, *Cal. State Paps., Dom.*, 1547-1580, p. 554; petition of Nicholas Vincent, 1547, *ibid., Addenda*, 1580-1625, p. 230; Bernard de Laude to the council, June 6, 1592; documents relating to the seizure of the *Holy Ghost* of St. Jean de Luz, July?, 1592; Privy Council to the mayor, etc., of Bristol, May, 1593, *ibid.*, 1591-1594, pp. 231, 248-251, 351.

⁴⁸ Privy Council to Sir John Gilbert, Richard Campernone, etc., July 29, 1591, *Acts, Priv. Coun., Dom.*, 1591, p. 345; the same to Sir Francis Drake, Sir John Gilbert, etc., June 10, 1593, *ibid.*, 1592-1593, pp. 303-304.

⁴⁹ Privy Council to Sir John Gilbert, Aug. 15, 1591, *Acts, Priv. Coun., Dom.*, 1591, p. 390.

⁵⁰ Parkhurst to Hakluyt, Nov. 13, 1578, Hakluyt, Extra Series, VIII, 9-16; his account of advantages arising from encouraging traffic to Newfoundland [1578], Lansdowne MSS., 100, no. 10, fols. 95-97.

THE BEGINNINGS OF THE FISHERY

latter part of the sixteenth century the English were gradually excluded from their own coastal waters, and the Dutch not only sold a great deal of fish in the English market but also bought English-caught fish there and shipped it abroad in foreign bottoms. England made several attempts to check a competition which threatened the ruin of an important domestic industry, but most of the legislative expedients devised to oust the Dutch proved utterly ineffective.[51] The English industry was so crippled during the sixteenth century owing to the competition of Holland and to the decline in the consumption of fish after the Reformation, that parliament enacted laws requiring the eating of fish in Lent and appointing certain days of the week as fish days.[52] There was no religious incentive behind the passage of this legislation, but the consumption of fish was to be encouraged for the purpose of strengthening the economic position of the state. The importance of the fisheries in the national economy is well illustrated by an early act of the reign of James I which set forth the principle that the fishing industry was the basis of the national mercantile and naval marine, being an important training school for sailors. This statute also emphasized the decline of England's domestic

---

[51] "The bill concernyng bying fishe upon the See," 1541-1542, 33 Henry VIII, c. 2; "An Acte for the Shipping in Englishe bottoms," 1558-1559, 1 Eliz., c. 13; "An Acte touching certayne Politique Constitutions for the maintenance of the Navye," 1562-1563, 5 Eliz., c. 5. This last act was repealed by "An Act for the mayntenance of the Navygation," 1571, 13 Eliz., c. 11. In 1580-1581, "An Acte for the encrease of Mariners & for the Maintenance of Navigation" was passed. It prohibited the importation of foreign cured fish, but specifically exempted that from Iceland, Shetland, Newfoundland, etc., 23 Eliz., c. 7. This law was succeeded by "An Acte for the encrease of Mariners and maintenance of Navigation . . .," 1597-1598, 39 Eliz., c. 10. Cf. Stock, *Proceedings and Debates*, I, 1, 3-4, 7.

[52] "An Acte touching the abstynence from Fleshe in Lent . . .," 2 & 3 Edward VI, c. 19; see also 5 Eliz., c. 6, clause 23; Capt. Robert Hitchcock to Cecil [1595], *Salisbury MSS.*, XIII, 556.

fishery in recent years.[53] The idea that the fishery was a
"nursery" or "seminary" for the training of sailors,
was accepted, and henceforward became a cardinal principle of English policy with respect to Newfoundland.

In the meantime, none of the legislation designed to
strengthen the declining industry proved effective. The
Dutch continued to be active in the coastal fisheries of
England, and English merchants found it more profitable
to sell Newfoundland fish to foreigners than to dispose of
it at home. During Elizabethan times there does not seem
to have been any extensive demand for Newfoundland
fish in England. "Newland" fish was sometimes used to
provision the fleet and the army in Ireland or to feed the
poor in England.[54] The greater part of the annual catch
of "Poor John" or "Poor Jack" was sold to foreigners
for shipment to southern Europe. Dutch, Flemish, and
French vessels called regularly at the ports of the West
of England after the return of the Newfoundland fleet.
Most of the shipments went first to France, whence they
were reëxported to Spain.[55] Irish ships also carried Newfoundland cod from the West Country to Ireland where it
was consumed by the Roman Catholic population. On at
least one occasion an Irish vessel carried a cargo to An-

[53] "An Acte to encourage the Seamen of England to take Fishe . . .,"
1603-1604, 1 James I, c. 29.

[54] References to the use of Newfoundland fish to provision the fleet during Elizabeth's reign are found in the following: *Cal. State Paps., Dom.*,
1581-1590, p. 593; *ibid., Addenda*, 1580-1625, p. 360; *Salisbury MSS.*, V,
387, 418; VI, 337; VIII, 415; XI, 297; Sir Walter Raleigh to Sir John Gilbert, Dec. 30, 1591, Huntington Library, HM 21,227. References to its use
in Ireland are found in the *Calendar of State Papers, Ireland*, 1574-1585,
pp. 256, 260, 268, 270, 281, 284; 1592-1596, pp. 450-451; 1599-1600, p. 12;
1600-1601, pp. 209-210, 217; 1601-1603, p. 388. Cited hereafter as *Cal. State
Paps., Ire.* See also *Cal. State Paps., Dom.*, 1601-1603, p. 84; and *Salisbury
MSS.*, IX, 345.

[55] [William Stallenge, mayor of Plymouth] to Cecil, Sept. 23, Oct. 15,
1595, Aug. 15, 1596; John Jefferey, mayor of Southampton to Cecil, Oct. 30,
1698, *Salisbury MSS.*, V, 386-387, 418; VI, 337; VIII, 415.

dalusia and the Basque Provinces. Nicholas Weston, mayor of Dublin, was the leader in this enterprise, as he was also the first to send Irish ships on fishing voyages to Newfoundland about 1600.[56]

Although the merchants of the West of England had been concerned with the Newfoundland fishery from the earliest times, they did not become interested in carrying their fish to foreign markets in English bottoms until after 1610. As long as they were content to permit foreigners to buy fish and ship it to southwestern Europe in alien vessels, the trade was necessarily roundabout and the profits accruing to English traders were small. However, after 1610 direct voyages were undertaken from Newfoundland to the foreign markets in increasing numbers. The development of this new method of handling the products of the fishery had a profound effect upon the fortunes of the western adventurers, as well as upon the development of the general carrying trade in the North Atlantic. During the period previous to 1610 the English fishery at Newfoundland was in a formative state. The nation was too preoccupied with other and more pressing matters to pay much attention to the development of these distant fishing grounds. Aside from the premature scheme of Parkhurst for the reorganization of the fishery and the establishment of an English colony in the island, and Gilbert's unfortunate attempt to carry out such a project, there were no organized efforts made previous to 1610 seeking to benefit the nation as a whole. During the early years what work was accomplished by English-

[56] Robert Eastfield to Burghley or Cecil, Dec. 20, 1596, *Salisbury MSS.*, VI, 530; Sir G. Fenton to ———, Jan. 12, 1596, *Cal. State Paps., Ire.*, 1592-1596, pp. 450-451; "Copy of a plot for furnishing the provant apparel, by Nicholas Weston, then mayor of Dublin and five others . . .," *ibid.*, 1598-1599, p. 296; an open warrant to the mayor, etc. of Poole, Jan. 28, 1601, *Acts, Priv. Coun., Dom.*, 1600-1601, pp. 132-133; Gerald Young, mayor of Dublin, to Cecil, Sept. 10, 1600, *Cal. State Paps., Ire.*, 1600, p. 418.

men in Newfoundland was done by the individual initiative of the merchants, shipmasters, and fishermen of the counties of Devonshire and Cornwall and a few nearby places such as Bristol, Southampton, and small ports in Dorsetshire. The localization of the industry in southwestern England was probably the result of the oversea extension of the coastal fisheries of that region. The regional character of the Newfoundland fishery played an important part in the events which followed during the seventeenth and eighteenth centuries.

The first definite project looking toward the exploitation of the fishery along other lines than those followed by the West Country groups was undertaken in 1610, when the London and Bristol Company was chartered for the purpose of carrying on trade and establishing a plantation in Newfoundland. The possibility of improving England's trade with Spain was the incentive which brought the transatlantic fisheries into greater prominence and interested a wider circle of Englishmen in them. The necessity of developing trade in Newfoundland and New England fish was one of the reasons urged during the reign of James I for concluding peace with Spain.[57] English merchants, then under the influence of bullionist theories, sought an opportunity to augment the wealth of the nation and of themselves by increasing the sale of English commodities in Spain and by controlling the carrying trade to that country. Codfish, which was an important staple of diet in the Iberian peninsula, offered possibilities in that direction. The merchants of London particularly were imbued with the idea of controlling the carrying trade by setting up monopolies in the several branches of foreign commerce in which the English people were interested at the time. They were not always successful, but their attempts to put into effect plans for

[57] Stock, *Proceedings and Debates*, I, 27.

THE BEGINNINGS OF THE FISHERY   37

controlling the Newfoundland and New England fisheries and the associated carrying trades profoundly influenced the early development of those regions. Especially in the case of Newfoundland the attempts to erect monopolies had far-reaching consequences.

After 1610 there was a marked tendency among the merchants to send fish directly from the fishing grounds to the markets of southwestern Europe. For some time a considerable part of the annual catch still continued to be carried away in foreign bottoms. Whereas previously foreign vessels had called at the ports of southwestern England after the return of the fishing fleet, they now went directly to Newfoundland. Besides the English vessels, French, Flemish, Dutch, and Hamburg ships called there for cargoes at the end of the fishing season.[58] The West Country fishing interests were indifferent as to the nationality of the carriers employed, and disposed of their catch to compatriots and foreigners alike. It is likely that the Dutch, always alert in shipping matters, introduced direct voyages from Newfoundland to Spain and the Mediterranean as early as 1593.[59] In England, the London, Bristol, Exeter, and Southampton merchants were more keenly alive to the advantages of direct trade with Spain in English ships than were most of the western adventurers. The merchants of these ports took the lead in urging a more systematic organization of the nation's carrying trade and the reorganization of the transatlantic fisheries.

[58] Astrid Friis, *Alderman Cockayne's Project and the Cloth Trade. The Commercial Policy of England in its Main Aspects, 1603-1625*, p. 163. Cited hereafter as Friis, *Cockayne's Project*. See also the petition of Robert Barker to the Duke of Buckingham, May 24, 1627, *Cal. State Paps., Dom.*, 1627-1628, p. 188; note of seven Hamburg ships that went to Newfoundland, 1627 [Jan. 19, 1628], *ibid.*, p. 522.

[59] Stock, *Proceedings and Debates*, I, 7; Friis, *Cockayne's Project*, pp. 178-179.

The first attempt of the Londoners and their associates to gain control of the trade to and from Spain was inaugurated when Sir Edwin Sandys, William Cockayne, and others obtained a charter for a company trading to France, Spain, and Portugal. They hoped to establish a monopoly of England's commerce with these countries, but were chiefly interested in Spain and her dominions. Because of the powerful opposition of the independent traders, particularly those of the West of England, the Spanish Company did not secure the desired monopoly, and the trade remained open to all. A monopoly would have obliged English importers and exporters dealing with southwestern Europe to receive and dispose of their goods through the company, and also to transport their articles of commerce in ships either belonging to the company or chartered by it. No longer would it have been possible to dispose of Newfoundland fish to foreigners calling at West Country ports and the monopoly would have prevented the western adventurers from making direct market voyages.[60]

Unsuccessful though they were in obtaining the desired monopoly of the Spanish trade, the London merchants, cognizant of the fact that New England and Newfoundland fish were important items in England's exports to Spain, sought to capture the transatlantic fisheries. During the first half of the seventeenth century they made persistent efforts to dominate them. The first enterprise of this character which affected Newfoundland was the plantation scheme undertaken by the London and Bristol, or Newfoundland, Company, which was chartered in 1610.[61] This joint-stock organization, composed of merchants of London and Bristol, with a sprinkling of

[60] Friis, *Cockayne's Project*, pp. 131-223, *passim*, but especially p. 163.
[61] Charter of the Newfoundland Company, May 2, 1610, Pat. Rolls, 8 James I, pt. VIII, no. 6; T. Carr, *Select Charters of Trading Companies, 1530-1707* (Selden Society), XXVIII, 51-62.

nobles and gentry, sent out a number of settlers under the leadership of John Guy, a Bristol merchant, in 1611. Guy had been the active organizer of the company, and he was to become the aggressive governor of the new settlement and the bitter enemy of the western adventurers.

Ostensibly the objects of the company were the same as those of the many other trading companies organized at the time. Actually, the intention of the group was from the first to secure a foothold at Newfoundland, and with the plantation as a base, eventually to bring the fishery completely under its control. This early attempt at settlement, although in itself to prove none too successful, introduced serious complications. The arrival of the settlers under Guy's leadership precipitated the first of many difficulties between the planters and the fishing interests of the West of England. Henceforward each group strove to control the fishery and to oust the other, bequeathing to future generations of planters and fishermen a heritage of hatred, violence, and mutual jealousy.

The possibility of conflict between the two groups had been anticipated to some extent. A clause was introduced into the company's charter which recognized the ancient fishing privileges of all nations and expressly excluded the patentees from jurisdiction over the fishery.[62] This provision was probably introduced not so much out of consideration for the feelings of the western adventurers as for the purpose of preventing Guy and his associates from becoming involved with the French and other foreign fishermen. In spite of the reservation, Guy issued orders shortly after his arrival which he expected the fishermen to observe, wherein he asserted his right as governor to jurisdiction over the fishery.[63] There was

---

[62] Charter of the Newfoundland Company, clause 7.
[63] Certain orders for the fishermen to observe . . ., published by John Guy, Aug. 13, 1611, C.O. 1:1, no. 40 (i).

some legal justification for this attitude because a good deal of the work was carried on ashore. The land was used for cleaning and curing the fish and making train oil. Stages, cookrooms, storerooms, train-vats, and other necessary buildings occupied the shore front in all the principal harbors. Timber and firewood were procured in the nearby woods, and sea-birds for use as bait were hunted ashore. Even the caplin, also used as bait, was seined in the shallow inshore waters. Extra boats, provisions, and supplies of all sorts were stored on the island at the end of the season to await the return of the fishing ships in the spring. Sometimes caretakers wintered there, charged with protecting this property. Consequently, with the arrival of a governor in charge of a group of permanent settlers, there were ample grounds for defining and limiting the activities of the fishermen, who heretofore had freely enjoyed all the facilities which the island offered. The jealousies of landsman and seaman cannot account altogether for the continual bickering and violence which resulted after permanent settlements had been made on the island. The antagonism was largely due to the mutual recognition by both groups of the fact that the other was an industrial rival operating under a totally different type of organization and employing different methods in the production of fish.

Guy's laws struck at the heart of the vested interests of the western adventurers. The West Country fishermen were forced out of some of the best fishing grounds by the new settlers. The governor ignored the old custom that permitted the first ship arriving in a harbor to have first choice of the fishing grounds, stages, and other equipment, and, like Gilbert, disposed of the best places to his favorites. The settlers prevented the fishermen from taking sea-birds for bait, and appropriated for their own use the provisions and equipment left during the

THE BEGINNINGS OF THE FISHERY 41

winter. The western adventurers complained that during the best part of the fishing season the planters summoned the fishermen to a court of admiralty, and fined them in fish or oil if they failed to appear. The fishermen also accused the planters of harboring and encouraging pirates, who attacked them and stole their fish and train oil.[64] Occupancy of the land during twelve months of the year gave the settlers an advantage over the West Countrymen who were only on the spot for about two or three months. The conflict was inevitable because the two groups were competitors from the start.

The colonists were not the only ones at fault, for according to complaints which the Newfoundland Company laid before the Privy Council, the West Country fishermen used violent methods to oust their rivals. In 1613 the planters complained of losses estimated at over £20,000, which they had suffered as the result of piracy. Anxious to restore order in its colony the Newfoundland Company asked for permission to send out an armed vessel to guard the coasts, and proposed that its expenses should be paid by the fishing fleet. The company also asked to be granted a commission from the High Court of Admiralty for the trial of pirates.[65] The complaints and mutual recriminations led to an official investigation, and in 1615 Captain Richard Whitbourne was commissioned by the Admiralty to investigate the abuses and disorders committed at Newfoundland. He was authorized to impanel juries, and was instructed to make recommendations as to the improvement of conditions there. He carried on the investigation at his own cost, and submitted certain pro-

---

[64] Articles of grievance of the western ports, no date, C.O. 1:1, no. 39; the Earl of Bath to the Privy Council, Oct. 19, 1618, inclosing petition of the merchants of Devon, *Cal. State Paps., Dom.*, 1611-1618, p. 586; order of the Star Chamber, Nov. 4, 1618, *Acts, Priv. Coun., Col.*, 1613-1680, §31.

[65] Privy Council to the judge of the admiralty, July 22, 1613, *Acts, Priv. Coun., Col.*, 1613-1680, §4.

posals which he later asserted were overlooked, so that his inquiry did not produce the good effects anticipated. As a matter of fact Whitbourne was not an impartial investigator. He had already been employed by the company and was committed to the project of colonizing the island in the interests of improving the fishery.[66] His partiality probably accounts for the failure of the crown officers to accept his recommendations.

The conflict between the two interests continued to rage for several years. In 1618 and 1619 the whole question was reviewed by the Privy Council. The principal point of dispute concerned the right of occupying fishing places. The planters considered it lawful to choose their own locations "and not leave the benefit thereof to the uncertain comers thither." They denied that they had ousted any fishing ships from the harbors they occupied, appropriated property belonging to the fishermen, or laid exactions upon them. The company assured the western adventurers of its willingness to permit the fishermen to hunt birds, and stated that it would issue orders to that effect to the planters. The Newfoundland Company accused the fishing interests of making heavy demands upon the planters, denied that its settlers had aided pirates, but asserted that their colony had been badly damaged by marauders, whom they insisted were furnished with provisions by the fishermen. The patentees maintained that they were anxious to coöperate with the western adventurers in suppressing piracy and in maintaining order, and protested that it was for this purpose that Guy had issued laws for the government of the fishermen,

---

[66] Richard Whitbourne, *A Discourse and Discovery of the Newfoundland* (London, 1620), p. 15; his "Voyages to Newfoundland, etc. . . .," in Samuel Purchas, *Hakluytus Posthumus, or Purchas His Pilgrimes* (Maclehose Edition), XIX, 427; and his "A Relation of Newfoundland," *ibid.*, p. 436. Cited hereafter as *Purchas his Pilgrimes*.

THE BEGINNINGS OF THE FISHERY

but they had refused to obey him.[67] In rebuttal the fishing interests insisted that the company's charter gave the planters no jurisdiction over the fishery, but on the contrary expressly reserved the ancient fishing rights to themselves. The colonists had no right to assume any precedence in the fishery, but if any such privilege was allowed it should be granted to the adventurers as they were the first Englishmen to discover Newfoundland and to engage in fishing there. The adventurers opposed abandoning the old custom whereby the ships took up their fishing grounds in order of arrival in the various harbors, as a discontinuance of the practice would lead to further complications. The West Country people, therefore, asked the Privy Council to confirm their old privileges as expressly reserved in the charter of 1610, and demanded that anyone violating the rules of the fishery should be sent to England for trial. The adventurers felt that they understood the management of the business better than did the planters and were therefore unwilling to submit to the jurisdiction of the colony. The documents which they submitted to the council show that the West Countrymen were perfectly aware that the plantation was being used by the patentees in an effort to gain the mastery of the fishery.[68]

These grievances were taken into consideration by the Privy Council in December 1618. Its attitude was on the whole sympathetic to the western adventurers, and the Newfoundland Company was enjoined from doing anything disadvantageous to the fishery "upon pain of fit punishment."[69] As the council failed to provide any gov-

[67] The reply of the Newfoundland Company to the charges of the fishermen, 1618?, C.O. 1:1, no. 40.

[68] The reply of the [West Country] petitioners to the answer of the governors of the plantation, Dec. 1618, C.O. 1:1, no. 41.

[69] Order in council, Dec. 13, 1618, *Acts Priv. Coun., Col.*, 1613-1680, §33.

44   BRITISH FISHERY AT NEWFOUNDLAND

ernmental machinery for enforcing this order, and as none existed at the time, the rivalry of the two groups went on unchecked. This order in council is the first of many similarly inadequate and unenforcible rulings with respect to Newfoundland.

About the time that the Privy Council gave its lukewarm support to the cause of the western adventurers, the Newfoundland Company, through the efforts of Captain John Mason, obtained the financial assistance of a group of Scottish merchants. Reinforced by new financial backing, in March 1620 it again protested against the disorderly conduct of the fishermen and the injuries inflicted upon its settlement by pirates. At this time the company showed its hand more openly than before, and besides asserting that a very profitable national trade would be ruined entirely if the abuses committed by the fishermen did not cease, it made a bid for complete control over the fishery by requesting that its patent be amended so as to give it jurisdiction over both English and foreign fishermen. Furthermore, it asked that Captain John Mason, Guy's successor as governor of the plantation, be created a lieutenant of the king and furnished with two or more ships to guard the coast and preserve order. If such powers were conferred upon the company and its governor, the monopolists maintained that the island could be settled upon a substantial basis and a catastrophe in the fishery averted.[70]

The company's request was referred to a committee of council, composed of the duke of Richmond and Lenox, the earl of Arundell, Sir Henry Carey, afterward Vis-

---

[70] Petition of the treasurer and company, with the Scottish undertakers [Mar. 16, 1620], C.O. 1: 1, no. 54; reasons to move the king to order that a lieutenant be sent yearly to Newfoundland, etc., no date, C.O. 1: 1, no. 54 (i). For a discussion of Scottish participation see George Pratt Insh, *Scottish Colonial Schemes, 1620-1686*, pp. 27-39. See also John Ward Dean, ed., *Captain John Mason* (Prince Society), *passim*.

THE BEGINNINGS OF THE FISHERY 45

count Falkland, Sir George Calvert, Henry, Lord Hay, afterward the earl of Carlisle, and Sir Thomas Edmunds or any four of them. Whitbourne's *Discourse and Discovery of the Newfoundland,* wherein he advocated the settlement of the island, was also submitted for consideration to this group. The committee was instructed to consider the company's petition and to call representatives of the two interests for consultation.[71] The committee failed to approve the radical changes which the company sought to obtain in its charter, but otherwise its recommendation to the Privy Council was favorable to the plantation. Acting upon this recommendation, the council sent letters to the mayors of the principal western ports charging them to caution the shipmasters not to interfere in any way with the planters. The council also expressed the opinion that encouragement of the plantation would prove beneficial to the fishing interests, and furthermore, informed the western adventurers that the fishermen would be held strictly accountable if they disregarded the order.[72] Thus, within a few years the crown had reversed its attitude from one of lukewarm support of the western adventurers to an equally half-hearted stand in favor of the company. The fact that Calvert and Falkland, both members of the committee of council which made a recommendation favoring the company, were purchasers of parcels of the original grant of the Newfoundland Company at about this time is not without significance.[73]

[71] Order of the Star Chamber, Feb. 14, 1620, *Acts, Priv. Coun., Col.,* 1613-1680, §43.

[72] J. D. Rogers, *Historical Geography of Newfoundland,* being Part IV of Vol. V of *The Historical Geography of the British Colonies,* edited by Sir Charles Lucas (Oxford, 1911), pp. 59-68. The account in D. W. Prowse, *History of Newfoundland* (2d ed., 1893), pp. 109-121, is inaccurate.

[73] Privy Council to the mayors of the western ports, Mar. 18, 1620, C.O. 1:1, no. 49.

The struggle with the West Country group and the difficulties of founding a new colony kept the London and Bristol Company continually in need of financial assistance. When it failed to obtain the desired changes in its charter which would give it control of the fishery, the company decided to raise new capital by disposing of large tracts of land. The policy which it adopted after 1620 was directly in line with that followed by Sir Ferdinando Gorges and the New England Council during the same period. Sir William Alexander and his Scottish supporters also sought to establish a colony in Nova Scotia along similar lines. In all cases the scheme of creating feudal domains and disposing of them to persons favorable to their cause was followed. Although not identical in personnel, all these associations had a common purpose, the control of the fishery adjacent to their particular field of colonization. In both New England and Newfoundland the policy of excluding independent fishermen was tried, but Gorges' group was more successful in keeping out "interlopers" than was the Newfoundland Company and its grantees, who had the well-established western adventurers to oppose them. The plan of making extensive land grants to private individuals, besides being in keeping with the feudal spirit of the age, relieved the patentees of the responsibility of making settlements, except as they too were privately interested in grants, and also brought in added capital wherewith the companies sought to improve their business by concentrating exclusively on fishing and trade.

The Newfoundland Company disposed of large tracts of land to Sir George Calvert, Viscount Falkland, and Sir William Vaughan. Calvert and Vaughan were more seriously interested than Falkland, and actually founded colonies in the island. Vaughan, a Welsh bard, established a plantation at Trepassey called "Cambrioll Col-

THE BEGINNINGS OF THE FISHERY 47

chos" where he spent a good deal of time in singing the praises of Newfoundland and the glories of English trade in verse, and in hurling poetic anathemas at the West Country fishermen who annoyed his settlement. Vaughan was too much of a dreamer to become a successful leader of a colonization enterprise and did not remain long in Newfoundland. In 1631 Robert Hyman, governor of Vaughan's plantation, attempted to arouse new interest in the place, and it is possible that his writings about Newfoundland interested the group which several years later backed Sir David Kirke in his attempted settlement of the island. Falkland advertised for settlers in Ireland but never went to Newfoundland himself. At the same time Sir William Alexander tried to found a Scottish colony in Acadia. This was but another link in the chain which the London merchants and their associates in Edinburgh, Bristol, and Plymouth sought to forge around the North Atlantic fisheries.[74]

Sir George Calvert was the most conscientious of all the grantees of the Newfoundland Company. Not content to remain a mere tenant, he secured royal confirmation of his holdings, receiving a grant by royal letters patent in 1622 and a regrant in 1623 of a strip of land extending across the southeastern peninsula from Ferryland to Placentia Bay, which he called the Province of Avalon. Calvert was much more concerned with establishing a colony than with assisting the Newfoundland Company to extend its control over the fishery. The Avalon charter contained a clause recognizing the right of all English subjects to fish at sea and in the ports and creeks of the province, and permitting them to salt and dry fish on the shore. These privileges were to be exercised by the fishermen without inflicting any injury or loss upon the pro-

[74] See note 72 above. A more recent account is found in *The Cambridge History of the British Empire*, VI, *Canada and Newfoundland*, pp. 127-128.

prietor or the inhabitants of Avalon.[75] Calvert was earnest in his desire to colonize Avalon. While he was still only a tenant of the Newfoundland Company he appointed Captain Edward Wynn governor of his grant in 1621. The proprietor went to Newfoundland in 1627, but returned for his family in the following year. He resided at Ferryland until 1629, when he left for Virginia.[76] His people there enjoyed more pleasant relations with the fishermen, probably because they did not attempt to compete in the fishery. Calvert took the fishermen's part against pirates and French raiders, but his religion always made him suspect.[77] It is significant that he quarrelled with the London merchants regarding prizes taken from the French at Trepassey.[78] The climate and soil of Newfoundland were probably too inhospitable for the type of colony which Calvert sought to establish.

If the London and Bristol Company expected to realize very much from the grants made to Falkland, Vaughan, and Calvert, it was doomed to disappointment.

[75] Grant of the province of Avalon to Sir George Calvert and his heirs, Dec. 13, 1622, Pat. Rolls 20 James I, pt. 15, no. 13. Copies of the regrant of Apr. 7, 1623, C.O. 1: 2, nos. 20, 23; B.M., Sloane MSS., 170, fols. 1-14. For the changes and additions made in the charter between December, 1622, and April, 1623, see *Cal. State Paps., Dom.*, 1619-1620, p. 543.

[76] Account of the relation of Sir George Calvert with Newfoundland, 1670, B.M., Sloane MSS., 3663, fols. 24-26. Baltimore intended to go to Newfoundland in 1625, *Cal. State Paps., Col.*, 1675-1676, §138. His religious beliefs got him into trouble after he reached Ferryland. Examination of Erasmus Sturton, late preacher to the colony at Ferryland, Oct., 1628, C.O. 1: 4, no. 59.

[77] Baltimore to Charles I, Aug. 19, 1629, C.O. 1: 5, no. 27; Charles to Baltimore, Nov. 22, 1629, C.O. 1: 5, no. 39.

[78] State of the case between Lord Baltimore and the merchants of London, 1628, C.O. 1: 4, no. 64. Cf. Sir Francis Cottington to Lord Treasurer Weston, Dec. 19, 1629, *Cal. State Paps., Col.*, 1574-1660, p. 94; petition of William Peasley on behalf of Baltimore to the Admiralty, Dec.? 1628, *ibid.*, p. 94; warrant for a privy seal to deliver one of the prize ships to [Leonard Calvert], [Dec.], 1628, *ibid.*, p. 95; memorandum by Sir Joseph Williamson that Baltimore was at Ferryland in 1628, *ibid.*, p. 95.

THE BEGINNINGS OF THE FISHERY 49

Not one of these proprietary projects was either successful or permanent. The net result in colonization was to leave a few more permanent residents in Newfoundland, whose descendants remained in conflict with the West Country fishing interests for several generations. Moreover, during the second decade of its existence the company's activities were much reduced, but it continued to function as late as 1628, though after its failure to secure the desired monopoly of the fishery and carrying trade, it was too weak to maintain its former aggressive attitude toward the western adventurers. During its last years it attempted to recoup some of its losses through the sale of land to other proprietors, and tried to develop the manufacture of bar iron by seeking to import ore from England. Some hope was expressed that iron and silver might be found, the exploitation of which would supplement the income derived from fishing, trapping, and the making of sarsaparilla.[79] Apparently no serious attempt was ever begun to prospect for metals, though curiously enough the largest single deposit of iron ore in Newfoundland—the Wabana field—lies on Bell Island in Conception Bay only a short distance from the settlements founded by Guy. After 1628 nothing more is heard of the London and Bristol Company, and the enterprise gradually dwindled away.

During the years when the Newfoundland Company was most active, the West Country interests became alarmed at the persistence of the Londoners and their associates in attempting to monopolize the North Atlantic fisheries, and sought relief in legislation. The counties of Dorset, Somerset, Devon, and Cornwall which were

[79] Reasons for the exportation of iron ore, etc., into Newfoundland, Apr. 11, 1620, C.O. 1:1, no. 50; *Acts, Priv. Coun., Col.*, 1613-1680, §53; Dr. James Meddus to Viscountess Conway, June 27, 1628, *Cal. State Paps., Dom.*, 1628-1629, p. 180; *Cal. State Paps., Col.*, 1574-1660, p. 92.

most actively interested in the transatlantic fisheries had a large and active representation in the House of Commons. In the session of February-March, 1621, a bill was introduced proposing to relieve the western fishermen of the tithes exacted by the ecclesiastical authorities. Apparently the bill died in committee, but it is significant as being the first of a series of attempts to improve and strengthen the position of the western adventurers.[80] The West Country people tried continuously from 1621 to 1628 to obtain the passage of legislation which would establish the "free" fishery on a statutory basis and prohibit monopolies such as that which Sir Ferdinando Gorges was attempting to enforce on the coast of New England or that which Guy's group desired at Newfoundland. On April 17, 1621, "An Act for the freer liberty of fishing and fishing voyages to be made and performed in the seacoast and places of Newfoundland, Virginia, New England, and other seacoasts and parts of America" was introduced in the House of Commons. The bill proclaimed that the free right of engaging in the fishery tended greatly to increase the number of ships and seamen, but that owing to recent attempts to interfere with the freedom of the fishermen, it should hereafter be provided that all English subjects should have the free right to fish on the shores of America, to select places for curing fish according to priority of arrival, and to take wood for fuel and repairs.[81] The bill passed the third reading on December 1 of the same year, but it was never considered by the House of Lords.[82] Undaunted by this failure the West Country interests tried again in 1624, a similar bill

[80] Stock, *Proceedings and Debates*, I, 25.
[81] Draft of "An Act for the freer liberty of fishing . . . in . . . Newfoundland, Virginia, New England, and other . . . parts of America," Apr. 17, 1621, Historical MSS. Commission, *House of Lords Manuscripts*, IV, 21. Cited hereafter as *House of Lords MSS.*
[82] Stock, *Proceedings and Debates*, I, 30-56, *passim*.

THE BEGINNINGS OF THE FISHERY 51

being introduced in the Commons on February 25. It was amended in committee, and brought up for the third reading on April 29, but consideration was deferred until May 4, when it was finally read and debated, though no vote appears to have been taken. The identical bill seems to have been resubmitted in 1625, when it passed the lower chamber, but again failed to receive consideration from the Lords.[83] The West Country members of the House of Commons persisted in their attempts to pass such legislation, the bill being reintroduced regularly in each session of Parliament until 1628.[84] Friends of the "free" fishery never secured the desired protection by statute, and confirmation of the old fishing customs had to be secured from the crown during the years when Charles I ruled without parliament.

Records of the discussions in parliament relative to these bills are scant, although they reveal the attitudes of the partisans in the fisheries disputes. The major attack was directed against the monopolists in the New England fishery, but the London and Bristol Company was subjected to considerable criticism. In the debates in the House of Commons, John Guy, who was a member of that chamber, aligned himself with the Gorges group. The West Country's usual complaints against the policies and methods of the Newfoundland Company were brought to the attention of the Commons, and the jealousy also of London for the outports manifested itself. The western interests asserted that the London merchants sought to monopolize all the shipping trades, and that they had been particularly annoying to the masters of the fishing ships at Newfoundland. Guy contended that the London

[83] Draft of "An Act for the maintenance and increase of shipping . . . and for the freer liberty of fishing . . . in . . . Newfoundland, Virginia, New England, and other . . . parts of America," May 4, 1624, *House of Lords MSS.*, IV, 123; Stock, *Proceedings and Debates*, I, 57-94, *passim*.

[84] Stock, *Proceedings and Debates*, I, 80-94, *passim*.

merchants were to be commended for their attitude toward the fishing monopoly, "whatsoever their greediness in other things, justly found fault with."[85] He argued that the legislation asked for by the western adventurers would deprive the planters of their right to fish, destroy the settlement, and leave the island open to occupation by foreigners. Both Guy and Calvert felt that this legislation was improper for the Commons to consider because it concerned America. Guy maintained "that the king hath already done by his great seal as much as can be done here by this act." Calvert thought the plantation worth encouraging and doubted whether the bill would receive the royal assent unless the planters were given priority.[86] Guy sought to minimize the effect of the plantation upon the fishery, stating that there were but three "real" settlements in Newfoundland and "never will be above twenty or thirty," while there was room for 500 fishing ships. Thomas Sherwell opposed the plantation as tending to create a monopoly, and expressed the opinion that the independent fishery should be encouraged because it was more important than the colony, bringing in £120,000 annually to the nation and had no exports except provisions. As a result of this discussion, Guy's proposal to amend the bill of 1621 in favor of the plantation was rejected, and the bill passed on December 1. The later discussions down to 1628 were concerned chiefly with the Gorges patent, and the Newfoundland fishery seldom entered into the debates. Supporters of the London and Bristol Company do not appear to have taken active parts in the discussions on bills of this nature introduced later than 1624.[87] The years from 1618 to 1621 mark the period of the greatest aggressiveness on the part of the company, but in parliament as well as out, its influence rapidly declined after 1621.

[85] *Ibid.*, I, 36.   [86] *Ibid.*, I, 55.   [87] *Ibid.*, I, 57-73, *passim*.

THE BEGINNINGS OF THE FISHERY 53

The reason for the stir created over the New England fishery arose not only from the more favorable situation of that region as a place of permanent settlement, but also from the fact that the monopolists were able to entrench themselves more firmly in the more southerly region than was possible at Newfoundland. The failure of the London and Bristol Company was in large part due to its attempt to drive out a well established vested interest, the beginnings of which date far back into the sixteenth century. On the other hand, the patentees of New England developed the fishery there concurrently with the establishment of permanent settlements, and the western adventurers who attempted to fish in that region were the interlopers.[88]

It should be clearly understood that there was no generally unified action on the part of the monopolists to secure control of these fisheries and no attempt to operate them in concert. Though both the New England and Newfoundland patentees had certain ideas in common, the two regions were never confused or regarded as forming part of one transatlantic fishery. The Newfoundland fishery had been organized in the sixteenth century under the auspices of the West Country ports. The managers of this industry were deeply influenced by localistic and medieval conceptions of commerce and based their operations upon ideas which were no longer acceptable in the century to come. On the other hand, the New England fishery was developed in a later age when localism in commerce and industry was gradually giving way to a national conception of economic activity. As yet the government had not adopted a fixed commercial policy, and, therefore, control of the various branches of commerce by private monopolies was the only way in which the new interests could express themselves. More-

[88] *Ibid.*, I, 74-96, *passim.*

over, the Newfoundland fishery had been established by Englishmen in the days before colonization had come to be readily acceptable as a means of controlling the nation's trade. Consequently, there was no tradition in the West of England which looked to the ultimate establishment of colonies overseas. Gilbert's unfortunate attempt to erect a plantation and impose civil government on Newfoundland had created an actual tradition of hostility to all such enterprises, and the Newfoundland Company, Calvert, Vaughan, and Falkland, as well as the planters whom they sent out, reaped a harvest of hatred at the hands of the western adventurers. The suitability of the two regions for colonization was another factor in deciding that the two fisheries should develop along divergent lines. The harsher climate of the northern island and the relative poverty of its soil even in comparison with rocky New England, were elements which played a large part in the history of these regions. It is impossible, except for a brief time, to consider the two fishing regions together, and in later years the differences become so marked as to make comparisons impossible. The failure of the London and Bristol Company to secure control over Newfoundland marks the end of the first phase of the struggle of the western adventurers to maintain themselves in an age when economic ideas and governmental policies with respect to trade and colonization were being revolutionized.

CHAPTER II

THE FIRST WESTERN CHARTER

TWO varieties of codfishery were carried on at Newfoundland from the sixteenth century onward: one for "wet" fish which were caught on the banks; the other for "dry" fish which were caught on the ledges near shore or within the extensive bays and harbors. During the early years the French, Portuguese, and Spaniards fished upon the banks, but from the first the English were principally engaged in the inshore fishery, wherein their rivals, the French also participated. During the sixteenth century ships of various nationalities fished side by side on the same grounds, but by the middle of the following century, nationalism had asserted itself sufficiently to produce a division of the inshore fishery into two zones one occupied by the French and the other by the English. The French zone was by far the larger, consisting of all of the southern, western, and northern coasts from Trepassey or Cape Race around the island to Cape Bonavista. The English occupied a comparatively restricted shoreline extending approximately from Trepassey, or roughly Cape Race, along the eastern shore northward to Cape Bonavista. Within the English zone were to be encountered not only the ships from such West Country ports as Bristol, Bideford, Barnstaple, Falmouth, Truro, Fowey, Plymouth, Dartmouth, Exeter, Weymouth, Southampton, and many others, but also the settlements established by Guy, Calvert, and Vaughan, whose settlers remained as permanent residents even after the collapse of the organized enterprises. Both the fishermen from England and the planters engaged in catching and curing cod generally employed similar methods.

The methods employed in the bank and inshore fisheries were quite different. Wet fish was caught on the banks, cleaned, salted, and packed in casks of brine, and then carried directly to market. As long as the Spaniards and Portuguese were active, they carried the pickled fish home and dried it before placing it on the market. The French carried the wet fish home in casks and barrels and disposed of it still in its pickled or "green" condition. The English do not appear to have engaged extensively in making wet fish, but because they carried on their operations close to the land, they brought the catch ashore, landing it on stages built out over the water, and after cleaning it and salting it, set it in the sun to dry until it was hard and firm. At the end of the season it was loaded into casks and barrels or carried to market in bulk. The French also engaged in this type of fishery along the coasts of Newfoundland as well as on the shores of Acadia and in the Gulf of St. Lawrence. Train oil or codliver oil was an important by-product which brought a good price in European markets. It was obtained from both wet and dry fish. The livers were separated from the rest of the offal and thrown into barrels on shipboard or into vats ashore. When thoroughly decomposed the oil was refined by removing the foreign material, and then poured into hogsheads in which it was shipped home.

The fishing season began early in June and lasted until late August, varying somewhat from place to place according to local conditions. The West Country shipping usually tried to sail from England about the middle of March or the first of April in order to arrive in time to have two or three weeks for preparations. Arriving sometime in May, the first two or three weeks were spent in

THE FIRST WESTERN CHARTER 57

choosing "fishing rooms," making repairs to boats, stages, and other equipment, procuring bait, and gathering supplies of salt for the season. Some ships made directly for favorite or previously chosen fishing grounds, while others spent some time in selecting new locations. In the latter case the ship would find a safe anchorage and send exploring parties in small boats up and down the coast in search of a promising situation for the summer's work. Sometimes the boats of the exploring parties were crushed by the ice or were driven upon the rocky coast, or the men perished from cold, starvation, or thirst. Once a suitable fishing ground had been reached, the ship was anchored upon good holding ground where it would be protected from sudden squalls, and left in charge of a few men. Most of the crew were landed and took up temporary quarters near the stages and cookrooms for the summer. The planters were able to begin fishing earlier than the West Countrymen because the long winter gave them ample time to do the preliminary work. In those harbors which were permanently inhabited and also visited by the English ships the homes of the planters and the temporary huts of the crews were side by side and located as close to their work as possible. Most of the fishing communities ranged along the shore for considerable distances but did not penetrate far inland. The rowboats used by both English fishermen and planters accommodated three men who fished with hook and line. Small sail boats do not appear to have been used in the early years, nor were trawls employed to catch fish. Seines were used in shoal water not only to obtain caplin and other small fish for bait but also to catch the young cod. The flesh of sea-birds was sometimes used for bait, but the English did not follow the French custom of baiting their hooks with the entrails of the cod but instead threw the offal into the harbors. The catch was measured by the "quintal," technically one hundred fish, but in order to allow for a wrong count, difference in size, and spoiling, it was customary to consider it as 120 fish. This measure was employed not only in reckoning the catch of fresh fish, but also the amount cured for market. Each producer culled his own fish, the "refuse" fish or "culls" being set aside and sold to less discriminating buyers than those who took "merchantable" fish. Train oil was measured either by the tun or by the hogshead. For every three men employed in fishing it was necessary to have three or four more at work ashore, engaged in cleaning, salting and curing the fish, making oil, and performing other necessary duties. The West Country vessels carried a considerable number of boats from England, but a good many shipmasters also kept extra ones in Newfoundland which were stored through the winter along with left-over salt and casks for the oil. Although it was possible to fish after the third week in August the season usually ended then in order that the fish and oil might be loaded, and the ships made ready for their market or homeward voyages before the autumn storms began. Ordinarily by mid-September all the ships had departed, but seldom did any remain until October. During the early years the fishermen lived upon hard biscuits, salt meat brought from England, and fresh-caught cod. Sometimes small quantities of "wet" fish or "cor" fish were pickled to serve as victuals on the homeward voyage. Scurvy was general, and the monotonous diet was only relieved by recourse to malt beverages, "spruce beer," wines, and ardent spirits. Drunkenness was a problem from the first. The wooden huts and work buildings of the fishermen and planters frequently caught fire, and general conflagrations which swept entire communities or destroyed the standing timber near the coast occurred from time to time. Very early in its history the Newfoundland voyage earned a reputation of being ardu-

THE FIRST WESTERN CHARTER 59

ous and dangerous, for besides the ever present danger to life and limb there was the constant menace of ill health from the sameness of the food and continued exposure to the elements.

During the period when the West Country fishermen and the London shipping interests were struggling to control the fishery and the carrying trade, the English merchants who were directly interested in the industry were faced with a decline in their business. This decline was the result of the wars of Charles I with France and Spain, the attacks of the Barbary pirates upon the Newfoundland shipping, and the increase of foreign competition in the carrying trade. Shipping was handicapped by the impressment of mariners and the seizure of Newfoundland fish to be used as provisions for the troops. Because their ships were detained, the western adventurers were unable to sell their fish at a good price, and they remonstrated against the seizures with such good effect that they frequently obtained releases. Ships bound for the fishery in the spring were held in order that their men might be impressed, but many vessels slipped away earlier than usual in order to avoid the embargoes laid upon shipping, and a great many sailors escaped service in the royal navy by fleeing to the fishery. The records of the reign of Charles I are filled with references to the arrest of ships bound for Newfoundland, the seizure of their crews, and the protests of the merchants and shipmasters against the pressmasters. Many ships were released, and it would appear that the Newfoundland fleet fared better in this respect than the ships in some of the other trades. As the Civil Wars approached there was an appreciably increasing tendency to impress men, and the number of protests is considerable.[1]

[1] There are many references to shipping embargoes, impressment, protests from merchants, and releases of vessels for the reign of Charles I in

The wars kept the English from their usual markets and enabled the French, Flemings, Dutch, and Hamburgers to extend their trade. Although Spain usually prohibited Dutch ships from entering her ports, an exception was made in the case of vessels laden with provisions. Taking advantage of this exception the Hollanders carried Newfoundland fish and other provisions to Spanish ports.[2] The English resented this extension of Dutch shipping in view of the hold which the Hollanders had upon the fisheries of the British Isles. The feeling was particularly acute in London where the merchants held more nationalistic views and felt the effects of Dutch competition more keenly than was the case with the outports. Although jealous of their special prerogatives, the western adventurers were indifferent as to who carried their fish to market. There are evidences, however, that the Dutch were not content to remain buyers only of fish from the French and English, but desired to establish their own fishery in North America.[3] Had this taken place, no doubt the West Country would have been as

Cal. State Paps., Dom., 1623-1639, and *Acts, Priv. Coun., Col.*, 1613-1680, *passim.* See also *Cal. State Paps., Col.*, 1574-1660, pp. 236, 263; and Historical MSS. Commission, *Cowper (Coke) Manuscripts*, I, 179. Cited hereafter as *Cowper (Coke) MSS.*

[2] Girolamo Lando, Venetian ambassador, to the Doge and Senate, Nov. 13, 1620, *Calendar of State Papers, Venetian*, 1619-1621, p. 473. Cited hereafter as *Cal. State Paps., Ven.* Petition of Robert Barker to Buckingham, May 24, 1627, *Cal. State Paps., Dom.*, 1627-1628, p. 188; note of seven Hamburg ships that went to Newfoundland in 1627 [Jan. 19, 1628], *ibid.*, p. 522; paper concerning a fleet from St. Malo that went to northern Newfoundland in 1627, C.O. 1: 32, no. 2; mayor, etc., of Plymouth to the Privy Council, Dec. 10, 1633, *Cal. State Paps., Dom.*, 1633-1634, pp. 318-319; paper on the breeding of seamen [1636], *ibid., Addenda*, 1625-1649, pp. 543-544.

[3] Project of the intended voyage . . . by Captain Andrews and Jacob Braems, Oct. [?] 1618, *Cal. State Paps., Col.*, 1574-1660, p. 19; mayor of Plymouth, etc., to the Privy Council, Dec. 10, 1633, *Cal. State Paps., Dom.*, 1633-1634, pp. 318-319.

THE FIRST WESTERN CHARTER 61

loud in its denunciation of the foreigners as it was in its complaints against the Londoners.

The war with France intensified the rivalry of the two nations in the North American fisheries. The various projects of colonization undertaken by the French and English in northeastern America were the direct result of increasing commercial competition. In not a few cases plantations were projected as much for the protection of the fisheries as for the exploitation of the territories occupied. Sir William Alexander's attempt in Nova Scotia, the Kirke project in Canada, and later the grant to the marquis of Hamilton, the earl of Pembroke, and Sir David Kirke in Newfoundland were designed partly to offset French expansion in the fisheries. The French likewise strove to extend the bounds of their control over land and water. Their fisheries were closed to all but subjects of the French king, and they planned to create a great training school for seamen in the North Atlantic fisheries. They developed a sedentary fishery along the coasts of Acadia and Canada. In the process of extending their fishing industry they drove Lord Ochiltree's colony out of Cape Breton Island.[4] All of these activities were but the prologue to the great struggle which was to take place after the Civil Wars and in the eighteenth century for control of the new continent.

During the French war the English lost all their markets in France, except the Huguenot ports. Although they preyed upon French ships returning from Newfoundland, they themselves suffered heavily at the hands of the French and their allies the Barbary pirates. Before the war English vessels had been attacked by the "Turks"

[4] The French king assumed the sole privilege of fishing in New England and Nova Scotia, Jan. [?] 1630, *Cal. State Paps., Col.*, 1574-1660, pp. 105-106; Coke's "Propositions for fishing," July [?] 1633, *ibid.*, pp. 170-171; Lord Ochiltree's information, Jan. [?] 1630, *ibid.*, p. 106. See Insh, *Scottish Colonial Schemes*, pp. 91-112.

in the Mediterranean, but between 1624 and 1652 the rovers became bolder and penetrated into the North Atlantic. The Salee pirates raided the coasts of western England, lay in wait for the outgoing and incoming Newfoundland fleets at the entrance to the English Channel, and captured many ships, killing or making slaves of their crews. They were daring enough to coöperate with the French in attacks upon the shipping along the New England coast and also threatened to raid Newfoundland.[5] During the reign of Charles I the navy was inadequate to give the rovers a decisive beating. Although the western adventurers were anxious to be rid of this menace to their shipping, the Newfoundland fishermen were disinclined to arm their vessels and rejected proposals aimed to furnish naval protection for their fleet or for the coasts of Newfoundland.[6] Their attitude was undoubtedly dictated partly by stubborn local pride and partly by their unwillingness to witness an extension of the royal authority over their affairs.

[5] There are numerous references to the depredations of the French upon the English fishing fleet. C.O. 1: 5, no. 3; C.O. 1: 32, no. 2; *Acts, Priv. Coun., Col.,* 1613-1680, pp. 3-4, 8, 11, 112-113; *Cal. State Paps., Dom.,* 1627-1628, pp. 25, 173, 184, 271, 297, 299, 362; 1628-1629, p. 62; 1629-1631, p. 43; 1636-1637, pp. 341, 511; *ibid., Addenda,* 1625-1649, pp. 319, 360. There are a great many references to the depredations of the Barbary pirates scattered throughout the *Calendar of State Papers,* both *Domestic* and *Colonial,* which may be found by consulting the indexes. See especially the following: John Delbridge to the mayor of Barnstaple, Feb., 1632, Historical MSS. Commission, *Manuscripts of the Corporation of Plymouth,* pp. 269-270a. Cited hereafter as *Plymouth MSS.* Kenelm Edisbury to Sec. Nicholas, Aug. 5, 1636, *Cal. State Paps., Col.,* 1574-1660, p. 239.

[6] Convoys were employed occasionally to protect the fishing fleet against the French and Barbary pirates. *Cal. State Paps., Dom., Addenda,* 1625-1649, p. 217; 1627-1628, p. 173. Usually the warships did not venture far beyond Land's End. *Ibid.,* 1629-1631, pp. 333, 342; *Cowper (Coke) MSS.,* I, 137; II, 440. The western adventurers objected to the convoy on the ground that it would harm their voyages and maintained that many of their ships did not sail direct for Newfoundland. Report of the Master, Wardens, etc., of Trinity House to the Privy Council, Feb. 23, 1637, C.O. 1: 9, no. 41.

THE FIRST WESTERN CHARTER 63

As a result of the wars, increased foreign competition, and the raids of the Barbary pirates, the whole fishery enterprise from Bristol to Southampton was considerably altered. Some of the more important fishing ports declined, while others improved their position at the expense of their neighbors. Poole, which was accustomed to send out twenty vessels annually lost that number (valued at £13,000) in four years. By 1628 the port had only sixteen ships left with a total burden of 838 tons, and only three planned to sail for the fishery in that year. Even as late as 1634 Poole had not recovered its former position in the fishing trade. At the same time others had gained. In 1631 Alderman Clement of Plymouth stated that during his lifetime there had been a considerable increase in Newfoundland shipping. Plymouth and Dartmouth, which formerly had sent only two or three ships each year, were now sending sixty and eighty respectively, and such other ports as Barnstaple and Topsham were also active in the trade. In spite of the losses occasioned by the wars and the raids of the Salee rovers, English traffic with Newfoundland actually improved. Before the war there had been about 300 ships in the fishing trade; by 1637 there were about 200 more of varying tonnage engaged.[7] Probably such places as Poole, the trade of which had been seriously injured, had increasing difficulty in competing with those ports the business of which had been unharmed or whose recovery had been more rapid. Better management and greater capital for investment in fishing ventures tended to concentrate the

[7] Mayor of Poole, etc., to the Privy Council, Aug. 23, 1634, *Cal. State Paps., Dom.*, 1634-1635, pp. 194-195; Alderman Clement of Plymouth to ———, Sept. 12 and 14, 1631, Historical MSS. Commission, *Manuscripts of the Duke of Northumberland*, p. 71. Cited hereafter as *Northumberland MSS*. Robert Lewes, *The Merchants Mappe of Commerce* (London, 1637), pp. 57-58.

industry in certain ports and to leave less fortunate places out of the running.

In 1634 circumstances combined to secure for the western adventurers legal recognition of their ancient privileges. This opportunity was due not only to the general depression in the carrying trade which made the London shipping interests willing to coöperate with the West Country merchants, but also partly to the monopoly conferred upon the Greenland Company which gave it the exclusive right to import train oil into England, and partly to the fear lest another attempt should be made to establish a fishing monopoly supported by the crown. The Greenland Company's privilege struck directly at an important feature of the fishery at Newfoundland, for train oil was the only product of the fishery that found a ready market in England. The western adventurers had called the matter to the attention of parliament in its last session in 1628, when the bill restraining fishing monopolies was introduced for the last time.[8] After the dissolution of parliament the Newfoundland traders complained to the Privy Council of the injury which the train oil monopoly inflicted upon them.[9] In 1633 the project of establishing a fishing monopoly in the New World was revived by some London merchants.[10] Until then the serious and long continued dispute between the West Country fishing interests and the London shipping group had continued unabated, but the general depression in these years of political uncertainty finally brought the antagonists closer together.

During the reign of Charles I various expedients were proposed to improve conditions in the carrying trade. In

[8] Stock, *Proceedings and Debates*, I, 82-83.

[9] Petition of divers merchants to the Privy Council [1634], *Cal. State Paps., Dom.*, 1634-1635, p. 393.

[10] Propositions for fishing, July? 1633, *Cal. State Paps., Col.*, 1574-1660, pp. 170-171.

THE FIRST WESTERN CHARTER   65

1624 consideration was given to a plan to exclude all foreigners. All imports were to be made in English ships, and aliens were to be forbidden to fish on the southern coast of England. On the other hand English fishermen were to be free from purveyance and were to enjoy a free market in London. All fish and other commodities were to be exported in English, Scottish, or Irish ships provided their crews should be subject to impressment. This measure was intended to strike at the Dutch, and following this suggestion an order in council was issued in 1626 which required that all herring or other fish caught by Englishmen on the coasts of England, Newfoundland, and New England should be exported only in English bottoms.[11] The order was not strictly enforced, for in 1629 one of the reasons given for the decline of English commerce and navigation was the transportation of sea coal, red and white herring, and Newfoundland fish to markets abroad in foreign-built ships.[12] So far attempts to regulate the fishing and carrying trades by orders in council had proved ineffective.

A more determined effort to carry out these regulations was made in the summer of 1631 when Lord Treasurer Weston ordered the customs officers to enforce the rules, pursuant to an order in council of September 29, 1630.[13] The action was taken at the behest of the master and wardens of Trinity House, who represented that English shipping lacked cargoes. The western adventurers had protested against the regulation in 1631, asserting that the outports were already overstocked with

[11] Consideration for employment of shipping and men, Jan. 30, 1624, *Cal. State Paps., Dom., Addenda,* 1580-1625, pp. 661-662; order in council, Nov. 10, 1626, *Acts, Priv. Coun., Col.,* 1613-1680, §181.

[12] Particulars concerning the decay of navigation, etc. [Dec. 13, 1629], *Cowper (Coke) MSS.,* I, 375.

[13] Lord Treasurer Weston to customs officers at Plymouth, etc., July 28, 1631, *Cal. State Paps., Dom.,* 1631-1633, p. 123.

pilchards and Newfoundland cod which they could not sell in England.[14] The probable reason for their desiring the withdrawal of this order was their commitments to foreign buyers. Contracts for future delivery to the carriers were customarily made before the beginning of the fishing season, and even before the departure of the fishing ships for Newfoundland. Moreover, the foreigners offered better terms, and usually paid cash on delivery. The western merchants admitted their willingness to employ some of the idle English cargo vessels, but desired to do so only on their own terms.[15] As a result of this protest and offer of partial coöperation with the owners of freighters, they obtained a temporary release from the order, and in 1631 were permitted to sell their fish to aliens and export it in foreign vessels.[16]

Although the order was relaxed only temporarily, to cover an alleged emergency in 1631, the western adventurers continued to use foreign bottoms. Two years later the Privy Council complained of the general disregard of the order in the western ports, particularly at Plymouth.[17] The council was inspired by the protest of Trinity House, which asserted that it had been trying to persuade the council to bar foreign shipping from the export trade in Newfoundland fish for eighteen or nineteen years. Although the association approved of the several orders in council intended to prevent foreign participa-

[14] Alderman Clement of Plymouth to ———, Sept. 12 and 14, 1631, *Northumberland MSS.*, p. 71.

[15] Mayor, etc., of Plymouth to the Privy Council, Dec. 10, 1633, *Cal. State Paps., Dom.*, 1633-1634, pp. 318-319; reasons tendered to the Privy Council in behalf of the West Country merchants, etc. (*circa*, 1631-1633?), *Plymouth MSS.*, p. 271.

[16] Order in council, Nov. 4, 1631, *Acts, Priv. Coun., Col.*, 1613-1680, §282; reasons tendered in behalf of the West Country merchants, etc., *Plymouth MSS.*, p. 271.

[17] Privy Council to the mayor of Plymouth, Nov. 23, 1633, *Plymouth MSS.*, p. 271.

THE FIRST WESTERN CHARTER 67

tion, the regulations had never been enforced. In the meantime foreigners had bought up all the fish and carried it away in their own vessels. As a result there were about 10,000 tons of idle shipping lying in the Thames. The Trinity House complained that there had been twenty-six or twenty-eight foreign ships buying fish at Newfoundland in one year, while sixteen or eighteen others of 300 tons each had loaded pilchards and Newfoundland fish in the western ports, all bound for Spain and Italy. Besides these there were eight Dutch ships waiting to load herring at London, twelve at Plymouth, and six others had already sailed.[18] This protest from Trinity House precipitated a general discussion of all matters relating to the Newfoundland fishery and the carrying trade which arose therefrom.

The council determined to find a way to exclude foreigners from the carrying trade. The West Country ports were directed to send representatives to present their views to the council in order that some means might be found to enforce the regulations satisfactorily.[19] Plymouth, Dartmouth, and Barnstaple were the only outports to send agents to Whitehall, but they had been instructed to work together in the interests of the entire region.[20] Hearings were held in December, 1633, and throughout January and February of the following year. The London shipping group was also represented. Correspondence was also received from the West Country corporations and considered along with the arguments presented orally. There were three principal questions taken into

[18] Petition of the Masters, etc., of Trinity House to the Privy Council, 1633?, *Cal. State Paps., Dom.*, 1633-1634, p. 367.

[19] Order of the Star Chamber, Jan. 29, 1634, *Acts, Priv. Coun., Col.*, 1613-1680, pp. 197-198.

[20] Alexander Staplehill, mayor of Dartmouth, to Roger Trelawnie, mayor of Plymouth, Dec. 3, 1633, *Plymouth MSS.*, p. 271; order of Star Chamber, Jan. 29, 1634, *Acts, Priv. Coun., Col.*, 1613-1680, pp. 197-198.

consideration by the council: the exclusion of the Dutch and French from the English carrying trade; the exclusion of foreigners from the carrying trade between Newfoundland and the markets of southwestern Europe; and the establishment of regulations for the Newfoundland fishery.

The western adventurers denied that they transported fish in foreign bottoms, but sold it after the return of the Newfoundland fleet to Dutch and French ships for re-export. They said that the foreigners paid the export tax on all the fish they carried away, and that the merchants, having an assured outlet for their fish, had doubled the number of ships and seamen at Newfoundland within a few years. Except in years when the fish failed, as had been the case in the summer of 1633, the West Country merchants were in a position to dispose of their goods to both Londoners and foreigners. They maintained further that they would be handicapped if foreigners were prohibited from buying fish, as there would be no other means of disposing of it except through the Londoners who were not in a position to accept more than half of the annual catch. The London group, they said, insisted on the right to fix the price, claiming that the fishermen had suffered unfortunate experiences when they had attempted to do so, and claimed that the West Countrymen were prevented from sending their fishing ships directly to Spain, because most of their vessels were Dutch built, which Spain prohibited from entering her ports. The western adventurers felt that if foreigners were prohibited from buying and carrying away their fish, it would cause the decay of the fishery, "the nursery of many thousand seamen"; would deprive the customs of £40,000 annually in export duties; would cause retaliation by foreigners; and would force the Dutch to engage

THE FIRST WESTERN CHARTER 69

in the Newfoundland fishery.[21] They informed the Privy Council that there were about 27,000 tons of shipping and 10,680 seamen engaged annually in the Newfoundland fishery, the annual gross income from the voyages of which was about £178,880. Train oil was the most profitable product, yielding £17,600 annually out of a net profit of £24,000.[22]

During the summer of 1633 the proposal was made that settlements be started in Canada and elsewhere to protect the New England and Nova Scotia fisheries, and that a company be chartered to carry on both the domestic and overseas fisheries.[23] Although at the time this plan was not considered the Privy Council continued to hold its hearings during the following winter, and the old threat of local supervision of the fishery was revived when the suggestion was brought forward that the foreign shipping trade from Newfoundland to Europe might be controlled by imposing customs duties in the island.[24] As such a suggestion implied the presence of officers authorized to collect the duties, those who presented it may have wished to establish some sort of government there. This was one of the important features of the Kirke plantation established three years later, and it would not be surprising if the London merchants were paving the way for their second attempt to control the fishery and the carrying trade.

[21] Mayor, etc., of Plymouth to the Privy Council, Dec. 10, 1633, *Cal. State Paps., Dom.*, 1633-1634, pp. 318-319. Spain needed food and, therefore, permitted the Dutch to bring it in, but prohibited other forms of commerce with Holland.

[22] Petition of divers merchants trading to Newfoundland to the Privy Council [1634?], *Cal. State Paps., Dom.*, 1634-1635, p. 393.

[23] "Propositions for fishing," by Sec. Coke, July? 1633, *Cal. State Paps., Col.*, 1574-1660, pp. 170-171.

[24] Note by Sec. Windebank of proceedings before the Privy Council, Dec. 14, 1633, *Cal. State Paps., Dom.*, 1633-1634, p. 322.

The discussions before the council resulted in a compromise. The West Country adventurers secured confirmation of their ancient fishing customs and recognition of their right to regulate not only the fishery at Newfoundland, but also the shipping going out and returning to England. On the other hand, they promised in the future to contract with the Londoners to take away their fish. However, because of the nearness of time for the departure of the fishing fleet when the contracts with the carriers were usually made, the requirement that only English ships be used as freighters was again temporarily suspended.[25] The Privy Council appointed a committee composed of thirty-eight persons, representing Trinity House, the London merchants, and the western adventurers to work out the details of a permanent plan for using English bottoms only, and directed the interested parties to submit suggestions in writing.[26] The committee held two meetings, probably in February or March, 1634. The West Country group presented written proposals according to instructions, but the committee could come to no decision because most of the fishing ports of England and Ireland were unrepresented. The committee members from the west of England asked for permission to return home because of the approaching assizes of Devon and the impending departure of the Newfoundland fleet.[27] Evidently permission was granted them to leave, for there are no records of further meetings of the committee of thirty-eight. No definite recom-

[25] Order in council, Feb. 10, 1634, *Acts, Priv. Coun., Col.*, 1613-1680, pp. 198-199.

[26] Copy of order in council, Feb. 10, 1634, appointing a committee of thirty-eight persons, *Plymouth MSS.*, p. 271; another of the same date, directing the West Country merchants to submit written proposals relative to carrying fish in English bottoms, *Plymouth MSS.*, p. 271. Cf. order in council of that date, *Acts, Priv. Coun., Col.*, 1613-1680, §325.

[27] Petition of Edmund Fowell and Roger Mathews to the Privy Council, March? 1634, *Cal. State Paps., Dom.*, 1633-1634, pp. 532-533.

mendations were made relating to the perennial question of the carrying trade, but the western agents had obtained what they wanted—another temporary suspension of the order regulating shipping; and more important still, confirmation of their fishing rights at Newfoundland. They remained as opposed as before to the permanent prohibition against selling to foreigners and shipping their catch in alien vessels. Evidently they made good their bargain with the Londoners, for they afterward contracted for the delivery of some fish in ships owned there.[28] The question of foreign shipping, however, was not settled by being ignored, and came to the fore again in 1637 when Kirke's proposed settlement of Newfoundland was under consideration.

For the western adventurers confirmation of their fishing rights at Newfoundland and the embodiment of regulations governing the fishermen while on voyages there were the most important outcome of the meetings which the Privy Council held in the winter of 1633-1634. During the period from 1621 to 1628 they had tried unsuccessfully but persistently to secure protective legislation from parliament, but now that Charles I was ruling without a parliament they turned to the crown for support, in the hope that henceforward they could look to the royal prerogative to defend their ancient privileges.

The charter of 1634 is known as "the first Western Charter" to distinguish it from a similar patent issued by Charles II in 1661 and the amended patent of 1676. Just when the West Country agents applied for it is uncertain, but their request was probably brought to the attention of the officials at Whitehall in December, 1633, or early in January, 1634. The agents of Plymouth, Dart-

---

[28] Petition of Anthony Hooper, Daniel Farvacks, Isaac Legaye, etc., of London to the Admiralty [June 2], 1636, *Cal. State Paps., Col.*, 1574-1660, p. 236; petition of Thomas Williams to the same, 1636?, *ibid.*, p. 236.

mouth, and Barnstaple were the only western representatives at the London hearings, but the communities which they represented had probably furnished them with instructions to try to procure royal confirmation of their old privileges, and possibly with an outline of the desired charter before they left for the capital.[29] Some doubt was entertained in court circles as to the king's power to legislate for Newfoundland, and the question was referred to William Noye, the attorney general, who rendered an affirmative opinion and submitted the proposed rules for the fishery to the Privy Council on January 24, 1634.[30] Noye's opinion is worth quoting: "In this acquired dominion, I do conceive his Majesty may give laws, and some that may serve for the present I have presumed to present to your honors to stand until it be otherwise ordained, with power to certain mayors of towns to execute them, and a command that they be published there."[31]

From the above opinion it will be seen that royal confirmation of the fishing regulations was intended to be only temporary, an intention which was to make the triumph of the West Country of short duration. The powers granted to the West Country mayors are significant features of the patent. Pursuant to Noye's recommendation

[29] The merchants, etc., of Plymouth, Dartmouth, and Barnstaple, etc., to the Star Chamber, Jan. 24, 1634, *Acts, Priv. Coun., Col.*, 1613-1680, p. 193.

[30] Older writers refer to these early fishery regulations as the "Star Chamber Rules." The author does not believe that they were promulgated by that court, but that they were approved by an order of the Privy Council at a session held in the Star Chamber. The personnel of the two bodies was almost identical, but the fact that the fishery regulations were entered in the Privy Council Register indicates that the action was a normal function of that body rather than of the Court of Star Chamber. For a discussion of this point see Edward R. Turner, *The Privy Council of England in the Seventeenth and Eighteenth Centuries, 1603-1784*, I, 91-92.

[31] [Order in council], Star Chamber, Jan. 24, 1634, *Acts, Priv. Coun., Col.*, 1613-1680, p. 193.

THE FIRST WESTERN CHARTER 73

the council approved of the rules, ordered them to be engrossed for the royal signature, "in regard the fishing season is now at hand," and directed that they be made patent under the great seal of England. The Privy Council also directed that copies of the letters patent be circulated throughout western England and in Newfoundland.[32] Two forms of the regulations are to be found among the records in England. One of these comprises copies of the order in council of January 24, and the other copies of the letters patent issued on February 10, 1634. In view of the imminent departure of the fishing fleet copies of the order in council were circulated in advance of the formal patent.[33] Strangely enough the "First Western Charter" does not appear on the patent rolls although it is to be found among the chancery warrants of the reign of Charles I.[34] What purports to be a copy of the letters patent under the great seal is said to be in the archives of Plymouth.[35] In view of the opinion of Secretary Noye that the regulations were expected to be only temporary, one wonders whether the omission of the so-called charter from the patent rolls was not intentional.

[32] [Order in council], Star Chamber, Jan. 24, 1634, *ibid.*, p. 197.

[33] The regulations are incorporated in the order in council, Star Chamber, Jan. 24, 1634, *ibid.*, §323; copies are also found in the *Plymouth MSS.*, pp. 271b, 289b; B.M., Additional MSS., 13,972, a.

[34] The charter is not to be found on the patent rolls, which the author caused to be searched unsuccessfully. Others have been no more fortunate, but the Quebec-Labrador Boundary Commission found the rules among the chancery warrants. "Regulations for the Newfoundland Fishery," Feb. 10, 1634, Chancery Warrants, series II, file 2106, no. 525; printed in *Documents Relating to the History of Newfoundland* in the *Joint Appendix*, IV, to the cause *In the Privy Council. In the Matter of the Boundary between the Dominion of Canada and the Colony of Newfoundland in the Labrador Peninsula*, §712. Cited hereafter as *Labrador Boundary Case, Newfoundland Documents.*

[35] "Letters Patent under the Great Seal of eleven Ordinances for the government of the King's people in Newfoundland . . .," Feb. 10, [1634], *Plymouth MSS.*, p. 280.

The charter of 1634 was designed to regulate the fishery according to the customs of the West Country. It consists of a preamble and eleven rules for governing the shipping trade to Newfoundland and the fishery there, besides a final clause authorizing the mayors of certain specified western ports to execute the regulations, and ordering the vice-admirals of the counties of Hants, Dorset, Devon, and Cornwall to proceed against offenders. The preamble reasserts the claims of the kings of England to dominion over Newfoundland and declares that the fishery is valuable to the nation because "a great number of our people have been set on work, and the navigation and mariners of our realm increased." It goes on to say that the fishery had originally been well conducted and the natives well treated, but that recently the English planters had inflicted serious injuries upon the fishery and the Indians, for which they could not be brought to justice in England. It concludes by announcing that the regulations are intended to protect the interests of the fishermen and to prevent the planters from interfering in their affairs.[36] These rules, though originally designed to be only temporary and though altered and amended from time to time, endured for many years as the fundamental law of the Newfoundland fishery.

The regulations provided that trials for murder and theft should be placed under the jurisdiction of the court of the earl marshal in England, but a majority of the articles were concerned with the management of the fishery and the relations of the fishermen and planters. They were designed to insure the prosperity of the fishery by preventing such practices as throwing ballast and press-

[36] Comparison of the preamble of the charter of 1634 with Guy's regulations of 1611, emphasizes the differing views of the West Country and proprietary groups. Draft of the Western Charter of 1634, *Acts, Priv. Coun., Col.*, 1613-1680, §323; Guy's rules for the fishery, 1611, C.O. 1:1, no. 40 (i).

THE FIRST WESTERN CHARTER   75

stones into the harbors, casting anchor where it might interfere with seining, defacement or injury to the equipment used in the catching and curing of fish and the making of train oil. The use or appropriation of fish, train oil, salt, provisions, or equipment belonging to others was forbidden except in emergencies. Conservation of the limited supply of timber and firewood near the coast was insisted upon, and even the morals of the fishermen were to be safeguarded and their efficiency maintained by a prohibition against taverns conducted by the inhabitants. Moreover all were enjoined to observe the Sabbath.[37]

One of the most important regulations was that regarding the choice of fishing grounds and working places ashore. Dispute over this right had been one of the principal causes for the antagonism between the fishermen and the planters. According to the charter the master of the first fishing ship to arrive in a harbor was to be considered admiral of that harbor. The admiral had first choice of the fishing grounds, stages, and boats' rooms, or of beaches and flakes in proportion to the number of fishing boats which he had. In addition, as a reward for his early arrival, he was granted an additional boat's room. All ships coming later were permitted to occupy enough space to accommodate the exact number of boats which they employed. Shipmasters who reserved space in more than one harbor were to decide within forty-eight hours after the arrival of other ships in those places, which fishing room they intended to retain, in order that none might be excluded or injured by delays. This rule was the written recognition of the old custom which the

[37] Draft of the charter of 1634, clauses 1-11, *Acts, Priv. Coun., Col.*, 1613-1680, §323. There is an unnumbered paragraph following that styled clause 4, but clause 3 was omitted in the draft and also in the Chancery Warrants. See *ibid.*, §323; *Labrador Boundary Case, Newfoundland Documents*, §712.

fishing interests had guarded so jealously in their struggle with the London and Bristol Company. It was an extremely important rule because it determined the whole West Country fishery administration in Newfoundland. Thereafter the fishing admirals were not only recognized as having special privileges, but were officially regarded as the persons responsible for the enforcement of the rules. Although the admirals were responsible for good order in their respective harbors they had no power to lay fines or to prescribe punishments for violations of the rules. This was a manifest weakness of the first charter, which the western adventurers sought to correct in later years.[38]

Right to enforce the regulations was vested in the mayors of Southampton, Weymouth and Melcombe Regis, Lyme, Plymouth, Dartmouth, East Loo, Fowey, and Barnstaple. These magistrates had cognizance of all complaints against violators of the rules. They were authorized to investigate the truth or falsity of accusations, to call witnesses under oath, to give judgment, and to prescribe awards or amends to the aggrieved parties. They could punish offenders by either fine or imprisonment or both, or confiscate their property either at Newfoundland or at sea. Furthermore, the vice-admirals of the counties of Hants, Dorset, Devon, and Cornwall also were authorized to proceed against offenses committed at sea. The judgments of the West Country magistrates and the vice-admirals were to be published by the fishing admirals in Newfoundland during the next fishing season.[39] At last to all appearances the West Country had complete control over the fishery, and had the planters at its mercy.

[38] Guy's rules provided for fines of from £5 to £10 and satisfaction to the aggrieved party. The opposition of the western adventurers to these exactions probably accounts for the omission of fines, etc., from the charter of 1634.

[39] The Western Charter of 1634, clause 12.

THE FIRST WESTERN CHARTER 77

This setting up of an administration in the west of England to govern a fishery nearly 2,000 miles away, an administration which could consider infractions of the rules, no matter how slight, months after they had been committed, constituted a clumsy and ineffective machine for controlling Newfoundland.

The charter confirmed the vested interest of the western adventurers in the fishery, gave them the legal recognition they had long desired, and placed the planters in a subordinate position. The imperfections of the instrument are obvious. The rules applied to the fishery in Newfoundland, but their enforcement was in the hands of officials living in the west of England. The fishing admirals were made responsible for good order but were clothed with no power to insure it, except as far as moral suasion or physical strength might be applied. The planters who remained in the island after the collapse of the early colonization schemes were left to the tender mercies of their fishing rivals. In no sense of the word did the charter provide a "government" for the fishery. It merely recognized the existence of a sort of law merchant for the Newfoundland trade, similar to the laws of Oleron and other medieval customary rules relative to navigation. From the point of view of the adventurers its greatest weakness was the indefiniteness of its duration. It made no mention of any specific number of years during which it was to be effective, nor did the crown grant the western ports their privileges in perpetuity. Within three years after the charter was issued the patentees were obliged to defend it before the Privy Council, when that body was considering the application of Sir David Kirke and his friends for a charter of colonization in Newfoundland.

In 1637 the western adventurers were faced with another and more serious attempt to establish a North Atlantic fishing monopoly. Favored by Charles I, the marquis of Hamilton, the earl of Pembroke, the earl of Holland, and Sir David Kirke applied for a charter granting them proprietary rights not only over Newfoundland but over practically the entire fishing area of northeastern America, including Cape Breton, Nova Scotia, and the banks.[40] Such a grant was of momentous consequence to the West Country. The movement to secure a monopoly in America was contemporary with a similar effort to establish a fishing monopoly in the seas surrounding the British Isles, and some of the persons interested in the British project were concerned also with the overseas plan.[41] The extent of the proposed grant in America promised difficulties with the French, for any attempt to exercise their authority in the vast region granted the patentees was bound to bring them eventually into conflict with England's chief rival in the fishery.[42] A group of London merchants was behind the movement, for in 1636, Edward Cason and others of that city had applied for permission to engage in bank fishing among the French.[43] Moreover, Sir David Kirke and his brother Lewis, a London merchant, had been associated with other business men there in their Canadian enterprise.[44] From the events which followed it is evident that the Londoners were again seeking to control the fisheries with

[40] A grant of Newfoundland to the marquis of Hamilton, the earl of Pembroke, and others, Nov. 13, 1637, Pat. Rolls, 13 Charles I, pt. 32; copy: C.O. 1:9, no. 76.

[41] Insh, *Scottish Colonial Schemes*, pp. 40-112, esp. pp. 40-90; Henry Kirke, *The First English Conquest of Canada* (London, 1871, reprinted 1908), *passim*.

[42] Anzolo Correr, Venetian ambassador, to the Doge and Senate, Apr. 17, 1637, *Cal. State Paps., Ven.*, 1636-1639, §199.

[43] Orders in council, June 25, July 9, 1637, *Acts, Priv. Coun., Col.*, 1613-1680, pp. 218-219, 220.

[44] Petition of Edward Cason, etc., of London to the Admiralty [Apr. 21, 1633], *Cal. State Paps., Col.*, 1574-1660, pp. 232-233.

THE FIRST WESTERN CHARTER 79

the object of improving their carrying trade. Involved with the Dutch in the home fisheries, the English were about to intensify their competition with the French in northeastern America.

Immediately upon receipt of the news that the king intended to issue letters patent to the applicants, the western adventurers prepared to resist, and were sufficiently influential to secure the attention of the crown. The king and council were willing to conciliate them, and the solicitor general was ordered to prepare a draft patent for the new plantation, to which the deputies representing the western ports might make exception.[45] Both sides were heard by the Privy Council and also by the solicitor general. The law officer was ordered to report to the council any points upon which the two parties could not agree in order that it might decide the disputed matters.[46] Lord Baltimore joined the opposition on the ground that his proprietary interest in the province of Avalon would be infringed by the new grant, but the king promised never to issue a *quo warranto* against his patents.[47] From the first the prospective proprietors showed an unusual willingness to come to terms with the West Country mer-

---

[45] The exceptions were heard by the council, Mar. 10, 1637. The solicitor general was ordered to prepare a draft patent for submission to the king in council, and to communicate its provisions to the West Country agents, in order that they might have an opportunity to take exception to any clauses affecting the fishery. Order in council, Mar. 10, 1637, *Acts, Priv. Coun., Col.*, 1613-1680, §362.

[46] Order in council, Apr. 30, 1637, *ibid.*, pp. 216-217.

[47] Lord Baltimore objected to the proposed grant early in 1637. Memorial of Lord Baltimore touching his right to part of Newfoundland, Feb., 1637, C.O. 1: 9, no. 43. He followed this with a petition partly upholding the West Country cause. Baltimore to the king, May 7, 1637, *Cal. State Paps., Col.*, 1574-1660, p. 252. See also Charles I to the Commissioners for Foreign Plantations, May, 1637, *ibid.*, p. 253. Meanwhile, John Crewkerne and other West Country agents petitioned for a hearing, Apr. 19, 1637, and the council decided to listen to their objections to the Kirke patent on Apr. 26, 1637, *Acts, Priv. Coun., Col.*, 1613-1680, §364.

chants. They declared that they had no intention of handicapping the fishermen or preventing them from exercising their liberties and privileges at Newfoundland, but that they were anxious to improve the fishery.[48] How far the two interests were able to coöperate in improving the industry was demonstrated within a short time after the new applicants received their patent and sent Kirke to Newfoundland.

Through the intervention of the Privy Council several important points upon which the two parties could not agree were determined, the negotiations between the groups being carried on through the solicitor general.[49] Among the disputed points was the proposal of the new patentees to levy an impost of five fish per quintal upon foreigners buying from either the fishermen or the planters. The West Country merchants felt that this would discourage foreign "sack ships," as the carriers were called, from touching at Newfoundland. To meet this objection the proprietors agreed to contract to carry away the seasonal catch on the basis of the average annual yield for the preceding seven years. They also promised to pay the same rate for fish as was usually paid by English carriers buying there, and to do so upon the same terms. These arrangements were approved by order in council on June 25, 1637, and the applicants were assured of the prompt passage of their patent through the seals.[50] Afterward the agreement was made more specific. The West Country merchants agreed to inform the patentees of the quantity of fish they expected to deliver, notification to be given before the departure of the fishing fleet. In order to be certain of the acceptance of their fish,

[48] Order in council, Mar. 10, 1637, *Acts, Priv. Coun., Col.*, 1613-1680, p. 214.
[49] Order in council, Apr. 30, 1637, *ibid.*, 1613-1680, pp. 216-217.
[50] A clause imposing a duty of five per cent on foreigners fishing, drying fish, or purchasing it at Newfoundland was inserted in the charter. Order in council, Star Chamber, June 25, 1637, *ibid.*, 1613-1680, pp. 218-219.

THE FIRST WESTERN CHARTER 81

the adventurers demanded of the patentees security for the fulfilment of the contract. The proprietors promised that this would be furnished "by sufficient merchants of London," but in order to make assurance doubly sure, the Privy Council ordered that the proprietors' ships should not be permitted to sail until security had been given. The agreement thus concluded was to be carried out from year to year.[51] Thus, at least on paper, the two groups had finally come to an understanding.

Kirke and his associates obtained their patent, but with such serious limitations that the western adventurers were the real victors. The charter as finally issued severely restricted the proprietors in a number of important ways, and prevented in many particulars their interference with the fishery. According to its provisions neither the proprietors nor the settlers were to be permitted to dwell within six miles of the shore between Cape Race and Cape Bonavista, the zone most frequented by English fishermen. They were forbidden to build houses or to cut down trees within the six-mile coastal area, and prohibited from appropriating any of the best or most convenient fishing places or beaches before the arrival of the fishing ships from England. The inhabitants were forbidden to destroy any property belonging to the fishermen, but were to permit them to store it safely in Newfoundland until the return of the fishing fleet in the following spring. The patent specifically stated that all the king's subjects were to have freedom to fish at Newfoundland, as well as to dry, salt, and store their catch there, and to have the right to cut wood, build stages, boats, and other conveniences, to make arrangements for the care of shipmasters, servants, and sailors, and to undertake all necessary things, according to custom. The fishermen were specifically exempted from the

[51] Order in council, July 9, 1637, *ibid.*, 1613-1680, p. 220.

civil and military jurisdiction of the proprietors, not being subject to the courts of the plantation nor required to bear arms in case of war or rebellion. No tax, other than the usual customs duties, was to be levied upon the products of Newfoundland exported to England, Ireland, or any royal domain. The proprietors were forbidden to build forts within six miles of the coast, since this might interfere with the fishery. Moreover, laws and penalties enacted for the government of the plantation were not to extend to anyone coming to Newfoundland on a fishing voyage. This last provision was a reaffirmation of the liberty of free fishing granted previously in 1634. Most important of all was a clause inserted in Kirke's charter which exempted the fishery forever from interference by either the proprietors or the planters, but placed the industry directly under the jurisdiction of the crown. Thus, at last, the West Country people had attained their objective and had curbed a threatening competitor by their willingness to compromise with respect to the carrying trade, but they had so restricted the prospective plantation at the start that no one in the future was likely to attempt to colonize Newfoundland and establish a civil government there under such onerous conditions. Moreover, the restrictions in the charter of 1637 guaranteed the inviolability of the Western Charter of 1634 for the lifetime of Charles I.[52]

In spite of the many restrictions thus placed upon the proprietors and planters, the old antagonism between them and the West Country fishermen broke out afresh. The adventurers complained of the activities of Sir David Kirke, who went out as governor in 1638, and of the planters under his rule. In general the objections were similar to those made against John Guy and the set-

[52] The Kirke Charter, Pat. Rolls 13 Charles I, pt. 32; copy: C.O. 1: 9, no. 76.

tlers sent out by the Newfoundland Company a quarter of a century before. The justices of the peace of Devonshire informed the Privy Council in 1640 that the planters had destroyed property belonging to the fishermen and had appropriated the best fishing places in the harbors where they resided, which they disposed of to foreigners or to anyone offering a good price. Sir David Kirke was criticised for licensing taverns for the sale of spirits, wine, beer, and tobacco in violation of his charter, thus rendering the fishermen unfit for their work. The West Countrymen declared that if these abuses were allowed to continue they would cause the decay of the Newfoundland fishery and shipping trade, with a consequent decline in the "seminary for seamen," and injure the fishermen and their families, who were dependent upon the trade for a living. They had nothing to say against the charter of 1637 nor any of the proprietors except Kirke, who, with the planters, was largely responsible for the violations. The Privy Council took the matter into consideration and directed the attorney and solicitor general to consider the charges. Meanwhile the council instructed Kirke to conduct himself and his colony in conformity with the charters of 1634 and 1637, and made one concession to the planters, which virtually admitted the impossibility of requiring the settlers to live inland and remain aloof from the fishery, by permitting them to reserve a fair amount of fishing room in each of the four harbors frequented by the fishing ships belonging to the proprietors.[53] From this it will be seen that the patentees had intended from the first to participate in the fishery.

The law officers reported that the charges of the western adventurers were denied by the defendants. Conse-

[53] Justices of the peace of Devon to the Privy Council, Jan. 10, 1640, C.O. 1: 10, no. 46; Robert Gabbes, mayor of Plymouth, etc., to Archbishop Laud, Jan. 22, 1640, *Cal. State Paps., Dom.*, 1639-1640, p. 363.

quently the Privy Council ordered a commission to take evidence in the West Country and report its findings.[54] Kirke wrote the council that he had complied with its orders by sending warrants to all the planters and fishermen calling upon them to observe the Western Charter, maintaining that he was innocent of the charges brought against him in the West Country, and laying countercharges against some of the shipmasters who had destroyed stages and cook-rooms. He asserted that some of "the most civil and wisest masters amongst them" had complained to him of these outrages, and considered anyone foolish who would presume to interrupt the valuable fishery at Newfoundland.[55] The council took no further action to settle the dispute, probably owing to the disturbed political condition in England, and the question of the desirability of exploiting Newfoundland along the lines favored by the western adventurers or those desired by the proprietors and planters was postponed, to be decided many years later.

In 1637 the Venetian ambassador in England prophesied that the Hamilton-Kirke project would bring the proprietors into conflict with the French.[56] Some of the English merchants trading to France became so alarmed that they coöperated with the fishing interests in opposing the patent. The Company of French Merchants of London, and a similar organization in Exeter, helped to pay the expenses of one of the West Country agents dur-

[54] Order in council, Sept. 9, 1640, *Acts, Priv. Coun., Col.*, 1613-1680, p. 290.

[55] Sir David Kirke to the Privy Council, Sept. 12, 1640, C.O. 1: 10, no. 77, in reply to the council's letter, Mar. 11, 1640, *Acts, Priv. Coun., Col.*, 1613-1680, pp. 279-280. Cf. John Harrison to John Winthrop, 1639, *Winthrop Papers* (*Collections* of the Massachusetts Historical Society, 5th Series), I, p. 120. Cited hereafter as *Winthrop Papers*.

[56] Anzolo Correr, Venetian ambassador, to the Doge and Senate, Apr. 17, 1637, *Cal. State Paps., Ven.*, 1636-1639, §199.

THE FIRST WESTERN CHARTER 85

ing the hearings of that year.[57] Sir David Kirke, after his arrival at Newfoundland, tried to impose a duty upon French ships fishing there, in addition to the levy he was entitled to make upon foreign carriers. He sent a fleet of three ships and a pinnace to the southern coast in 1639 to collect the duty from some Frenchmen there, but their fleet of fourteen sail was too strong for the proprietor's little squadron, and he was obliged to abandon the attempt.[58] He must have been successful in collecting from some of the French fishermen, because the French ambassador complained of the exactions. Secretary Windebank was directed to be firm in his reply to the French envoy and to explain Kirke's action as a retaliation against recent impositions placed upon English merchants in France.[59] If English authority were to be asserted at Newfoundland in the interest of developing the nation's commerce, the imposition of duties upon foreigners catching or buying fish would be a practicable way to accomplish the purpose. It would have the effect of driving away the aliens and leaving England sole master of the fishery. A memorandum, made probably during the investigation of 1640, indicates the persistence of those anxious to eliminate foreign competition, who claimed that English interests would be less injured if both the fishermen and planters had less to do with foreigners. The nation's shipping, they said, would be advanced if the Dutch were prevented from buying fish at Newfoundland, "which they buy in great abundance to the hurt of English merchants, taking the prime of the market," and

[57] Governor [of the Company of French merchants] of Exeter to the mayor of Barnstaple, 1637, Historical MSS. Commission, *Manuscripts of the Corporation of the Borough of Barnstaple*, p. 215. Cited hereafter as *Barnstaple MSS*.
[58] Harrison to Winthrop, 1639, *Winthrop Papers*, I, 120.
[59] Sec. Coke to Sec. Windebank, May 16, 1639, *Cal. State Paps., Dom.*, 1639, p. 179; *Cal. State Paps., Col.*, 1574-1660, p. 294.

the crown would be the gainer on the same account, because the fish, if carried to England, would pay a subsidy and customs duty. If the crown did not see fit to prohibit the Dutch and other aliens from trading altogether at Newfoundland, the foreigners ought at least to pay the same export duties there that were required in England, which would amount to over a thousand marks a year. The writer of the memorandum notes that a scheme for levying these duties at Newfoundland could be easily worked out.[60] In spite of the desire of the English to eliminate their competitors from the fishery and carrying trade, Kirke's attempt to levy taxes on the French only succeeded in intensifying the ill-feeling already existing between the two nations. The Dutch problem was not settled until the close of the third war with Holland, although after 1652 the Dutch seem to have become less important than the French as competitors at Newfoundland. The governments of the Commonwealth and Protectorate tried to effect the collection of an impost of five fish per quintal. They issued orders to shipmasters and others to accomplish it but their plans were not carried out. As late as 1660 the Kirke family asked the government to instruct the convoy commanders to assist them in collecting the duties.[61] The request was denied, and except for an occasional attempt to revive it the impost on foreigners was never seriously considered in later years.

The disputes between the western adventurers and the proprietary interests from 1610 to 1640 show that the two groups entertained fundamental differences of opin-

[60] Memorandum concerning "the bill" to be preferred by the West Countrymen [1640], C.O. 1:10, no. 80; hearing by the council of the complaints of Fowell of Plymouth, representing the western adventurers, Mar. 8, 1640, *Acts, Priv. Coun., Col.*, 1613-1680, §460; Richard Breton to [Windebank], 1640?, *Cal. State Paps., Col.*, 1574-1660, p. 315.

[61] Coke to Windebank, May 16, 1639, *Cal. State Paps., Dom.*, 1639, p. 179; *Cal. State Paps., Col.*, 1574-1660, p. 294.

ion relative to the conduct of the Newfoundland fishery and the carrying trade, based on very divergent economic views. Aside from the profits accruing to the individuals concerned in which all were naturally interested, the proprietors and their allies, the London merchants and those of some of the outports, had as their chief objective the increase of the national wealth. They desired to conduct the fishery as a national industry by creating a monopoly that would strengthen efficiency and reduce foreign competition. They were attempting to regulate a public trade by means of private control at a time when commercial legislation was still rudimentary and undeveloped and when the government had not yet accepted a position of responsibility with respect to the formulation and execution of a national commercial policy. The proprietary group sought to apply their theories through the agency of corporate enterprise. In order to accomplish this they had to obtain governmental regulations restricting the activities of foreigners in the carrying trade, and to provide a group of dependents to assist in the industrialization of the fishery, and to provide not only a government capable of controlling the fishery in Newfoundland, but also one that would be forceful enough to eliminate England's competitors. Adequate governmental regulation failed to materialize for two reasons: first, because of the determined opposition of the western adventurers; and secondly, because the government under the early Stuarts lacked a definitely settled commercial policy. The crown wavered in its support from the monopolists to the West Country fishing interests and back again. Little could be expected from parliament during the years that it was in session because the House of Commons was influenced by the western adventurers and the House of Lords was indifferent to legislation affecting trade. Successful colonization in Newfoundland was hampered not only by the unfavorable climate and the unsatisfactory character of the soil for agriculture, but also by the restrictions placed upon the proprietors and planters in the charters of 1610, 1622-1623, 1634, and 1637. Moreover, the patentees had difficulty in obtaining sufficient money to carry on their projects. The London and Bristol Company was in poor financial straits by 1620; Calvert, Vaughan, and Falkland, either invested very little or else lost more than they gained; while the Hamilton-Kirke group, although better organized and evidently possessing funds at the start, was handicapped by the disturbed political conditions which existed in England after 1640 which naturally had a depressing effect on all commercial and colonizing enterprises at that time.

Of all the groups which sought to colonize Newfoundland or to control the fishery and carrying trade there, the Hamilton-Kirke association had the most definite plans. They were more conservative than the others in their expectations and in the policies which they adopted. They had profited by the mistakes made by Guy's company, and tried to avoid some of the pitfalls into which that enterprise had fallen. Then, too, they were outwardly much more conciliatory in their attitude toward the West Countrymen than were the merchants. They did not seek to eliminate the western adventurers at one stroke, but expected to undertake their own fishery in such an improved manner that the fishing interests would eventually be driven from the field or accept their control. Their willingness to accept the severe restrictions contained in the patent of 1637 shows that they did not expect immediate results, but sought rather to attain their ends first by making their establishment an accomplished fact and then by beginning a process of boring into the fishery. The whole question of proprietary control at Newfoundland challenged the western adven-

THE FIRST WESTERN CHARTER 89

turers in more respects than the one matter of administering the regulations governing the fishery. It raised the issue as to whether the new policy of conducting fisheries by permanent residents under civil government throughout the entire year was not more efficient and profitable to the nation economically than the very brief fishing season conducted by non-residents. The profits of the latter were eaten into by the cost of maintaining the fishing fleet and their economic ideas were so localistic that they were willing to permit the foreign carrying trade to obtain the benefits which otherwise would have assisted England in building up her national wealth.

Selfish though they may have been, the proprietary interests and their London associates had a much larger vision than did the West Country group. They felt that under the existing system the profits of the individual merchant and of the nation were smaller than they ought to be. They believed that the carrying trade could be encouraged by the impost on fish, and by contracts between the shipping interests and the western adventurers which would call for the transportation of Newfoundland fish in English bottoms. The Kirke group in particular was convinced that the fishery could be made more efficient by substituting the wage system for the old West Country practice of fishing on the share basis, while all proprietors considered the old coöperative method wasteful. True capitalists in the modern sense of the word, they were certain that the individual merchant was deprived of a considerable profit because of the payment of labor on a share basis, and, therefore, was unable to obtain additional funds to expand his business. They thought that the payment of wages would eliminate a good deal of the loss and that other expenses would be cut by the manufacture of salt in Newfoundland and by furnishing provisions there for the return voyage. The western adventurers bought foreign salt, and also had a good deal of capital tied up in provisions for the entire fishing voyage. The patentees intended to engage in brewing beer, evaporating salt, and making bread, in carrying on the manufacture of potash, and in exploiting the iron deposits of the island.[62] All of these activities were intended to assist in the more efficient operation of the fishery. From the modern point of view the Hamilton-Kirke group had a wider perspective than their antagonists, but their project had in it too much of the narrow economic nationalism and industrialism of later times to make one very sympathetic with their aims.

The western adventurers naturally resented and resisted these innovations. They objected particularly to the introduction of the wage system. The old system of fishing on shares which had been practiced for many years had made the Newfoundland fishery an industry wherein the entire community of the West of England participated either directly or indirectly. In good fishing seasons the merchants, shipmasters, and crews all shared proportionately in the profits of the voyage. In poor seasons the losses were borne by all concerned. The vessels used in the trade were fitted out, supplied with provisions and other necessaries, and their crews recruited in the West Country. Local farmers and tradesmen were interested in the prosperity of the fishery because the profits were spent at home. Is it any wonder then that the western adventurers were alarmed at any proposal which threatened to alter the economic structure of the fishery? Any radical change would seriously affect the fundamen-

[62] Sir David Kirke's reply to the "answer to the descrip: a of Newfoundland," Sept. 29, 1639, C.O. 1:10, no. 38. His "Description of Newfoundland" is listed among the papers deposited with the council in Jan 1637, *Cal. State Paps., Dom., Addenda*, 1625-1649, p. 549. See also Kirke to Laud, Oct. 2, 1639, C.O. 1:10, no. 40; the same to the Privy Council, Sept 12, 1640, C.O. 1:10, no. 77.

THE FIRST WESTERN CHARTER 91

tal basis of life in most of southwestern England. The introduction of the wage system would reduce the interest of the fishermen in their work, and throw out of employment those who were unwilling to lose their economic independence by condescending to work for hire. Under the plan for a highly organized industry the capitalist would look after himself and disregard the rights of his employee.[63] With all their faults, chief among them being a narrow localism which refused to permit them to look beyond the confines of the southwestern counties and an extreme conservatism which prevented their accepting with ease any new practices, the western adventurers were too strongly entrenched to be excluded by the monopolistic movement of the early seventeenth century. The coöperative spirit and sectional solidarity extant in the West Country made it impossible to oust them and difficult to control them. They were deaf to the appeal to the spirit of economic nationalism which the London merchants and the proprietors attempted to arouse. That both sides were selfish cannot be gainsaid, but the tenacity of the western adventurers in supporting the old medieval conception of trade was to affect the history of Newfoundland and the fishery for many years after the early attempts to found organized plantations in Newfoundland had failed.

[63] In spite of the continued quarrel between Kirke and the western adventurers, and in spite of attempts to end the dispute, no definite action was taken during the remaining years of the reign of Charles I. The last reference to any action is the order in council, Sept. 9, 1640, *Acts, Priv. Coun., Col.*, 1613-1680, §415.

CHAPTER III

THE FISHERY UNDER CROMWELL
AND CHARLES II

THE Interregnum and the Restoration were significant epochs in the development of English commerce. The foundations of mercantilist policy were laid at this time and serious attempts were made to apply the doctrine to all trades. During the days of the Commonwealth and Protectorate the merchant came into his own. Thereafter he was to participate increasingly and more actively in the government of his country, and the government in turn was to heed his requests and harken to his advice much more than had been the case in the early seventeenth century. At this time the state became definitely aware of the importance of foreign trade in the economic life of the nation. The first war with Holland was dictated in part by the demands of the merchants for the elimination of the Dutch from the fisheries surrounding the British Isles and for their exclusion from the carrying trade. Cromwell's war with Spain was undertaken in order to extend England's plantation system and to break down the Spanish commercial monopoly in America. The merchants demanded the extension of national trade in order that greater wealth might accrue to England and also to the traders themselves. In spite of the attempt of Charles II to set the clock back to 1649, too many changes had occurred in the interim to warrant a return to the *laissez-faire* attitude of the early Stuarts. The restored monarch was under obligations to a con-

UNDER CROMWELL AND CHARLES II 93

siderable body of merchants for his return to the throne. In spite of his personal leanings toward France, his administration was obliged to heed the desires of the merchants. Consequently, a good many of the commercial policies inaugurated during the Interregnum were continued after 1660, being altered and amplified to meet the new circumstances. Although in some respects a new order of things grew up in this period, a good deal of the old remained to hamper the full development of a governmental commercial program. In spite of such innovations as the various acts of trade and navigation, and in spite of the government's assumption of responsibility for the regulation of trade in accordance with those statutes, the persons accountable for the administration of these laws were hampered by the revival of interest in the private control of national commerce and colonization. The government continued to charter great companies and to make extensive grants to feudal proprietors for the purpose of widening the bounds of the plantation system and expanding English foreign trade. Both the attempt to regulate commerce by statute and the recrudescence of the proprietary movement had considerable effect upon the Newfoundland fishery during the period from 1650 to 1675.

With respect to Newfoundland the Interregnum was a period of transition. The fall of the monarchy left the fishery under the dominance of the West Country, and in large measure the policies of the western adventurers were adopted by the new government. When the Commonwealth established the Council of Trade in 1650, the importance of the fishery in the economic life of the nation was recognized and its value as a naval reserve was appreciated. The act of August 1, 1650, "for the Advancing and Regulating of the Trade of this Commonwealth" contained in its eleventh clause directions to be followed by the Council of Trade relative to the fisheries. The council was to consider the entire fishing trade, not only that of England and Ireland, but also the overseas fisheries of Iceland, Greenland, Newfoundland, New England, and elsewhere, in order that "the fishermen may be encouraged to go on in their labors, to the increase of shipping and mariners."[1] The problems of Newfoundland occupied the attention of this council to some extent, and the Council of State was also interested. Parliament, too, was concerned for the prosperity of the fisheries, and besides paying special attention to the activities of Sir David Kirke and the planters, passed several laws which affected the commerce and industry of Newfoundland.

Among these laws were several, enacted between 1647 and 1657, which applied either generally or specifically to the Newfoundland trade. That of January, 1647, freed goods designed for the American plantations of all duties except excise for a period of three years. Goods for the plantation at Newfoundland were explicitly denied this privilege. The act also provided that goods which were the growth of the plantations were not to be shipped to foreign countries except in English bottoms.[2] It is doubtful whether this requirement was intended or not to apply to the products of the island, but the reservation specifically exempting the plantation at Newfoundland from the benefits of the law indicates not only that the West Country interests were still influential, but also that the government did not regard Newfoundland as being a true plantation, a definition which continued to be accepted

[1] "An Act for the Advancing and Regulating of the Trade of this Commonwealth," Aug. 1, 1650, clause 11, C. H. Firth and R. S. Rait, eds., *The Acts and Ordinances of the Interregnum*, 1642-1660 (3 vols., London, 1911), II, 405. Cited hereafter as *Acts and Ordinances, Interregnum*.

[2] "An ordinance for encouragement of adventurers to the several plantations . . .," Jan. 23, 1647, *Acts and Ordinances, Interregnum*, I, 912, 913.

UNDER CROMWELL AND CHARLES II 95

for many years.[3] The Navigation Act of 1651 contained three provisions affecting Newfoundland. Besides a general clause confining imports from Asia, Africa, and America to English vessels, there were two others pertaining directly to the fisheries. One of these was designed to prevent the importation of foreign-produced fish and train oil, which could not be introduced into England, Ireland, or any English possession unless they had been produced by citizens of the Commonwealth and unless the shipping employed was owned by them. Another clause provided that after February 1, 1653, no cod, ling, herring, pilchard, or any other salt fish could be exported except in ships owned by citizens of the Commonwealth, the master and a majority of the crew of each of which were English. Penalties and forfeitures were provided for violations of the act.[4] Although this law was obviously intended to strengthen English commerce by excluding the Dutch and was applicable to the entire fishing industry and carrying trade, it was modified, at the behest of the western adventurers, during the war with Spain (1656-1659). Then an emergency measure was enacted by parliament which again threw open the carrying trade to foreigners and for this reason. Owing to hostilities, the West Country was unable to dispose of its fish in the best market, while at the same time there was a great demand for it in Spain. Unless they could work through the agency of neutrals, the western merchants feared that the fish would rot before they could sell it. George Downing, always interested in matters relating to trade and colonization, introduced a bill, November

[3] There is no specific statement excluding the Newfoundland carrying trade from the provisions of the act, but exemption from the payment of all duties except excise specifically omits Newfoundland, and also provides for the carriage of plantation goods in English bottoms.

[4] "An Act for the increase of Shipping, etc.," Oct. 9, 1651, *Acts and Ordinances, Interregnum*, II, 559-560.

1656, to relax the restrictions on foreign carriers, with Captain Henry Hatsell and John Fowell, representing Devonshire constituencies, its chief advocates. There was some opposition, but the bill finally passed and received the approval of the Lord Protector on June 9, 1657. It permitted certain specified fish to be exported in either domestic or foreign bottoms from England, Scotland, or Ireland, the ships' crews to be of any nationality. Export duties were to be paid, and the duty on Newfoundland fish was fixed at 3*d.* per quintal. Aliens were permitted to purchase fish in Newfoundland and New England and to carry it from those possessions to foreign markets free of export or other duties. The act was to continue in force until December 25, 1659.[5] From the position assumed by Hatsell and Fowell it is clear that the western adventurers were again behind the movement to modify the restrictions upon foreigners in the carrying trade, because any Anglo-Spanish differences were bound to have an adverse effect upon the Newfoundland trade, and consequently upon West Country prosperity.

The Puritan authorities were decidedly favorable to the western adventurers. At their behest an official investigation was undertaken of the activities of Sir David Kirke. In 1646, before the establishment of the Commonwealth, the merchants and shipowners of Plymouth had petitioned Parliament for action against Kirke, accusing this "notorious malignant" of occupying the best fishing places, destroying the property of the adventurers, and debauching the fishermen and seamen in taverns. The charges were not new, but as a supporter of the royalist cause Kirke had threatened to seize the fishing ships, alleging that their operators were rebels. The Plymouth

[5] "An Act for giving License for Transporting of Fish in Forein Bottoms" [June 9, 1657], *Acts and Ordinances, Interregnum,* II, 1099-1100. Cf. Stock, *Proceedings and Debates,* I, 237-242, *passim.*

people asked that Kirke be brought home to answer their charges, and demanded that the fishing trade should have adequate protection in the future. The petition was referred to the Committee for Foreign Affairs but no action was taken at the time.[6] Kirke, meanwhile continued to make the plantation a center for royalist activities. In 1649, Prince Rupert was expected to use Newfoundland as a base for his operations against the commerce of the Commonwealth, and the government was sufficiently alarmed to provide a convoy for the Newfoundland fishing fleet.[7] Kirke recruited several hundred able seamen in the West Country, by offering them high wages ostensibly for work in the fishery, but the Puritan authorities suspected that the men were destined to join Rupert's fleet. The press-masters were dismayed at the prospect of losing so many skilled sailors to the royalists, and steps were taken to prevent the departure of the ships upon which the men had embarked.[8] Although reports and rumors of this character were received by the officers of the Commonwealth at Whitehall, it was some time before the long-suspected Kirke family were subjected to a serious investigation of their activities in the royalist cause. When Kirke was finally examined, it was done at the request of the western adventurers.

The plea of 1647 having failed to produce any results,

[6] Petition of the merchants, etc., of Plymouth to Parliament [Mar. 24, 1646], Stock, *Proceedings and Debates*, I, 177, 178.

[7] Order of the Council of State to the generals at sea, May 7, 1649, *Cal. State Paps., Dom.*, 1649-1650, p. 128; *Cal. State Paps., Col.*, 1574-1660, p. 329; Council of State to Vice Admiral Moulton, Feb. 24, 1649, *Cal. State Paps., Dom.*, 1649-1650, p. 18; order of Parliament to the Council of State, Feb. 26, 1649, *ibid.*, p. 19; proceedings of the council, Apr. 13, 1649, *ibid.*, p. 83.

[8] Council of State to Moulton, Feb. 20, 1649, *Cal. State Paps., Dom.*, 1649-1650, pp. 9-10; the same to the customer, comptroller, and searcher of the ports of Plymouth, Dartmouth, and Barnstaple, Feb. 23, 1649, *ibid.*, p. 16.

98   BRITISH FISHERY AT NEWFOUNDLAND

the Plymouth merchants and shipowners again petitioned in 1650. Now that the king had gone to the block there was more chance that the government would heed their request. The charges against him were that he was partial to the planters, favorable to the French, and unfriendly to the Commonwealth.⁹ No definite action was taken at the time, although when Lewis Kirke proposed a voyage to Newfoundland in 1650, he was obliged to give a bond of £2,000 not to injure the interests of the Commonwealth while in the island.¹⁰ Busy with other and more pressing matters, the Council of State did not move to settle the affairs of Newfoundland until in 1651 and the two years following the whole question was reviewed by its committees.¹¹ Commissioners were appointed to go to Newfoundland in 1651 to protect the interests of the Commonwealth, to take possession of all arms, ammunition and fishing equipment belonging to Kirke, and to turn these materials over to the western adventurers for their use. The commissioners were also instructed to collect the imposts upon foreigners "until Parliament declare their future pleasure."¹² Captain Thomas Thoroughgood was commissioned to arrest Sir David and bring him back to England.¹³ He arrived home some time in the autumn of 1651 with Kirke in his custody. Sir

---

⁹ Petition of the merchant adventurers of Plymouth to the Council of State [1650], *Winthrop Papers*, I, 499-501.

¹⁰ List of recognizances to the Council of State, May 25, 1650, *Cal. State Paps., Dom.*, 1650, p. 516.

¹¹ Orders of the Council of State, Mar. 17, 1651, Apr. 2, 1652, *Cal. State Paps., Col.*, 1574-1660, pp. 353, 376. Kirke's petition was referred to the Newfoundland committee, Apr. 12, 1652, *ibid.*, p. 377.

¹² Warrant to John Littlebury, John Treworgie, Walter Sikes, Capt. Thomas Thoroughgood, Capt. Thomas Jones, and Capt. William Haddock, Apr. 8, 1651, authorizing any two of them to take possession of Kirke's property, etc., for the use of the western adventurers, *ibid.*, pp. 354-355.

¹³ Council of State to Capt. Thomas Thoroughgood, Apr. 8, 1651, *ibid.*, p. 354.

UNDER CROMWELL AND CHARLES II 99

David set about immediately to defend himself, but the government took its time before investigating his affairs and those of the fishery.

The investigation begun in October 1651 was not finished until the midsummer of 1653. During the autumn of 1651 Kirke presented a written account of current conditions in Newfoundland to the Council of State.[14] During November and December his accounts were being audited at Worcester House, but not until January 1652 did his examination really get under way.[15] On January 2, the council ordered Kirke to appear and directed that the complaints against him be presented.[16] A committee was appointed to conduct the investigation, to peruse the papers describing Sir David's conduct at Newfoundland, to call for an account of the shares and profits which he had forfeited to the state, and to report upon the entire matter to the council.[17] Kirke was held under bond not to leave England and to remain in readiness to attend the committee whenever he should be called.[18] The committee made its first report in June 1652, which was distinctly favorable to the West Country. It recommended that the best interests of the Commonwealth would be served and the fishery best encouraged were the proprietary interests entirely eliminated, and that the fishery should be conducted by the western adventurers as before, but it added the significant recommendation that both the fishery and the plantation should be supervised

[14] Order of the Council of State, Oct. 31, 1651, *ibid.*, p. 363.
[15] Order of the Council of State to ——— Rowe, Nov. 7, 1651, *ibid.*, p. 364.
[16] Order of the Council of State, Jan. 2, 1652, *ibid.*, p. 364.
[17] The personnel of the committee was specified in an order of the Council of State, Jan. 12, 1652, *ibid.*, p. 376. The group named were identical with the Committee for Foreign Affairs. Order of the Council of State, June 7, 1652, *ibid.*, p. 381; minutes of a Committee for Foreign Affairs, June 11, 1652, *ibid.*, p. 381.
[18] Order of the Council of State, Jan. 30, 1652, *ibid.*, p. 373.

by commissioners appointed by the government. Kirke or a person representing him was to be permitted to go to Newfoundland to settle the proprietor's affairs there.[19] As a result of this recommendation, the Council of State commissioned Walter Sikes, Captain William Pyle, John Treworgie, and probably Robert Street to administer Newfoundland in the interests of the Commonwealth, and directed them to seize the estate of Sir David Kirke.[20] Treworgie remained in the island for several years, probably from 1652 to 1659 or 1660, but was finally obliged to return to England for want of funds, his salary as commissioner being six years in arrears. Convinced by his experience there that some sort of governmental regulation was necessary, he requested that he be given another commission, and asked for two or three ships of war to assist him in collecting the imposts from foreigners and to protect English shipping.[21] The impending Restoration altered the situation entirely, and Treworgie failed to receive any further recognition as a government agent. He did, however, revive the old question of adequate government for the island fishery, which was to receive much more serious consideration than it had ever had before in the years which were to follow.

[19] Order of the Council of State, June 7, 1652, *Cal. State Paps., Col.*, 1574-1660, p. 381; minutes of a Committee for Foreign Affairs, June 11, 1652, *ibid.*, p. 381.

[20] Petition of Walter Sikes, Capt. William Pyle, and John Treworgie to the Lord Protector, Apr. 24, 1654, C.O. 1: 12, no. 20; item referring to the salary for Robert Street, employed as a commissioner for Newfoundland in 1652, Dec. 5, 1655, *Cal. State Paps., Col.*, 1574-1660, p. 433; see also p. 437.

[21] Petition of John Treworgie, commander of the English colony in Newfoundland, appointed in 1653, begging a commission to order affairs there [1659?], Egerton MSS., 2385, fol. 262. He was originally told to return home at the end of the fishing season, but according to the above he appears to have remained in Newfoundland for several years or else made several successive voyages. His instructions, June 3, 1653, *Cal. State Paps., Col.*, 1574-1660, p. 403. See the order of the Committee of Council for Plantations, Apr. 27, 1660, on Treworgie's petition, C.O. 1: 33, no. 73.

UNDER CROMWELL AND CHARLES II 101

The establishment of the Commonwealth in 1649 necessitated the issuance of new laws and ordinances regulating the fishery. That the new government was favorably inclined to the fishing interests of western England is obvious from its treatment of Kirke. The temporary rules issued by the Council of State in 1652 and again in 1653 were intended to apply only until parliament should establish some form of permanent regulation. The predominance of the western adventurers was somewhat curtailed by the presence of the resident commissioners in Newfoundland, and by the omission from the temporary rules of those clauses of the old charter of 1634 which required the trial of criminals in England and gave the mayors of the western ports supervision over the discipline of the fishery. On the other hand two clauses were added which were distinctly favorable to the West Country. One of these limited the planters in their possession of stages and fishing rooms, in the construction or maintenance of houses and buildings, and in keeping livestock near places used for storing or drying fish. The intent of these restrictions was the same as in the case of the limitations placed upon the planters by the Hamilton-Kirke charter of 1637—the elimination of the inhabitants from the fishery. Another clause, providing for the open sale of provisions, was intended to prevent engrossing, and was included as a result of the complaints of the fishermen against Kirke and the planters, whom they accused of monopolizing salt and other necessaries.[22] Though these rules were only temporary they remained in force throughout the period of the Interregnum. Thus, owing to the increasing interest in national commerce, the regulation of the Newfoundland fishery and carrying trade

[22] "Laws, rules and ordinances, whereby the affairs and fishery of the Newfoundland are to be governed until Parliament shall take further order," June 16, 1652, C.O. 1: 38, no. 33 (iii).

received a good deal of attention from the government of the Commonwealth and the Protectorate.

The two most important events of the period, with regard to Newfoundland, were the appointment of the commissioners to conserve the interests of the Commonwealth and the establishment of a regular system of naval protection for the fishing and market fleets. The use of commissioners, although discontinued after the return of Treworgie, brought officials of the government into direct contact with Newfoundland for the first time. The arrangement suggested the further possibility of erecting a permanent civil authority in Newfoundland under the direct control of the central government in England, and implied close supervision by the officials at home. This implication almost became a reality when, in 1656, the Council of State was on the point of establishing a permanent administration in Newfoundland.[23] About the same time a plan to carry on the fishery by reëstablishing the plantation was proposed.[24] Nothing came of either of these proposals, except that the direct interest of the government in the problem of Newfoundland from this time forward became more pronounced.

The most important and lasting contribution of the Interregnum to the development of the trade and fishery was the establishment of the convoy service. During the early Stuart period the western adventurers had resisted attempts to provide naval protection for the Newfoundland fleet beyond the waters immediately surrounding the British Isles. During the Civil Wars and the wars with Holland and Spain in Cromwell's time, the merchants and shipowners experienced a change of heart. Menaced

[23] Order of the Council of State, July 1, 1656, *Cal. State Paps., Col.*, 1574-1660, p. 433. Cf. order of June 3, 1656, *ibid.*, p. 441.

[24] Regulations for the Newfoundland trade and fishery were drawn up by Hugh Lamy, probably sometime during the period of the Interregnum, Additional MSS., 5489, fols. 39-40, 41.

by Prince Rupert's fleet, and threatened by Dutch and Spanish warships, their ships required protection. Therefore, when it could do so, the government furnished convoys.[25] In 1566, the Council of State ordered the preparation of instructions for the convoy commanders and the "governor" of Newfoundland. This step was antecedent to the practice, which was adopted regularly after the Restoration, of granting the convoy commanders supervisory powers over the fishery and plantation. This form of regulation was substituted for the supervision previously exercised by the mayors and magistrates of the West Country. By 1656, the government was well aware of the importance of Newfoundland and the other English fisheries. During this period some relief was given to the western adventurers by relaxing the requirements relative to the impressment of seamen in the Newfoundland trade.[26] In 1659, when the Council of Trade was established by act of parliament, one of its duties was to encourage the fishing on the coasts of England, Scotland, Ireland, and Newfoundland.[27] It is significant that the New England fishery was omitted from the act, probably because the people of that region were conducting the industry so successfully that they had no cause to bring pressure upon the home government. The Newfoundland fishery, however, was undergoing investigation by the Council of Trade when Charles II was restored to the throne of his fathers, and the work went on, apparently without much interruption. The Interregnum is a signifi-

[25] Admiralty Committee to the Navy Commissioners, Mar. 1, 1650, *Cal. State Paps., Dom.*, 1650, p. 17. There was no convoy in 1652. Minutes of a Committee of Foreign Affairs, Dec. 15, 1652, *Cal. State Paps., Col.*, 1574-1660, p. 394; order of the Council of State, June 5, 1655, *ibid.*, p. 425; the same, Apr. 14, 1656, *ibid.*, 1685-1688, §2003.

[26] Order of the Council of State, June 5, 1655, *ibid.*, 1574-1660, p. 425; the same, June 3, 1656, *ibid.*, p. 441; the same, July 1, 1656, *ibid.*, p. 433.

[27] Act of Parliament constituting a Council of Trade, May 19, 1659, *Cal. State Paps., Dom.*, 1659-1660, pp. 349-350. See especially clause 21.

cant period in the history of the Newfoundland fishery, because during that time, some of the policies which later became fixed practices were inaugurated. Although the Restoration brought with it some reactionary tendencies, the Newfoundland fishery and carrying trade were henceforward more directly the concern of the state, and the merchants, although they did not always agree among themselves or with the plans proposed by the authorities, became increasingly aware of the value of government support for this important branch of the national trade.

The encouragement given to the western adventurers by the Commonwealth and Protectorate suffered a temporary setback after the accession of Charles II. Cecilius Calvert, Lord Baltimore, was responsible for the reactionary step. In June 1660, he petitioned the king praying that the Province of Avalon might be confirmed to him, on the strength of the patent which his father had received from James I. Baltimore attacked the validity of the Hamilton-Kirke patent of 1637, which he maintained had been procured surreptitiously. He denied that his father had abandoned Avalon as the rival group of proprietors alleged, maintaining that Sir George Calvert had appointed Captain William Hill governor of the province when he left for Virginia. Hill was still in residence when the others obtained their grant to the entire island. Kirke had forcibly dispossessed Baltimore's colony, and although the proprietor had complained of this he had been unable to secure satisfaction owing to the outbreak of the Civil War. Baltimore made a serious charge against Kirke, which, if substantiated by other evidence, would go a long way toward explaining the interest of the Cromwellian government in Newfoundland. He alleged that in 1655, Kirke had obtained the support of the Protector for a revival of his enterprise at Newfoundland by making over five-sixths of his patent to

John Claypoole, Colonel Robert Rich, Cromwell's sons-in-law, Colonel Goffe, and others. On the death of Sir David, which occurred shortly before the return of Charles II, Sir Lewis Kirke and Sir James Kirke, his brothers, had endeavored to secure confirmation of the patent of 1637, in behalf of themselves and Sir David's sons who were resident in Newfoundland. Baltimore asked that the Kirke petition be rejected, and that he should be restored to his possession.[28]

Baltimore's plea for recognition of his rights not only reopened the controversy between the Kirke and Calvert proprietary groups, but also introduced once more the question of governmental supervision of the fishery. On June 17, 1660, the petition was referred to the lord chief baron of the Exchequer, Sir Orlando Bridgman, and to Sir Heneage Finch, the solicitor general. These law officers were instructed to investigate the rights and interests of the Kirkes in Newfoundland, and to determine to whom the rights properly belonged.[29] On July 14 they heard the two parties. Lord Baltimore presented his case, alleging that he had suffered damages to the extent of about £30,000 besides the loss of his father's initial investment of £20,000 at Ferryland. He attacked the Kirkes, and recounted his efforts during the Civil Wars and the Interregnum to secure satisfaction from them.[30] Baltimore's claim for damages was met by a counter claim from the Kirkes, demanding compensation for the upkeep of the province of Avalon during the time that it

[28] Petition of Lord Baltimore to the king [June, 1660], inclosed with an order in council, June 17, 1660, C.O. 1: 14, no. 9.

[29] Order in council, June 17, 1660, C.O. 1: 14, no. 9.

[30] Lord Baltimore's case concerning the Province of Avalon, no date, Egerton MSS., 2395, fol. 310; petition of Sir Lewis Kirke in behalf of himself and the sons of Sir David Kirke, deceased (1660), C.O. 1: 14, no. 8; Lady Kirke to Charles II, asking that the government of Newfoundland be conferred upon her eldest son, George, as governor, Egerton MSS., 2395, fol. 258; testimony of William Wrixon and others, Sept. 13, 1661, *ibid.*, fol. 309.

had been under their control. Baltimore naturally objected to this claim. Subsequently he submitted a second petition which was referred to a committee composed of Sir Heneage Finch, Sir James Ware, and Sir Maurice Eustace, who were instructed to report their findings to the king.[31] Consideration of the conflicting claims seems to have been undertaken by Finch and Bridgman, who was now chief justice of the Court of Common Pleas. After hearing the Kirkes and their counsel, both commissioners felt that the Avalon patent was still in force, not having been voided by the later patent to the Hamilton-Kirke group. Bridgman and Finch found it difficult to recommend a satisfactory way out of the difficulty, at the same time conserving the royal service and the interest of the public.[32] Finally royal letters of command and a proclamation were issued on March 30, 1661, ordering the Kirkes to vacate the province of Avalon, and directing all commanders of ships and other officers to assist Captain Robert Swanley, Baltimore's lieutenant at Ferryland, in the government of the province.[33]

Lord Baltimore immediately set about assuming control over his domain. He sent Captains Pease and Rayner as governor and deputy governor in the summer of 1661. The arrival of these officials caused confusion at Ferry-

[31] Lord Baltimore's second petition to Charles II [1660 or 1661?], C.O. 1: 15, no. 38. Baltimore had attempted to recover Avalon during the Interregnum. Baltimore's libel against Sir David Kirke, Dec. 1651, Admiralty Court, Instance and Prize Libels, &c., File 110, no. 329. Cf. "The Lord Baltimore's Case, concerning the Province of Avalon in New-found-land, an Island in America," Dec. 23, 1651, printed in L. D. Scisco, "Calvert's Proceedings against Kirke," *Canadian Historical Review*, VIII, 132-136, the petition being on pp. 133-135. See also the deposition of James Pratt of St. Saviour's Docke, Mar. 31, 1651, printed in *ibid.*, pp. 135-136.

[32] Answer by Sir Orlando Bridgman, chief justice of the Common Pleas, and Sir Heneage Finch, solicitor general, Feb. 28, 1661, C.O. 1: 14, no. 9 (i).

[33] Proclamation of Charles II recognizing Baltimore's claim to Avalon, and requiring the Kirkes to surrender the province, Mar. 20, 1661, C.O. 1: 14, no. 10.

UNDER CROMWELL AND CHARLES II 107

land. They threatened to dispossess the members of the Kirke family who were residing there. Finally an agreement was reached whereby the Kirkes became tenants of Baltimore for the house and land at Ferryland. The new régime was not popular. Captain Rayner was regarded as a desperado, though Pease, according to Charles Hill of Ferryland, "a beaten soldier," was persuaded to give up his position "upon honourable terms." Rayner was favorable to the Kirkes and unwilling to do anything in Newfoundland for which he would have to answer in England. His inclination to favor the Kirkes was probably due to his former service under Sir David in the navy. The governors pretended to have a commission from the king, which they asserted had been granted three days before their departure, and which authorized them to manage the affairs of all Newfoundland. The real cause of Rayner's unpopularity was probably his attempt to collect arrears of rents due Lord Baltimore, which few paid because few trusted him.[34] There is no evidence that the royal commission was ever issued to Rayner and Pease. It is not improbable, however, that Charles II may have continued for a short time the practice instituted by Cromwell of appointing commissioners to superintend the affairs of Newfoundland. This view is partly substantiated by Rayner's statement that he attempted to enforce the Navigation Act of 1661 against a Dutch trader.[35]

Within a short time Baltimore's authority over Avalon collapsed. Rayner and Pease departed, leaving no representative of the proprietor behind them. In 1666, several of the inhabitants of Avalon asked George Kirke, son of Sir David, to assume the proprietorship of the province.[36] Nothing came of this request and the unceremonious de-

[34] Charles Hill to George Kirke, Sept. 12, 1661, Egerton MSS., 2395, fol. 309.
[35] Petition of John Rayner to the king [1662], C.O. 1:16, no. 113.
[36] Petition of Robert Brouse, De Belleville, and others, "householders and

parture of Baltimore's representatives marked the end of proprietary control in the island. The elimination of the proprietary groups did not solve the problems of those interested in the trade and fishery. In fact, it left the planters without any strong backing in England and without any individual or group immediately responsible for the protection of their interests. The struggle henceforward was to be between the western adventurers, who advocated, as always, a minimum of governmental regulation, and the London merchants and their allies, the Newfoundland planters, who desired to see a stable civil government set up in the island. While the Kirkes and Calverts were arguing over the validity of their respective claims, the lines of battle were re-forming for the contest between London and the outports for supremacy in the trade and fishery of Newfoundland.

Commercial depression prevailed widely in England during the years immediately preceding and following the return of Charles II. The depression resulted partly from the internal political disorder, and partly from the wars with Holland and Spain. English merchants suffered heavily, particularly those of the West of England.[37] The merchants engaged in the Newfoundland trade experienced considerable capital losses. When they attempted to recover their former position they were faced with serious obstacles. Their difficulties appeared to be so insurmountable that they not only despaired of enjoying any future prosperity but they expected an even greater decline in the industry than they had already experienced. Competition had increased tremendously in the twenty years previous to 1660. The trade in dried codfish, in which the merchants of southwestern England

inhabitants of Avilonie,'' to George Kirke, Mar. 18, 1666, Egerton MSS., 2395, fol. 447.

[37] Report of Thomas Povey to the Committee of Council on Foreign Plantation, May 11, 1660, Egerton MSS., 2395, fols. 263-264.

had enjoyed an important place, was now threatened by the French and New Englanders. The West Country was faced with a curtailment of its markets in France, Spain, Portugal, and Italy. After the Restoration the western adventurers began to feel the effects of the new economic policy of France, as directed by Colbert. The French completely excluded the English from their market for "dry fish" and at the same time secured a firm foothold in southern Europe. The English merchants now had to pay more attention to price, quality, and delivery than ever before.

At the same time that they were confronted with an increased effort on the part of France, they were also faced with competition from the New Englanders. Since the death of Charles I the latter had been developing the fishery along their own coasts, which the colonists managed, controlled, and regulated themselves. They had also begun to operate their own mercantile marine, and were carrying fish and other commodities to foreign markets. The merchants of the West of England felt the effects of this new development very keenly. They resented the New England competition in dried fish, but they were especially disturbed by the trading activities of the colonists, who besides competing with them abroad were threatening to capture the entire trade in provisions and supplies there, a branch of commerce heretofore largely in the hands of the western adventurers. Competition was intensified by the curtailment of the market in Spain owing to the growing laxity of the Spaniards in observing Lent and other fasts. In order to meet these problems satisfactorily a reorganization of the trade and fishery was required. The western adventurers desired to retain control over Newfoundland, but in order to do so they had to develop a successful and prosperous business, and be in a position to make prompt and effective delivery of their commodity to foreign markets. Those who sought to nationalize the industry felt that the West Countrymen were incapable of accomplishing the needed reforms.

There were many obstacles in the way of improvement. Internally, the fishery was in no condition to offer the western adventurers much encouragement. Not only did their old rivalry with the planters continue to hamper the efficiency of the industry, but they were now faced with the competition of "byboatkeepers" or independent fishermen financed by merchants at home. The byboatkeepers went out each year in large numbers to Newfoundland, where they employed small vessels and fishing boats and numerous capable fishermen. They and their employees frequently remained in the island as residents, thus adding to the population and becoming permanent competitors of the adventurers. Byboatkeeping had been introduced by Sir David Kirke, but it had developed rapidly during the Interregnum. Its methods of catching and curing the fish were the same as those practiced by the planters and western fishermen, but the employees were paid in wages instead of shares. Thus the West Country not only had to meet the competition of the French and New Englanders, but they were also confronted with the necessity of preventing any further loss of prestige in the fishery and reëstablishing themselves in their former position of dominance at Newfoundland.

As before, the struggle for control of the fishery centered on the question of the proper method of regulation. Both the business methods and old rules of the western adventurers had become antiquated and patently inadequate. The West Country people sought to secure permanent safeguards recognizing their particular interest and they also sought to eliminate all rivals. Their opponents, the planters and byboatmen, aided and abetted by the London merchants, sought to reorganize the fishery entirely and to establish its regulation on a more firm

UNDER CROMWELL AND CHARLES II 111

and enduring foundation. The conflict between these two groups forms a large part of the history of Newfoundland from 1660 to 1763. The search for a way to improve the fishery, which took place between 1660 and 1675, laid the foundation for later developments and fixed the policy of the government so unalterably that the authorities were often in a dilemma as to how to accomplish even necessary changes.

The depressed condition of the Newfoundland trade is set forth clearly in the petitions and complaints of the merchants. In 1667 some of the West Country adventurers were so poor that they were unable to undertake the journey to London, even though the Privy Council was considering their business.[38] The Bristol merchants feared the complete loss of this trade to the French and Dutch.[39] Three years later the merchants complained that for many years the Newfoundland fishery had proved "a very slender and bare employment" to those adventuring in it. Few had earned more than ten per cent of the capital invested, while many had lost their principal. In 1669 Dartmouth merchants placed their figure at nearly a quarter of the capital they had invested in the trade, and similar amounts were reported from Plymouth and some of the other western ports.[40] The depression in the English fishery contrasted unfavorably with the reputed prosperity of the French. In 1670 it was estimated that France employed 400 ships and 18,000 seamen at Newfoundland, while England had only 300 ships and 15,000 men.[41] Although these numbers do not imply such a great

[38] Order in council, Aug. 28, 1667, *Acts, Priv. Coun., Col.*, 1613-1680, §713.

[39] Order in council, Dec. 6, 1667, *ibid.*, 1613-1680, §735.

[40] The reply of the merchants, etc., of the western parts of England to the allegations of Capt. Robert Robinson [1668?], C.O. 1: 22, no. 71.

[41] Reply of Capt. Robinson to the answer of the West Country gentlemen [1668], *Cal. State Paps., Col.*, 1661-1668, §1732.

difference in conditions as the adventurers believed existed, they were bad enough, and seemed to promise the loss of the entire trade.[42]

In spite of the depressed condition of the fishing industry and the impending collapse of the nation's trade, the West Country fishing interests were still sufficiently influential to make their wants heard in court circles. Shortly after Charles II was established upon his throne the western adventurers secured royal confirmation of their ancient fishing rights. The new king reissued the Western Charter of 1634, almost *verbatim*.[43] The only important change was the addition of a paragraph forbidding the transportation to Newfoundland of persons who were neither planters nor members of the crews of fishing vessels.[44] This provision was intended to restrict the byboatkeepers, whose competition was seriously injuring the prosperity of the West Country as well as threatening the ruin of the entire trade.[45] The merchants were also strong enough to secure consideration of their problems by the newly established Council of Trade. This

[42] Petition of the merchants, etc., of the western parts to the king in council, read Dec. 23, 1670, *Cal. State Paps., Col.*, 1669-1674, §362; order in council, Feb. 12, 1675, C.O. 1: 67, no. 15.

[43] The Western Charter, Jan. 26, 1661, Pat. Rolls, 12 Charles II, pt. 17, no. 30; copy: C.O. 1: 15, no. 3.

[44] Western Charter of 1661, clause 8 and the unnumbered clause following 11.

[45] The byboatmen paid from 40s. to 50s. for a passage. Most of the men were able seamen, although from a fifth to a quarter were green. The byboatkeepers made the voyage in small vessels, which carried enough men to provide crews for four or five of the larger ships operated by the adventurers. The use of smaller ships entailed a smaller initial investment and lower operating costs. Moreover, the byboatkeepers made profits from the passenger fares and the sale of fish to the sack ships, avoiding by the latter practice the risk and cost of carrying their catch to market. Notes in [Sir Joseph] Williamson's hand of the evidence of Gould and Parrett, Feb. 27, 1675, C.O. 1: 34, no. 16. The Western Charter of 1661, last clause, contains provisions designed to put a stop to the activities of the byboatkeepers, but

body, like its predecessors during the Interregnum, was charged with supervision of both the domestic and oversea fisheries in order that "they be most improved, and regulated to the greatest advantage of the stock and navigation of the nation, by excluding the intrusion of our neighbors in it."[46] Parliament was also called upon for assistance, and the first act applying specifically to Newfoundland was passed in 1663. One of its clauses provided that no tax should ever be levied on codfish caught by Englishmen at Newfoundland; another prohibited the use of seines or nets in procuring cod fry; while a third provided for the protection of property left behind by the fishermen at the end of the season.[47] These enactments were probably intended to protect the western adventurers against the possibility of the passage of unfavorable laws should a civil government be established in the island, and to correct some of the more glaring abuses committed in the fishery. This obscure law was

it was ineffective from the first, and the western adventurers sought to prohibit their rivals from participating in the fishery altogether. Petition of the merchants, etc., of the western parts to the king in council, read Dec. 23, 1670, *Cal. State Paps., Col.*, 1669-1674, §362. See the order in council, Dec. 4, 1663, *Acts, Priv. Coun., Col.*, 1613-1680, pp. 374-375. The proposed additions to the charter, made in 1671, contain definite provisions designed to eliminate the byboatkeepers. Order in council, Mar. 10, 1671, clauses 6, 10, 12, *ibid.*, 1613-1680, §915.

[46] Draft instructions for the Council of Trade, 1660, in Charles M. Andrews, *British Committees, Commissions, and Councils of Trade and Plantations, 1622-1675* (Johns Hopkins University Studies in Historical and Political Science, XXVI), pp. 1-151. Note especially clause 4, p. 73a. Cited hereafter as Andrews, *British Committees, etc.* Povey and Noell, who drafted the scheme, expanded clause 4 in the final instructions to cover all plantation trade. The draft instructions for the Council of Trade and Foreign Plantations, Sept. 27, 1672, contained in clause 3 instructions relative to the improvement of both home and overseas fisheries, *ibid.*, p. 127.

[47] "An Act for regulating the Herring and other Fisheries and for repeale of the Act concerning Madder," 15 Chas. II, c. 16. See especially clauses 3 and 4. The bill passed through parliament between February and July, 1663, Stock, *Proceedings and Debates*, I, 315-319, *passim.*

destined to have a tremendous effect upon the history of Newfoundland, because prohibiting the taxation of codfish, the only form of wealth in Newfoundland for many years, blocked all attempts to establish a civil government for a long time.

The Council of Trade was at first more concerned with the domestic fishery than with that of Newfoundland, but finally gave the overseas fishery some attention. Subsequently the affairs of Newfoundland were taken under advisement by the Privy Council, which conducted an investigation and took steps to enforce the Western Charter of 1661.[48] The council was particularly concerned with the new restriction placed upon the byboatkeepers. The mayors of the western ports were warned that the shipmasters disregarded the prohibitions and continued to transport others than *bona fide* planters or fishermen to Newfoundland.[49] The Privy Council advised the magistrates that if such actions were not checked they would injure the trade seriously by lessening the number of ships and seamen, and prejudice the interests of the crown by the complete destruction of the fishery.[50] Content with warnings of this nature, the government made no detailed investigation into the affairs of Newfoundland until after Lord Baltimore had failed to reëstablish his authority in the province of Avalon.

After the collapse of the proprietary movement, the London merchants and Newfoundland planters who had supported Kirke and his associates turned to the crown

[48] On Dec. 2, 1663, the Privy Council appointed a committee to consider the proposed letter to the West Country mayors and customs collectors. The committee was instructed to compare the letter with the charter of 1661. *Acts, Priv. Coun., Col.*, 1613-1680, §610.

[49] Meeting of the Council of Trade, Dec. 17, 1663, Egerton MSS., 2343, fol. 137.

[50] Privy Council to the mayors of Southampton, Weymouth, Lyme, Dartmouth, Plymouth, Fowey, and Barnstaple, Dec. 4, 1663, *Acts, Priv. Coun., Col.*, 1613-1680, §612.

UNDER CROMWELL AND CHARLES II 115

for assistance. They sought the establishment of civil government in the island and direct supervision by the royal authority. Becoming alarmed at this trend in favor of stronger control, the western merchants and shipowners protested to the Privy Council in 1667 that certain persons were secretly endeavoring to establish a government in Newfoundland for their own "sinister ends."[51] The West Country people were not entirely in accord in thus opposing the erection of civil government, for the Bristol merchants, following the tradition established in the days of John Guy, took their stand with the Londoners and suggested that some person of ability be sent out as governor. They believed that the trade would be lost unless some security were given to the fishery, partly by the creation of a local government and partly by assistance from the crown in the form of military defense.[52] These and others who were anxious to wrest control of the fishery from the grasp of the West Country put forward in 1668 several proposals looking to the establishment of civil authority. They laid special stress upon the growing activity of the French, who were expanding their fishery and fortifying and garrisoning the southern coast of Newfoundland at Placentia. They suggested that another attempt be made to impose a tax on foreigners, and pointed out that in the years when Sir David Kirke had collected taxes from foreign buyers and fishermen, the English fishery had been in a prosperous condition, yielding an annual profit to the merchants of about £500,000, and that in recent years the profits had declined to about one-third of their former level. They believed that a strong civil government was necessary,

[51] Petition of the merchants, etc., of Plymouth to the king, Aug. 28, 1667, *Cal. State Paps., Col.*, 1661-1668, §1561. Almost identical petitions were received from Totnes, Dartmouth, and other ports, *Acts, Priv. Coun., Col.*, 1613-1680, §716.
[52] Order in council, Dec. 6, 1667, *ibid.*, 1613-1680, §735.

116 BRITISH FISHERY AT NEWFOUNDLAND

not only to protect the fishery against pirates and enemies, but also to prevent the perennial violation of the fishing rules, and that the loss of the fishery would deal a heavy blow to the royal navy because recruits for it were trained in the fishery.[53] Should the French gain complete control of Newfoundland they would become a serious menace to New England, New York, and Virginia. Therefore, they proposed that the local administration in Newfoundland be supported by a tax upon fish and train oil.[54] In view of this proposal it is easily understood why the western adventurers had taken the precaution in 1663 to get parliamentary sanction against taxation of the products of Newfoundland.

The Londoners and their allies were so determined that their plans should receive consideration at the hands of the crown, that the Privy Council finally directed the Committee of Council for Trade and Plantations to investigate the matter. A beginning was made in January, 1670. The London merchants and the agents for the western ports were ordered to appear before the committee, which was directed to hear all sides of the question and report its findings to the council. With this report as a basis the Privy Council proposed to make recommendations to the king relative to any changes in "the present constitution of that place."[55]

[53] Reasons for the settlement of Newfoundland and the trade thereof under government [1668?], C.O. 1: 22, no. 69.

[54] Capt. Robinson's proposals to the king, 1668, C.O. 1: 22, no. 70. Robinson, who had visited the island as commander of the convoy, evidently aspired to the royal governorship. See his reply to the answer of the West Country gentlemen, 1670, *Cal. State Paps., Col.*, 1669-1674, §368. He submitted a great deal of written information. See certain arguments or reasons for a settled government. Robinson to the king in council, 1669, C.O. 1: 66, nos. 71, 71 (i); C.O. 1: 25, no. 111. He also submitted nearly identical proposals to William and Mary. Sir Robert Robinson's reasons for a settled government, received Jan. 25, 1694, C.O. 1: 68, no. 99.

[55] Order in council, Dec. 15, 1669, C.O. 1: 66, nos. 72, 73.

UNDER CROMWELL AND CHARLES II   117

During the subsequent hearings the western adventurers had an opportunity to present their views. They admitted that the fishery had been unprofitable for some years, but that it deserved the same encouragement from the crown that it had received in the past. They felt that the appointment of a governor would be harmful, and in contrast to the Londoners, minimized the danger of French encroachment. They emphasized the disadvantages arising from the presence of the inhabitants and desired their removal from the island, believing that the presence of a governor would encourage the planters and at the same time injure the fishery. They considered the fortification of important points in Newfoundland as futile, alleging that such projects would prove useless in view of the extent of the fishery and the dispersion of the settlements; and that in any case were fortifications erected the burden of defense would still rest upon the fishing ships. In fine, the western adventurers insisted that Newfoundland did not need a government because of its scanty population and unfitness for settlement, and they were convinced that the proposals to establish civil government were inspired by their competitors in London. They maintained that the proposed taxes to support the government, though seemingly inconsiderable, would increase the cost of production of fish and train oil and hinder the revival of the "decaying, dying trade" by adding a little more to the losses of the merchants during seasons of bad fishing voyages. Furthermore, they called attention to the illegality of the proposed tax, and played their trump card when they pointed out the danger that Newfoundland might develop along the same lines as recalcitrant New England were a governor sent there and the fishery surrendered to the residents. The inhabitants, they showed, were already receiving provisions from New England, and should the adventurers be excluded from control of the fishery all that would be left in English hands would be the carrying trade.[56] Although both contesting groups were given opportunity to air their views before the committee, no decision was reached in 1670. The outbreak of the third Dutch war postponed consideration of the problem of Newfoundland until 1675.

Meanwhile, the western adventurers continued to complain of their many discouragements. No improvement took place between 1671 and 1675 because of the general commercial uncertainty due to the war with Holland. The merchants continued to complain of the decline of the trade, and as before gave French competition and the destructiveness of the planters and fishermen as the reasons for its decay. The critics of West Country management, on the other hand, maintained that a considerable amount of the falling off in the industry was due to the time wasted on arrival in unproductive repairs, which occupied about twenty per cent of the fishing season as well as considerable outlay for materials and labor, and that the fishermen were hindered by those of the planters who built their houses and laid out their gardens so near the shore that they occupied places suitable for drying fish. The fishing was also injured by the general habit of throwing ballast into the harbors, and by using seines to take bait, which destroyed the cod fry.[57] Faced with these obstacles, it is not surprising that the New Englanders and foreigners could undersell them in world markets.

The decay of the industry manifested itself in a shipping decline and a falling off in demand with a consequent lowering of prices for dried codfish. Formerly about 200 or 300 English ships had voyaged annually to

[56] The reply of the merchants, etc., of the western parts to the allegations of Capt. Robinson [1668], C.O. 1:22, no. 71.

[57] Reasons for the settlement of Newfoundland and the trade thereof under government [1668], C.O. 1:22, no. 69. Thomas Farr, mayor of Southampton, to the Lords of Trade, Mar. 21, 1675, C.O. 1:34, no. 28.

UNDER CROMWELL AND CHARLES II 119

Newfoundland to fish.[58] By 1675 it was estimated that there were only 150 ships employing about 7,000 men.[59] Another estimate placed the number of fishing ships at one-third of the former fleet, and in some years only about thirty ships made the voyage.[60] Sometimes not more than fifty or sixty ships made the return voyage to England because an increasing number of people were remaining permanently in the island. The steady annual decline in the fishing fleet was matched by a similar reduction in the number of ships in the carrying trade to Spain. Previously about 500 vessels had been engaged in the voyage to Bilbao, but in 1675 there were not more than four or five bound for that market.[61] This was probably due to the decline in demand for dried fish in Spain. As has already been remarked, the Spaniards were becoming increasingly lax in the observance of Lent and other fasts when fish would ordinarily be consumed. This occasioned a fall in prices which affected the merchants adversely. Sir Josiah Child gave this as the primary cause of the decline of the trade.[62] Thus at a time when England was running the risk of losing all her Newfoundland fishery and trade to the French, her merchants were confronted with a falling market and keen competition from the byboatmen and planters. Trade was so bad that many of the West Country ships "lay by the wall."[63] Any further decline was certain to ruin the handicraftsmen,

[58] Pretended reasons against erecting the king's government [Feb. 23, 1675], C.O. 1: 67, no. 18.
[59] Notes by Williamson of the evidence of Gould and Parrett, Feb. 27, 1675, C.O. 1: 34, no. 16.
[60] Pretended reasons against erecting the king's government [Feb. 23, 1675], C.O. 1: 67, no. 18.
[61] Notes by Williamson, Feb. 27, 1675, C.O. 1: 34, no. 16.
[62] Sir Josiah Child, *New Discourse of Trade* (3d ed., London, 1718), pp. 205-212. J. R. Tanner, ed., *Samuel Pepys' Naval Minutes* (Naval Records Society), pp. 196-197 and note.
[63] Notes by Williamson, Feb. 27, 1675, C.O. 1: 34, no. 16.

120 BRITISH FISHERY AT NEWFOUNDLAND

farmers, and shopkeepers of the West of England who were dependent upon the trade for their support.[64]

There is a note of desperation running through all the complaints and petitions of the West Country people during these years. The decrease in the Newfoundland trade was a serious blow to the economic stability of southwestern England. In 1650, that region had so many seamen available that it had been possible to send out 200 fishing ships, manned by 10,000 men. The trade had brought in an annual customs revenue of £150,000, while about £100,000 more had been spent yearly in England for the products and manufactures consumed in the fishery. In a reasonably good fishing season the total returns to the nation from the trade in codfish and train oil amounted to nearly £300,000 yearly. After the return of the fleet to England in the autumn a great many of the fishermen engaged in farming for the rest of the year. They were, therefore, available in case of war. Besides the fishing ships there had been nearly a hundred merchantmen engaged in carrying salt to Newfoundland and in transporting fish and oil to Europe. Under the fishing methods prevailing in 1650, fish had cost about a quarter as much to produce as in 1675; buyers had been certain of delivery according to contract, and consequently had been able to sell more cheaply. The thousands of handicraftsmen who had formerly been engaged in producing materials used by the fishing fleet had become impoverished as a result of the decline in the trade. The decrease was attributed by the West Country interests to the development of the fishery carried on by the planters and byboatkeepers. These disturbing elements not only had caused an increase in the cost of production but also were responsible for the reduction in the number of fishing ships employed.

[64] Remonstrance of the merchants, etc., of Plymouth, Dartmouth, and other western ports, to the king, read Mar. 25, 1675, C.O. 1: 67, no. 26.

UNDER CROMWELL AND CHARLES II 121

By 1675, the inhabitants were threatening to develop their own carrying trade by sending fish direct to foreign markets in their own bottoms, thus following in the footsteps of New England.[65] These views were expressed by a London merchant named Parrett, who gave frequent testimony during these years when the Newfoundland trade was being investigated by the crown. Other Londoners, such as John Gould, felt that the situation was gradually growing worse, and that more and more trade was likely to be lost to France.[66]

The London merchants who were more concerned with the development of the port of London and England's carrying trade than with the waning prosperity of the West Country, rallied to the support of the planters and byboatkeepers. In 1675, James Houblon, an influential Londoner interested in trade with Spain, attributed the decline of the Newfoundland trade and fishery to improper armed protection against enemies. The vessels employed in fishing were of little defensive worth. During the war with Spain, 1656-1659, the English had lost about 1,000 ships, many of which were fishing vessels. This loss had impoverished Plymouth, Dartmouth, Lyme, Poole, and other western ports to such an extent that as late as 1675 they had not recovered from the blow. As a result of their losses in the Spanish and Dutch wars, the western adventurers had been obliged to mortgage their shipping in order to finance their fishing voyages. They were paying interest at the rate of twenty-one or twenty-two per cent, and hence handed over to the usurers more than the total profit from their voyages in bad years. Even in good years they found it difficult to improve their

[65] Parrett to Williamson, read Mar. 25, 1675, C.O. 1: 67, no. 27; Parrett's paper, read Mar. 20, 1675, C.O. 1: 67, no. 29.
[66] John Gould to [Sir Robert Southwell], Mar. 18, 25, 1675, C.O. 1: 67, nos. 22, 24.

position because the high interest charges and amortization reduced their profits. Houblon thought that in time, if the French continued their activity, the English fishing trade would dwindle away to nothing. He felt that the only way to improve the situation was to meet the French on their own terms by producing fish cheaper and more efficiently than was possible under the direction of the western adventurers. This could be accomplished only by encouraging the planters who were more advantageously situated for developing the industry along efficient lines. The extensive outlay of capital invested in fishing ships would be obviated were the planters to conduct the entire fishery. The fishing ships lost a good deal of time in transit to Newfoundland and after their arrival. Were the inhabitants to operate the fishery this waste of time would be avoided, while the ships released from their former employment in the fishery could be used in carrying salt and provisions. The inhabitants would have greater incentive to produce well-cured fish, because local competition would encourage them to take pains, and because of the fear that their product might be rejected by the buyers on the "sack ships." They would turn out a better quality at a lower price than did the West Country fishermen. Houblon was confident that such an arrangement would prove so attractive that the western adventurers would themselves become planters.[67]

Other Londoners added their approval to Houblon's suggestions. John Gould argued that the operation of the fishery by the planters was the only economical way to manage the trade. He maintained that this was not generally recognized because of the influence exerted by the operators of the English fishery, who received commissions on the chartering of fishing vessels and on the sale

[67] James Houblon to [Southwell], Mar. 20, 25, 1675, C.O. 1: 34, no. 27; C.O. 1: 67, no. 23.

of the fish, and who were consequently unwilling to be deprived of their profits. Gould felt, however, that the number of inhabitants could be increased to a point where they would be able to make fish at a rate of 16 to 20 *reales* per quintal, and insisted that England would benefit from this arrangement because the inhabitants would barter their fish to the merchantmen in exchange for English commodities.[68] These opinions were further reinforced by the arguments of an anonymous commentator on the inefficiency of West Country management. Writing under the pen-name, "An Englishman," he asserted that the planters could produce fish more cheaply than the adventurers for a number of reasons. In the first place, they were on the spot and lost no time in transit as did the fishing ships, or in preparing for business after their arrival. During this time the fishermen were an expense to the merchants financing the voyages. There was also a considerable loss through the idleness of the ship itself during the fishing season when the men were engaged in catching cod from fishing boats and curing it ashore. The inhabitants were in a position to begin fishing whenever they wished, provided their stages had not been destroyed by the West Country fishermen, and did not have to bear the losses resulting from idle ships, long voyages, and idle men. With the fishery operated by residents, the fishing vessels could be converted into merchantmen to engage in the carrying trade, and would remain at Newfoundland only long enough to procure their lading before proceeding on market voyages. Consequently the fish would be delivered earlier and the merchants would get quicker returns. The planters would also have the advantage of a fishing season of seven months, as compared with one of three months for the adventurers, and could, therefore, make their fish cheaper and in larger quanti-

[68] Gould to Southwell, Mar. 18, 25, 1675, C.O. 1: 67, nos. 22, 24.

ties. On the other hand the western adventurers claimed that the time wasted by the inhabitants in idleness offset the losses of the fishermen during the season. If the merchant vessels went to Newfoundland only to buy fish the inhabitants would be in a position to exact high prices, knowing that the sack ships could not afford to go away empty. They also urged that the quantity of English provisions consumed by the inhabitants would not be as great as was the case in the fishery managed from the West Country, and that such an arrangement as that upheld by the Londoners would not train as many sailors as did the old system. The Londoners in reply stated that the seven months' fishery would greatly reduce the idle time of the inhabitants, that their earnings would become greater than those of the fishing ships operating for only three months, and, as "An Englishman" pointed out, it would be impossible for the planters to exact any higher prices for fish from the merchants than the traders could demand from them for provisions imported from England. In the long run, therefore, there would be no lessening of the profits to the merchants, as the increase in population in Newfoundland would create a greater demand for salt and provisions. This would not only increase profits but also result in the employment of better ships in the carrying trade. Thus encouraging the inhabitants to operate the fishery would prove of advantage to the English merchants by preventing the trade from falling into the hands of France. If the merchants were to obtain greater profits and meet French competition, they must be able to make their own fish as cheaply as possible by reducing the overhead costs, disposing of the greatest quantity possible in foreign markets, and securing quicker returns, while at the same time deriving an additional income from the increased provision trade "An Englishman" felt that the adoption of such business methods would eliminate the principal grounds for complaint.[69]

From the foregoing it is readily seen that the West Country operators had not only lost heavily as the result of the wars, of increasing competition from France and New England, and of the curtailment of their markets abroad, but were being hard pressed also by the planters and byboatmen who were able to make fish at lower cost. Moreover, from the criticisms of the London merchants and from the admissions of the adventurers themselves, it is apparent that the financial condition of industry was unsound and that the economic situation in the West of England was generally most unsatisfactory. With due regard to the prejudice of their opponents one cannot help but feel that many of the improvements suggested by the Londoners merited a fair trial. The western adventurers were reactionary, uncompromising in their stiff-necked insistence upon the maintenance of the *status quo* in the fishery, and imperviously rigid in their attitude toward any suggestions for alterations in the management and regulation of the industry. In spite of the depression in their business, the western adventurers were as tenacious as ever in defending their vested interest, and were sufficiently influential, not only to procure a certain amount of attention from the crown, but also to obtain the support of parliament. Confirmation of the old regulations by Charles II in 1661 was important for the West of England, but the statute of 1663 which prohibited the levying of taxes upon fish caught by Englishmen at Newfoundland was the most effective bar which they could possibly have obtained against any threat to reëstablish proprietary control or to erect a local government in Newfoundland under the auspices of the royal authority.

[69] Some modest observations and queries by "An Englishman," Mar. 25, 1675, C.O. 1: 34, no. 32.

CHAPTER IV

THE PROBLEM OF REGULATION, 1660-1676

THE contest between the western adventurers and the planters assisted by their allies in London took the form of a discussion as to the most satisfactory type of regulation which could be applied in Newfoundland. The West Country merchants naturally sought to retain the Western Charter as the fundamental law of the fishery and settlement, while the planters in self-defense and the London merchants in self-interest desired the creation of an adequate government for the island and the placing of the fishery under the jurisdiction of responsible royal officers.

After Lord Baltimore's failure to reassert his rights to Avalon, the earliest proposal for the establishment of a civil government in Newfoundland was made to the crown by Captain Robert Robinson of the Royal Navy in 1669. Robinson was much interested in Newfoundland which he had visited as commander of the naval convoy during the second Dutch war. He evidently hoped to be appointed the first royal governor, a plan that received the support of the London merchants. Robinson's petition formed the basis for the discussion of the two conflicting points of view relative to the fishery and its regulation during the years from 1669 to 1676.[1] In January,

[1] Capt. Robinson's proposals, 1668, C.O. 1: 22, no. 70; his petition to the king in council, 1669, C.O. 1: 66, no. 71, annexing: Certain reasons for a settled government, C.O. 1: 66, no. 71 (i); his reply to the answer of the West Country gentlemen, 1670, *Cal. State Paps., Col.*, 1661-1668, §1732; *ibid.*, 1669-1674, §368; his proposals, Jan. 25, 1694, C.O. 1: 68, no. 99; order in council, Dec. 15, 1669, C.O. 1: 66, nos. 72, 73.

1670, the Privy Council Committee reported unfavorably upon the captain's proposals. The report was accepted by the Privy Council which also agreed to an alternative proposal of the committee that the commander of the convoy be given greater power to regulate abuses committed at Newfoundland, though the Western Charter of 1661 was still to remain the basis for the regulation of the fishery.[2] In spite of this recommendation and its acceptance by the council no steps were taken at this time to give the convoy commander, or "commodore," the proposed extension of his power. This plan to supervise the fishery through the agency of an officer of the Royal Navy was probably suggested in part by the practice adopted under Cromwell of sending commissioners to Newfoundland, and in part by the regularity with which convoy protection was now furnished to the Newfoundland shipping. It is interesting to note that the development of local government in Newfoundland was henceforward identified to a considerable extent with the activities and recommendations of the commodores of the convoy, and that when civil government was finally established in 1729 it was accomplished through the agency of a commodore. The first royal governor and his successors for many years were officers of the navy.

In December 1670, the Newfoundland question was reopened by the western adventurers who complained that the order in council of December 4, 1663, was continually violated by the byboatkeepers. The mayors and magistrates authorized to act under the Western Charter asked permission to depute persons to execute the order in council that the fishery might continue to be carried on in accordance with the ancient custom. Acting on this com-

[2] Report of the Committee of Trade and Plantations, Jan. 18, 1670, and order in council approving it, Feb. 4, 1670, C.O. 1: 66, no. 77.

128 BRITISH FISHERY AT NEWFOUNDLAND

plaint, the Privy Council ordered a hearing.[3] The question was referred to the select Council for Foreign Plantations in January 1671, which was instructed to consider the best means for regulating and advancing the trade and fishery, securing it against foreign encroachment, controlling it in order to increase the number of seamen, and managing it to the best advantage of the king and his subjects. The council was also directed to consider the charter of 1661, and the additional powers desired by the West Country traders. All papers relating to the question were referred to the council and persons who had been called by the Privy Council in its initial inquiry of January 11, 1671, were ordered to attend the meetings of the subordinate council and to assist that body in preparing a report.[4] The Council for Foreign Plantations took considerable time to prepare its report to the Privy Council, and it was not rendered until March 10, 1671. When completed it was a lengthy document recommending twenty-seven additions to the charter of 1661. Eighteen of these recommendations were approved by order in council on the day the report was received, and ordered incorporated in a bill confirming a new western charter.[5]

Again the privileges of the western adventurers were confirmed by action of the crown. The eighteen proposed additions to the Western Charter covered the points upon which the West Country interests had been most insistent. Their acceptance by the Privy Council was in the nature of a victory for the adventurers, and spelled a temporary setback for the planters and London merchants.

[3] Petition of the merchants, etc. of the western parts to the king in council, read Dec. 23, 1670, C.O. 1: 66, no. 85.

[4] Order in council, Jan. 11, 1671, C.O. 1: 26, no. 5.

[5] The additional rules were included in the report of the Council for Foreign Plantations, Mar. 2, 1671, C.O. 1: 26, no. 5.

THE PROBLEM OF REGULATION 129

According to the new rules English subjects were always to have freedom to fish and take bait at Newfoundland, liberty to go ashore on any part of the island or adjacent ones for the purpose of salting, drying, curing, or storing fish, and making train oil. Besides these privileges the fishermen were allowed to cut wood for the construction of stages, houses, vats, boats, or other fishing conveniences, provided they submitted to and observed all rules and orders "as now are or hereafter shall be established by his Majesty, his heirs or successors for the government of the said fishery in Newfoundland." No aliens or strangers were to be permitted to fish or take bait anywhere between Cape Race and Cape Bonavista. The inhabitants were prohibited from cutting, burning, or destroying any trees useful for timber, and denied the right to erect houses or other buildings or to lay out gardens within six miles of the seacoast between the two capes or upon any island lying within ten leagues of the shore in that zone. Moreover, no planter or inhabitant was to occupy any stages or other fishing conveniences before the arrival of the fishermen from England, in order that the latter might be first provided with fishing grounds and equipment. The masters and owners of fishing ships were prohibited from transporting to Newfoundland any persons other than members of their ships' crews or those engaged for the voyage on shares or for hire. The last two provisions were intended to restrict the expansion of the fishery carried on by the inhabitants and byboatkeepers.

In order to restrict further the development of the plantation the number of persons whom it was permissible to carry on any ship for a fishing voyage was limited to sixty for each hundred tons burden. All masters trading to Newfoundland were ordered to give bond each year before their departure to the mayors of the western or other English ports under penalty of £100 not to violate the above provision relative to the transportation of unauthorized persons, and to agree to bring all their men back to England at the end of the voyage, including those engaged for market voyages. The shares or wages of any person who should desert ship at Newfoundland were to be forfeited and paid to the mayor of the port from which the deserter's ship had sailed, before the master or owner would be released from his bond. In order to provide a more continuous supply of sailors, the proposed rules required that one out of every five men engaged for the Newfoundland voyage should be a green man, "that is to say not a Seaman." Before clearing port in England the master and owners of fishing ships were enjoined to provide all provisions and other necessaries, except salt, for the entire voyage. This clause was aimed at the practice of certain owners who permitted their ships to load provisions in Ireland where they could be procured more cheaply than at home. Dependence upon Irish provisions would prove injurious to England's trade, and was, therefore, to be discouraged. Definite rules were laid down for the time of departure of the Newfoundland shipping. No vessels bound there directly were to leave before March 1, while those going first to the Cape Verde Islands for salt were not to sail before January 15. Other clauses dealing specifically with abuses committed at Newfoundland were also included. Hereafter no staging already built was to be occupied by less than twenty-five men, all of whom were to be members of the same ship's company. No fishermen were to be permitted to remain in Newfoundland during the winter.

The new rules granted more power to the fishing admirals and the West Country magistrates. The admirals were required to preserve peace and order among the fishermen and seamen ashore as well as at sea, to execute

THE PROBLEM OF REGULATION 131

the royal charter concerning the fishery, and to apprehend and punish all offenders. According to the old custom, the fishing admirals were empowered to secure all felons and capital offenders and transport them to England for trial, and this obligation was reaffirmed. The admirals were also to publish the orders prohibiting the fishermen from remaining in the island during the winter, and were to keep journals of all their proceedings, copies of which were to be delivered to the Council for Foreign Plantations. They were also to furnish that body with information relative to the condition of the fishery in their respective harbors. In England the magisterial functions of the West Country officials were to be extended to include the recorders, deputy recorders, and justices of the peace, who were authorized to join with the mayors of towns in the same county in determining disputes among the fishermen. These officers were to take cognizance of all offenses against the laws, rules, and orders established by the king for the Newfoundland fishery, and they were to determine such cases as were described in the charter of 1661, thus restoring to the West Country its authority over the fishery. Fines, penalties, and forfeitures were to be imposed upon violators of the regulations.[6]

These tremendously important recommendations of the Council for Plantations were approved by the king in council on March 10, 1671. Sir Heneage Finch, the attorney general was ordered to prepare a bill confirming the charter of 1661 with the additions recommended. One other change was proposed. In the original charter the earl marshal of England had been given jurisdiction over capital crimes and felonies committed in Newfound-

[6] The rules were approved and recommended for incorporation in a new western charter, by order in council, Mar. 10, 1671, *Acts, Priv. Coun., Col.*, 1613-1680, §915.

land, but this office no longer existed. Finch was asked to propose a system of judicial procedure suitable for hearing and determining cases of treason, felony, murder, and all criminal offenses committed either ashore or afloat in the fishery. The Privy Council ordered that the procedure adopted should be in accord with law and equity, and that proper methods of administering the criminal law in the above mentioned cases should also be provided.[7]

Besides the tremendous increase in powers for the western adventurers and the fishing admirals, and the change in the method of conducting criminal cases, the Council for Foreign Plantations also recommended that the powers of the commanders of the naval convoys should be increased. It was intended that these rules also be included in the new Western Charter, but the Privy Council felt otherwise and decided to delegate these powers to the Admiralty. The lord high admiral was directed to issue orders to the commodores describing their duties while on the Newfoundland station. These rules extended the authority of the officers of the navy very considerably, but at the same time prevented them from interfering in matters connected with the fishery which were not strictly maritime in character. The commodores were to aid the fishing admirals in preserving peace and good order, and to assist in apprehending criminals. In order to make his supervision as thorough as possible the commander of the convoy was not to remain in one harbor for any great length of time, but was to ply from port to port, except when necessity or security required him to remain. Moreover, the commodore was forbidden to transport any passengers to Newfoundland, and neither the officers of the navy nor the men un-

[7] Order in council, Mar. 10, 1671, *Acts, Priv. Coun., Col.*, 1613-1680, §917.

THE PROBLEM OF REGULATION 133

der their command were to engage in fishing or curing fish, or in transporting it on their ships, except such as was necessary for their own use. The captains of the vessels forming the convoy were to make it their business to survey the fishing harbors from Cape Race to Cape Bonavista, gathering information relative to the fishery as carried on by the ships from England and the planters. Copies of their journals and observations on the state of the fishery were to be delivered by the commodore to the lord high admiral on the return of the convoy to England.[8]

The Privy Council enjoined the mayors of the western ports and the owners of ships to observe the new rules and orders carefully, "and govern themselves in every particular, as they will answer contrary at their peril."[9] Temporary orders to make the new regulations immediately effective were issued. These were identical with the additional clauses approved for incorporation in the new charter and in the directions given to the Admiralty, except that one extremely important clause was added to the commodore's orders. He was directed to encourage the inhabitants "to transport themselves and their Families to Jamaica, St. Christophers, or some others of your Majesty's Foreign Plantations."[10] This recommendation had been included in the representation of the Council for Foreign Plantations of March 2, 1671, unquestionably at the behest of the western adventurers.

During the investigations which had been under way ever since Captain Robinson had presented his petition in

[8] Order in council, Mar. 10, 1671, *ibid.*, 1613-1680, §916. The recommendations, included in the representation of the Council for Foreign Plantations, Mar. 2, 1671, clauses 13, [16], 17-22, were incorporated in this order.
[9] Order in council, Mar. 10, 1671, *ibid.*, 1613-1680, §917.
[10] *Ibid.*

1669, all the proposals calling for a radical readjustment of the Newfoundland trade in favor of the creation of a plantation there had been rejected. Influential though they may have been in other branches of commerce, the London merchants so far had been unable to make any headway against the West Countrymen. The entire program of reorganization as submitted by the Council for Foreign Plantations and approved by the Privy Council was exactly what the western adventurers had desired for many years. Some of the so-called new regulations were identical with the restrictions which the West Country merchants had forced Kirke and his associates to accept in the patent of 1637. The fishing interests of the West of England had altered their demands in only two respects: whereas before the Civil Wars they had been unwilling to consent to naval protection and supervision at Newfoundland, they were now ready to accept such assistance from the crown; and they were now committed to a policy of absolute extermination for the plantation, not only by restricting settlement to the inland parts of Newfoundland but also by demanding the complete removal of all the inhabitants and the adoption of measures which would prevent the resettlement of the island. Those who had fought monopoly now sought to maintain it, even though as yet it was but a paper victory for the West Country.

In spite of the triumph of the West Country in having its views accepted by the government, no immediate action was taken by the crown to make the new rules effective. Execution was deferred because the government had more pressing matters on its hands owing to the outbreak of the third war with Holland in 1672. The elaborate changes made in 1671 were still largely on paper. The temporary order confirming the new rules had been executed but the proposed additions to the Western Charter

THE PROBLEM OF REGULATION 135

still remained to be incorporated in a new patent. In most respects the regulation of the fishery was still in an unsettled state, and a great many alterations in the policy laid down by the Council for Foreign Plantations in 1671 were to be made before a definite series of rules could be finally established.

Meanwhile the situation in the fishery was even worse than before. By 1675, after the war with Holland had been concluded, French competition was felt more keenly than ever before. The rivalries within the English fishery itself, arising from the growing activity of the planters and byboatkeepers, were intensified by the increasing participation of Ireland and New England in the provision trade, which gave the western adventurers grave concern. The spectre of civil government reappeared to haunt the West Country merchants, for the planters and their friends in England resumed their agitation as soon as the war with the Dutch was over. By 1675 both sides were demanding another investigation and a permanent settlement of the question.

The investigation of 1675 was instigated by the planters who for the first time began to take a direct interest in their own fate. William Hinton, himself a resident of Newfoundland, presented the case of his fellow-countrymen. He petitioned to have the question reopened, and his plea was referred by the Privy Council to the Lords of Trade on February 12, 1675. Hinton asked the government to reconsider its decision of 1671. He pointed out that owing to the recent war no final arrangement had been made to execute the new regulations approved four years previously. The uncertainty as to whether the crown intended to carry out its earlier decision was producing a great decline in the trade, and England was in danger of losing all. Hinton proposed that a civil government be erected in Newfoundland and that the planters

136 BRITISH FISHERY AT NEWFOUNDLAND

be encouraged to carry on the fishery.[11] As directed the Lords of Trade undertook the investigation. At first they were favorably impressed with Hinton's proposals, but they decided to proceed cautiously, and notified the West Country mayors to transmit their opinions on the points at issue and to furnish the investigators with the latest information relative to conditions in the settlement and fishery. The Lords of Trade also decided to obtain information from other sources. They asked the customs commissioners and convoy commanders to provide them with data, with the result that they had much more knowledge upon which to base their final conclusions than had been the case in previous investigations.[12]

The Lords of Trade had to account for the decline in trade and the decay of the fishery, and to find a suitable form of regulation, either by following the precedent laid down by the existing charter, or by recommending the establishment of a royal government with both civil and military powers. The arguments put forward by both contending factions were approximately the same as those presented to the Council for Foreign Plantations four years before, but the value to the crown of a permanent settlement of the question was emphasized more than ever. The advocates for the Newfoundland planters were more aggressive and emphatic in their demands, probably owing to the presence of Hinton. On the other hand the stiff-backed conservatism of the western adventurers was even more apparent than in the previous investigation as was shown by their unwillingness to accept any alterations in the regulations except such as would benefit themselves. The contest was the most bitterly fought

---

[11] Order in council, Feb. 12, 1675, C.O. 1: 67, no. 15; Petition of William Hinton to the king, read Feb. 12, 1675, C.O. 1: 67, no. 16.

[12] Minutes of the Council of Trade and Plantations, Feb. 23 and 25, 1675, *Cal. State Paps., Col.*, 1675-1676, pp. 177-178; draft order from the Lords of Trade to the mayors of the western ports, Feb. 25, 1675, C.O. 1: 67, no. 20.

THE PROBLEM OF REGULATION 137

of all the disputes between the two groups up to this time.

Hinton and his London allies appealed for support on the ground that the establishment of civil government and the encouragement of a fishing industry controlled by the inhabitants were politically expedient and economically necessary. They explained that the government had a direct interest in settling the fishery regulations in a satisfactory manner because of its own claims to sovereignty over the island, and because of the strategic position of the island as the "next part of the West Indies" adjacent to the king's European dominions. Moreover the commercial prosperity of Newfoundland was of benefit to the crown, because the nursery for seamen was valuable for defense, and because the revenues derived from goods imported into England in exchange for the fish sold abroad went into the royal treasury. The encroachment of the French had been possible because of the absence of civil authority. France had extended her operations so that her fishermen were now making dry fish at Newfoundland, an enterprise in which they had not been previously engaged; she had occupied some of the best harbors on the southern coast, which her government was fortifying; and the latter was even persuading the English inhabitants to live under its protection. Thus the French, who had formerly paid the proprietary government for the right to fish, now paid no longer, and were threatening the very existence of the English fishery. The planters demanded equal justice with the fishermen, the elimination of the abuses which threatened the prosperity of the fishery, and protection against pirates and enemies.[13]

[13] Reasons for the settlement of Newfoundland under government, [1668?], C.O. 1:67, no. 16 (i), annexed to Hinton's petition. See C.O. 1:22, no. 69, another and evidently earlier copy. Gould to Southwell, Mar. 20 and 25, 1675, C.O. 1:67, nos. 22, 24; Houblon to the same, Mar. 20, 1675, C.O. 1:34, no. 27; 67, no. 23.

The advocates of royal intervention charged the western adventurers with insincerity. They alleged that the latter did not give their real reasons for opposing the changes. Although the adventurers maintained that the creation of a civil and military establishment and the encouragement of colonization would prove burdensome to the trade, their rivals denied the validity of these assertions by pointing out that military protection and civil order would lead to an increase in the capital invested in the fishery and would develop the training school for seamen to a greater extent than ever before, while the cold climate and barrenness of the soil would naturally prevent overpopulation.[14] The real reason why the West Country opposed these reforms, said the Londoners and the planters, was the financial predicament in which the operators found themselves. Many of them were heavily in debt to prominent merchants, having borrowed money to finance their fishing voyages "at bottomry," thus mortgaging their ships for considerable sums. These operators were afraid to favor the establishment of government, lest they antagonize their creditors and thus be unable to borrow more money to carry on their trade.[15] There is no doubt that the fishing industry of the West of England was badly organized and badly financed, as later events were to prove, but it is unreasonable to assume that the indebtedness of the adventurers was the only reason for their opposition to the proposed reorganization of the fishery. Their opposition had been too persistent and too long continued to be accounted for on such slender grounds as these. The West Country relied overwhelmingly on the Newfoundland fishery as one of the principal pillars of the economic structure of the re-

[14] Pretended reasons against erecting the king's government, with answers thereto, read Feb. 23, 1675, C.O. 1: 67, no. 18.
[15] True reasons why the West Country fishermen are against government, read [Feb. 23, 1675], C.O. 1: 67, no. 17.

THE PROBLEM OF REGULATION 139

gion, and the adventurers there could in no way afford to surrender control to other groups lest it bring disaster to them.

These western adventurers naturally objected to the establishment of civil government and the delivery of the fishery into the hands of the inhabitants in Newfoundland on a number of grounds. They minimized the activity of the French, whom they asserted were more concerned with protecting the fur trade than with the extension of their fishery; scoffed at the suggestion that additional protection, other than that offered by the ice in winter or the ships in summer, was necessary; and maintained as always that the trade of England would suffer were the planters allowed to control the fishery, for they would purchase provisions and supplies from New England and engage in illicit trade with the French and other foreigners to a greater extent than ever before. They questioned the practicability of maintaining a civil establishment in view of the destructiveness and debauchery of the inhabitants. They pointed out that the only feasible means of supporting the government would be by a tax on the fishing trade. This in turn would injure the business and be detrimental to the best interests of the crown because the revenues derived from imports from southern Europe would be reduced thereby, and the consequent shrinkage in shipping would in turn provide a smaller supply of seamen upon which the Royal Navy might draw in case of need.[16] Therefore, the West Country group felt that the continuation of the fishery under

[16] Williamson's notes of the evidence of Gould and Parrett, Feb. 25, 1675, C.O. 1:34, no. 16; Parrett to Williamson, read Mar. 25, 1675, C.O. 1:67, no. 27; the same to the same [Mar. 30, 1675], C.O. 1:67, no. 24; remonstrance of the merchants, etc., of Plymouth, Dartmouth, etc., to the king, read Mar. 25, 1675, C.O. 1:67, no. 25; reasons tendered by George Pley for a government, Mar. 17, 1675, C.O. 1:34, no. 25; letters from the mayors of the western ports, C.O. 1:34, nos. 19, 21, 22, 28, 38, 39.

the charter regulations was the only way in which the trade could be improved and the best interests of king and country be maintained.

The investigation occupied the Lords of Trade for the greater part of February, March, and April 1675. Representatives of the contending factions were heard frequently, and papers from the various outports were read and considered. The committee soon became convinced that the continuation of the fishery under the old form of regulation was the only proper method of controlling the trade. Lord Anglesey, the lord privy seal, was apparently the only member of the group who believed firmly in the establishment of civil government, but he was overruled by the majority of the committee, and on April 1, 1675, the Lords of Trade resolved to report to the king their full approval of the rules and orders settled in council in March 1671. The mayors of the western ports were to take out a new charter with additional powers covering crimes committed at Newfoundland. Only a few changes were made in the rules recommended in 1671 and these were of a more or less formal nature. Owing to the abolition of the Council for Foreign Plantations in 1674, the fishing admirals were directed to return their journals to the Lords of Trade at the end of the fishing season, and the inhabitants were to be prohibited from living within six miles of the coast between Cape Race and Cape Bonavista. The merchants were ordered to attend the committee on April 8, to give their opinions regarding the proposed amendments to the charter and the removal of impediments to the trade. The Lords of Trade also ordered that draft instructions for the commodore be prepared relative to the French trade and concerning the complete removal of the inhabitants from Newfoundland.[17] Thus

[17] Minutes and journals of the Lords of Trade, *Cal. State Paps., Col.*, 1675-1676, pp. 177-178, 179, 186, 197-198, 201, 204-205, 209.

THE PROBLEM OF REGULATION 141

again in spite of the efforts of the planters and their London associates, the West Country had scored another victory.

The final report of the Lords of Trade, drafted April 15, 1675, accepted entirely the point of view of the West Country interests. Besides containing a very comprehensive description of conditions in the Newfoundland trade since 1660, the report incorporated the recommendations of the committee for improving the fishery and insuring its stability. In general, the Lords of Trade adopted the West Country interpretation of the decline of the trade, which they attributed to three factors: intensified French commercial rivalry; the increased competition of the New England fishery; and the antagonisms existing within the English fishery at Newfoundland. They rejected the proposals for the creation of civil government and the encouragement of settlement on several important grounds, and recommended the enforcement of the regulations contained in the charter of 1661 together with the additional rules approved in 1671 but never put into effect. In many respects this report of the Lords of Trade is the high water mark of West Country influence in Newfoundland and the fishery.[18]

The Lords of Trade agreed that the adventurers had lost heavily through several poor fishing seasons and the destruction of their shipping in the wars with Holland, and attributed their inability to recover their former position to the intensified activity of the French, who had developed their fishery so industriously by means of government assistance that the English had not only been driven from their former market in France, but had suffered in other foreign markets as well. Recognition of in-

[18] Report of the Lords of Trade to the king, Apr. 15, 1675, C.O. 1: 67, no. 30 (i); approved by order in council, May 5, 1675, C.O. 1: 67, no. 30; C.O. 1: 34, no. 71; *Acts, Priv. Coun., Col.*, 1613-1680, pp. 621-626.

creased French competition did not prevent the Lords of Trade from following the West Country interpretation of French activity in southern Newfoundland, and they minimized the danger to the English fishery from that quarter. The New England fishery, which was carried on locally by the colonists, was recognized as a serious competitor, because its annual catch, which amounted to about 60,000 quintals, was sold in the same places as the Newfoundland fish, and consequently injured the business of the English merchants and fishermen. Within the Newfoundland fishery itself, the adventurers were faced with the competition of the planters who, in disregard of the restrictions placed upon them by the charter of 1637, lived near the shore and engaged actively in fishing.

In rejecting the pleas for civil government and local management of the fishery by the Newfoundlanders, the Lords of Trade were actuated by more than their prejudice in favor of the western adventurers. They considered the establishment of civil government, the erection of fortifications, the maintenance of garrisons, and the encouragement of colonization in Newfoundland as contrary to the best political and economic interests of England. They pointed out that civil government was not feasible because the population of the island was scanty, the settlements widely separated, and communication between them difficult particularly in winter when most of the disorder occurred. Although a governor might be resident in one community there would be forty or more places without any government at all. They also rejected the plea for additional military protection, accepting again the West Country argument that defense in winter was unnecessary because of the ice, and protection in summer was furnished by the English shipping, "for that place will always belong to him that is superior at Sea." Moreover, they felt that the fishing trade would be unable

THE PROBLEM OF REGULATION 143

to provide sufficient revenue to maintain a civil and military establishment. Fear that the proposed government would prove a financial burden to the crown was undoubtedly responsible for this feeling. Indeed, one cannot fail to realize that the ready acceptance which all these government commissions and committees gave to the ideas of the West Country was dictated by the dread lest the crown be involved in expenses which it could ill afford in a time of financial stringency.

The Lords of Trade were equally fearful lest any encouragement given to the erection of a government in Newfoundland might lead to the development of another region similar to New England, the economic, political, and even religious ideas of which would be antagonistic to the policies of the home government. The committee felt that Newfoundland was unsuited for colonization owing to the rigors of the climate and the infertility of the soil, disadvantages which not only inclined the population to idleness and debauchery, but also made them dependent upon New England for provisions and liquor. The Lords of Trade were apprehensive lest New England influence be extended in some way to Newfoundland, Even were the climate and soil of the island more favorable to colonization than was actually the case, they felt that it would be harmful to England to encourage settlement or local government, as Newfoundland would tend to adhere to New England, "and in time tread in the same steps."[19] The fear that another disaffected colonial area might be created at Newfoundland henceforward actuated the home government to a very considerable extent in determining its policies with respect to the plantation and the fishery. This fear accounts for the readiness of the crown to accept without qualification West Coun-

[19] *Acts, Priv. Coun., Col.*, 1613-1680, p. 623.

try control of the fishery, and to rely so persistently upon an otherwise weak and unsatisfactory form of regulation.

The Lords of Trade laid down an important principle with respect to the fishing industry, which explains why Newfoundland's economic position in the seventeenth and eighteenth centuries was looked upon as differing from that of the plantation colonies of the West Indies and the mainland of North America. They pointed out that the sort of regulations which were effective in controlling the trade of the plantation colonies were inapplicable to a commodity like fish. This product was unlike the "enumerated" commodities raised in the plantations because it could not bear the cost of first being transported to England, but had to be carried directly to its final destination, the foreign market where it was consumed. This opinion amounted to a declaration that the traffic in codfish was not a branch of the nation's domestic commerce as was the case with the plantation commodities, but that it was actually a part of England's *foreign* commerce. This differentiation made it impossible to include Newfoundland in the plantation system and impracticable to endow the island with the political privileges enjoyed by the true plantations. Judged by contemporary economic standards the island and its principal industry were no more conformable to the mercantilist conception of a true unit in the plantation system than were the New England and Middle Colonies. If this statement of the function of Newfoundland's trade is coupled with the other pronouncement of the Lords of Trade concerning the danger of creating another colonial region like that of New England, one can readily understand why they rejected the pleas of the planters and London merchants and left the fortunes of the fishing trade to the tender mercies of the West Countrymen. The foregoing interpretation does not imply that the Newfoundland

THE PROBLEM OF REGULATION 145

trade did not conform in some respects to the mercantilist standards of the day, but rather that its function was different from that of the true plantation and, therefore, that the regulations applied to it should be of a different kind.

The Lords of Trade concluded their report by recommending that the rules accepted in council on March 10, 1671, were generally suitable, although necessitating a few minor changes. The committee proposed that the commodore for the year 1675 should be instructed to publish at Newfoundland the royal intention that the inhabitants were to have a choice of removing six miles inland or of leaving the island altogether for England or one of the plantations. The withdrawal of the residents from the fishing zone was to be superintended by the convoy commander, who was to seize any planters remaining in the forbidden area and bring them to England for trial. The Lords of Trade held that the validity and good effect of the whole regulatory system depended upon the execution of this project. The convoy commander was to assist in carrying home those planters who should choose to return to England, and the governors of plantations were to be instructed to receive the former Newfoundlanders should they decide to emigrate to the other American possessions. Thus would the fishery at last be entirely in the hands of the merchant adventurers of the West of England, the planters could no longer threaten to disrupt it, and no new disturbing element would henceforward be permitted to establish itself in Newfoundland.

Aware of the increasing activities of the French, the Lords of Trade recommended that the commodore's orders should include directions for the more careful investigation of the rival fishery than had been customary and to report his findings to them on his return to England.[20]

[20] *Acts, Priv. Coun., Col.*, 1613-1680, pp. 623-624.

146 BRITISH FISHERY AT NEWFOUNDLAND

This suggestion is indicative of the change which had taken place in English foreign policy following the third war with Holland, and shows that with the Dutch safely out of the way, England was looking upon France with more jealous eyes than had been the case during the early part of the reign of Charles II, when that monarch had imposed a pro-French policy upon the country.

It will be recalled that no definite recommendations had been made in 1671 relative to the conduct of criminal actions in Newfoundland. In their report of 1675, the Lords of Trade proposed that the king's counsel should inquire into and review the powers given in the charter of 1661 with respect to trials for treason, felony, and murder committed at Newfoundland and report any deficiencies in the provisions of the patent. If deficiencies were found, the law officers were to recommend a suitable form of judicature for hearing and determining such offenses.[21] When the investigation of the judicial powers had been completed and approved in council, the mayors of the western ports were to be required to renew their charter, including the additional powers recommended in 1671. The new patent was then to be printed, and a proclamation issued in the king's name, calling the attention of all concerned to the necessity of enforcing its provisions.[22]

The approval of the report of the Lords of Trade was given by the Privy Council on May 5, 1675, and on the same day other orders were issued for carrying the program into effect. The new rules were to be enforced at once through a temporary order, until the western mayors could take out a new charter permanently incor-

---

[21] Order in council directing Sir William Jones, attorney general, to consider and revise the powers granted in the charter of 1661 relative to the trial and punishment of criminals, and to draft a new charter, May 5, 1675, C.O. 1: 67, no. 31; *Acts, Priv. Coun., Col.*, 1613-1680, p. 626.

[22] *Ibid.*, p. 626.

porating the additions suggested by the Lords of Trade. The new Western Charter was issued on January 27, 1676, and remained in force for twenty-three years, until superseded by the Newfoundland Act of 1699.[23] The charter contained the changes and additions which have already been discussed and was itself merely a confirmation of the concessions made to the western adventurers by the crown.

The favorable report of the Lords of Trade and the order in council of May 5, 1675, approving the recommendations of 1671, as well as the final promulgation of the new charter in 1676, were triumphs for the West Country and severe blows to the Newfoundland planters and their friends, the mercantilists of London. The fight which the adventurers had waged in defense of their ancient fishing rights ever since 1610 reached its climax in the promulgation of the charter of 1676. Not only did the crown ignore the claims of the planters and Londoners, but it went to the length of ordering the forcible removal of the population in order that the West Country fishermen might be left in complete and undisturbed control of the fishery. Though they were undoubtedly self-seeking, the crown felt that the exigencies of national prosperity and political safety demanded recognition of their privileges.

England in the period of Cromwell and Charles II had no colonial policy at all in the modern sense of the word, and no well defined commercial policy. The foundations of mercantilist policies and practices for the plantation trade were laid during this epoch, and at the same time the basis of a fishery policy for Newfoundland was established. The attitude of the government toward trade in general during the Restoration period was dictated by

[23] Charter granted to the West Country adventurers for the fishery at Newfoundland, Jan. 27, 1676, Pat. Rolls, 27 Charles II, pt. II, no. 12; copy: C.O. 1: 67, no. 36.

the necessity of overcoming the economic and financial handicaps under which the nation labored. The position assumed by the crown with respect to the peculiar circumstances existing in the Newfoundland fishery was determined by the conservatism and penuriousness of the royal administration as well as by the steadfastly reactionary attitude of the West Country toward any radical changes which might endanger its vested interests in the fishery. The crown was not anxious to introduce innovations of doubtful value or possible expense, and was fearful lest it create another colonial entity the future life of which might prove totally incompatible with the political and economic doctrines which prevailed in England.

CHAPTER V

THE COLLAPSE OF CHARTER REGULATION, 1676-1699

THOUGH the acceptance by the crown of the ideas of the western adventurers during the investigation of 1675 represents the high water mark of the success of West Country fishing interests in their attempts to control the fishery, the struggle between them and the Newfoundland planters did not abate with the adoption of the revised charter in 1676. Indeed, the rivalry increased rather than diminished, owing to the attempts of the western adventurers, assisted by the crown, to enforce the new regulations as well as the old, and because of the struggle of the planters to secure modifications of the rules and to obtain civil government in their stead. The campaign was waged intermittently from 1676 until the passage of the Newfoundland Act of 1699.

Scarcely were the additional rules governing the fishery approved than many of them were found to be inoperative. This discovery was made by Sir John Berry, R.N., who was sent to Newfoundland as commander of the convoy in the summer of 1675. He was charged with the duty of investigating conditions in the fishery and settlement and was instructed to enforce the restrictive measures upon the planters. Berry arrived at St. John's in July, and immediately proceeded to carry out his orders. He sent two subordinates northward and southward along the coast to take an exact census, and to notify the inhabitants that they were either to leave Newfoundland or else to withdraw six miles inland, while he remained at St. John's to execute these orders there. The Newfoundlanders "with all humility" promised to obey the royal will and pleasure, but Berry found that it would be impossible to remove them from their native island because many were too poor to pay their passages elsewhere. Unless the government sent a ship to carry them away, passage free, they would be obliged to remain in the island, where their existence would be impossible were they forced to move six miles inland. The commodore criticised the proposed transportation of the population to other parts of the English world on the ground that they were so poverty stricken they would become public charges wherever they went. Were they removed to England the planters would suffer considerable hardship owing to the lower earning power of the individual worker in the mother country. A laborer in Newfoundland could earn £20 during the summer because he obtained most of his food from the sea, while in England he would be fortunate if he could make £3 during the same period. Berry felt that the removal of the inhabitants was inadvisable and would create as many problems as it was intended to solve.[1]

Commodore Berry also devoted some time to investigating the charges which had been made against the Newfoundlanders by their rivals in England. He discovered that most of the accusations were false and that the English fishermen were as responsible as were the inhabitants for the destruction of stages, woods, and other conveniences used in fishing. He found that most of the English shipmasters at St. John's favored such practices, as

[1] Sir John Berry to Sir Joseph Williamson, July 24, 1675, C.O. 1: 34, no. 118; the same to Sir Robert Southwell, Sept. 12, 1675, and the inclosure, C.O. 1: 35, nos. 17, 17 (iii).

COLLAPSE OF CHARTER REGULATION 151

they found it profitable to destroy the stages in order to sell the wood to the sack ships for fuel. Berry was not convinced that the planters enticed the fishermen to remain in Newfoundland at the end of the fishing season, as the West Country interests had alleged, but on the contrary believed that the shipmasters were in the habit of persuading the planters to accept their men, presumably as indentured servants, in order to avoid paying them their wages for the return voyage to England. He felt that those who desired the removal of the planters were anxious to get the entire control of the fishery for themselves, but his experience at Newfoundland convinced him that the forcible depopulation of the coastal zone would drive the inhabitants to seek homes on the south coast where they could live under French protection.[2]

Sir John Berry deserves great credit for his fearless espousal of the cause of the unfortunate planters. He was the one man of this period associated with the administration of the fishery who showed a spark of human understanding and sympathy during the perennial squabbles of the Newfoundlanders and the West Countrymen. Perhaps the fact that he had risen from the ranks in the navy may have given him a sympathy for his fellow men which is strangely lacking in most of the commodores and other persons associated with the enforcement of the fishing regulations in the seventeenth and early eighteenth centuries. He expressed his opinions freely while at St. John's, for later some of the West Country merchants complained of his favorable attitude toward the planters, asserting that the latter had become so encouraged by Berry's outspoken interest that they had been led to expect a complete reversal of royal policy toward them, and that encouragement of colonization instead of discour-

[2] Berry to Williamson, July 24, 1675, C.O. 1: 34, no. 118.

agement would be the future program.³ In defense Berry replied that he could not help but pity the poor inhabitants about whom so much false information had been circulated.⁴ In reading his reports of conditions in Newfoundland one has the feeling that here at least is one man who held warm-hearted but honest opinions, based on his own observation and experience and the work of his subordinates who visited the outports above and below St. John's.

Commodore Berry's report caused the Lords of Trade to hold a lengthy debate on the Newfoundland question in December 1675.⁵ Nothing decisive was done at the time, and the matter of rescinding the order of removal was postponed until the summer of 1676, when further hearings were held.⁶ Meanwhile the new charter had passed the seals and become effective. During the investigation which followed, Berry was supported by Captain Davies, who had been with him at Newfoundland the year before, and upheld his position in defense of the planters. In spite of the merchants' allegations to the contrary, Berry insisted that he had obeyed his royal instructions and had enjoined the inhabitants to leave Newfoundland or withdraw from the coastal zone. He pointed out that forty-five of the more important West Country shipmasters had told him that the trade would be utterly destroyed if the inhabitants were removed, because they were of considerable assistance to the English fishermen, who were much more dependent upon the Newfoundlanders than some of them were willing to admit.⁷ Many of the West

---

³ Minutes of the Committee of Trade and Plantations, Dec. 4, 1675, *Cal. State Paps., Col.*, 1675-1676, pp. 310-311.

⁴ Berry to Williamson, Dec. 17, 1675, C.O. 1: 35, no. 60.

⁵ Minutes of the Committee of Trade and Plantations, Dec. 4, 1675, *Cal. State Paps., Col.*, 1675-1676, pp. 310-311.

⁶ Journal of the Lords of Trade, Aug. 8, 1676, *ibid.*, 1675-1676, p. 439.

⁷ Berry's observations, 1676, C.O. 1: 35, no. 82.

COLLAPSE OF CHARTER REGULATION

Country fishermen, who had set out in small boats at the beginning of the season to seek out desirable fishing places, would have perished from hunger and cold if the inhabitants had not been there to give them relief. Hundreds of men were cured of scurvy by the planters. If there were no help at Newfoundland in case of illness, many fishermen would be discouraged from shipping for the voyage there. The presence of the inhabitants was also valuable in relieving the distress of vessels engaged in the West Indian and other trans-Atlantic trades which were forced by stress of weather to take refuge in the ports of the island.

Berry also considered that the plantation was sufficiently large to consume large quantities of English goods, and deprecated the assertions of the western merchants that there was danger of the development of a local carrying trade by the Newfoundlanders, since the latter had "neither ship, bark, nor boat to create any trade whatever." Most important of all, he pointed out, was the fact that the sixteen hundred inhabitants made an annual catch of 70,000 quintals of fish, or about a third of the total seasonal production. These fish were sold to English merchants in exchange for manufactured goods, and were transported to foreign markets in English ships.[8] Thus he was the first independent investigator to point out not only the dependence of the West Country fishermen upon the inhabitants, but to demonstrate that

[8] There was a resident population of 1,655 between Cape Race and Cape Bonavista. The inhabitants cured over 69,000 quintals of merchantable fish, almost all of which was shipped in English vessels. Their total production, including wet fish and train oil, was worth about £47,000, about a third of the total for the entire fishery. Berry to Williamson, Sept. 12, 20, 1675, C.O. 1: 35, nos. 16, 21; the same to Southwell, Sept. 12, 1675, C.O. 1: 35, no. 17. Lists of the planters were inclosed in these letters. Cf. "An Account of H.M. the King of England's Subjects in Newfoundland," Sept., 1676, C.O. 1: 38, no. 82.

the planters themselves were assuming an important position in the fishing industry.

The western adventurers were correct in assuming that Berry's outspoken sympathy for the Newfoundlanders would hearten the planters immeasurably. John Downing, the planters' agent in London, asserted that the fishing interests of the West of England had tried ever since the days of Sir David Kirke to ruin the planters in order to get complete control of the fishery, and maintained that the adventurers had resorted to calumny and unfair business practices in order to force the planters out of the trade. Some of the West Country merchants refused to accept any fish other than those caught and cured by their own men, and were accustomed to buy from the inhabitants only at their own price and at their own convenience. Moreover, they insisted that all men employed in the fishery should work for them, serving for such food and wages as they pleased. Downing confirmed Berry's statement that the wrongs attributed to the planters were exaggerated, and that instead of working to the detriment of the fishing interests the Newfoundlanders were of considerable assistance to them. He also felt that the return of the planters to England would place a burden upon the landed gentry, "to whom the extreme poverty of distresed families will neither be pleasant, nor profitable."[9] Berry's report and Downing's representation produced in official circles such a favorable reaction toward the planters at home that the crown determined in 1677 to reconsider the Newfoundland question.

As a matter of fact the West Country was quite as dissatisfied with the new regulations as were the planters al-

[9] A brief narrative of Newfoundland, by John Downing, received Nov. 24, 1676, C.O. 1: 38, no. 70; Egerton MSS., 2395, fols. 560-563; names of the English inhabitants, C.O. 1: 38, no. 89; an account of the colony and fishery, C.O. 1: 38, nos. 73, 74; Additional MSS., 13,892, fols. 15-25.

though for different reasons, and they, too, petitioned for a reconsideration. The new rules had proved ineffective from the start not only with respect to the management of the fishery but also the shipping trade itself. The merchants complained that there were some ports engaged in the trade to Newfoundland which were outside the jurisdiction of the corporations the magistrates of which were authorized by the charter to enforce the regulations. Some of these communities were unincorporated and had no magistrates within their limits, while the magistrates of other nonsubscribing ports were engaged profitably in the trade. The West Country merchants admitted that the amended charter had not produced the desired effect, and asked that the mayors and magistrates of the various corporations as well as the custom-house officers be enjoined to observe the patent carefully and to execute it effectually.[10] Early in 1677 the Privy Council referred the petition to the Lords of Trade for examination and report. The order of reference contains this significant clause, "His Majesty in Council being willing to gratify the petitioners in anything that may cause the late Letters Patents touching the fishery of Newfoundland to be made effectual to them."[11] From this statement it is evident that the crown was definitely committed to the cause of the West Country, and desired to warn the Lords of Trade not to listen to any proposals for the modification of the rules that might be construed as favorable to the planters and their sympathizers in London and Bristol.

In spite of the warning from the Privy Council, the

[10] Petition of the merchants and traders to the king, received Feb. 16, 1676, C.O. 1: 67, no. 36 (a).

[11] The merchants' petition and one from John Downing were referred to the Lords of Trade by order in council, Feb. 21, 1677, C.O. 1: 39, no. 43; *Acts, Priv. Coun., Col.*, 1613-1680, §1125; Downing's petition, C.O. 1: 38, no. 43. Cf. Petition of Rider, recorder of Dartmouth, C.O. 1: 67, no. 36 (a); and draft letter, Blathwayt to Rider, Feb. 27, 1677, C.O. 1: 67, no. 36 (b).

156 BRITISH FISHERY AT NEWFOUNDLAND

Lords of Trade reconsidered the question very fully in a series of meetings during March 1677. They not only reviewed their own report of April 15, 1675, but considered the statements of Sir John Berry and examined sundry individuals representing the conflicting interests. The West of England was represented by Benjamin Scutt and John Pollexfen, who made the usual allegations against the planters, maintaining, in addition to the oft-repeated statement that the French menace was exaggerated, that the decline of their fishery had been caused by the encouragement given the Newfoundlanders by their English friends. This, they claimed, was clearly shown by the revival of trade which had occurred as a result of the execution of the new regulations, for in 1675 the West Country fishery had shown distinct improvement.[12] John Downing was the principal advocate of the planters. Besides making the usual counter-accusations against the fishermen, he asserted that the planters were legally entitled to dwell in Newfoundland by virtue of the early proprietary patents, which he maintained were not invalidated by the Western Charter. This attack upon the fishermen from the legal standpoint was the only new factor introduced into the quarrel. Otherwise Downing's statements were conventional, even to the emphasis which he placed on the growing danger from France.

Downing presented a better legal case than had any of the earlier supporters of the plantation. On March 14 1677, as agent for the planters, he petitioned the king to suspend the clause of the new charter which provided for

[12] An account of the colony and fishery [1677]. C.O. 1:38, no. 73; journal of the Lords of Trade, Mar. 3 and 4, 1677, *Cal. State Paps., Col.*, 1677-1680, pp. 76-78. Cf. report of the Lords of Trade, Apr. 15, 1675, C.O. 1:67, no. 3 (i); *Acts, Priv. Coun., Col.*, 1613-1680, pp. 621-625; Berry to Williamson Sept. 12, 1675, C.O. 1:35, no. 16; the same to Southwell, same date, C.O. 1:35, no. 17.

COLLAPSE OF CHARTER REGULATION 157

the forcible removal of the inhabitants.[13] Although Downing asked only for a temporary suspension during the approaching fishing season, he confounded his opponents very cleverly by raising the question whether the Newfoundlanders were not entitled to permanent residence in the island by virtue of the earlier proprietary patents. The western adventurers, represented before the Lords of Trade by merchants and members of the House of Commons, found themselves unprepared to answer this query. The investigation got under way during March and April, but was adjourned for a short time to give the adventurers an opportunity to obtain legal advice.[14] The West Countrymen felt sure that they could prove the validity of their patent, but were so evidently uneasy that they extracted a promise from the Lords of Trade that nothing would be done to injure their interests during the short recess. The planters were again supported by Lord Anglesey, who pointed out that the condition of the fishery was such that England was running the risk of losing all to France unless the existent abuses were speedily removed.[15] Downing's cleverness in questioning the legality of the removal clause of the charter of 1676 in the light of the earlier patents, not only confounded his enemies, but placed the whole matter of the regulation of the trade and fishery on a higher level than it had heretofore occupied.

Though the Lords of Trade agreed to postpone their final decision until the western adventurers could obtain

[13] Petition of John Downing to the king, received Mar. 14, 1677, C.O. 1: 39, no. 45. His earlier petitions: C.O. 1: 38, no. 33; C.O. 1: 39, no. 12.
[14] Report of the Lords of Trade, Mar. 26, 1677, received by the Privy Council and action taken thereon, Mar. 28, 1677, C.O. 1: 39, nos. 49, 50; the order in council, C.O. 1: 39, no. 53; *Acts, Priv. Coun., Col.*, 1613-1680, §1128, clause 3.
[15] Lord Anglesey to Williamson, Mar. 16, 1677, C.O. 1: 39, no. 46.

legal advice, the committee took immediate steps to suspend that portion of the new Western Charter which was intended to restrict the planters or require their removal. On the recommendation of the Lords of Trade, the Privy Council issued orders to the fishing admirals announcing the temporary suspension of the restrictive clauses, and directing them to forbid the fishermen from interfering with the planters and their fishing activities provided the inhabitants conformed to the other rules of the new charter.[16] The inhabitants were consequently permitted to continue in possession of their property along the shore and to reside in the coastal zone pending the report of the commodore for the summer of 1677, and until royal action had been taken thereon.[17] Although the restrictions were only suspended temporarily, the planters had gained an important victory in their struggle to break down West Country control and gain official recognition of their right to exist.

The information furnished by Sir John Berry and his determined stand in opposition to the removal of the planters taught the government the value of impartial official investigators at Newfoundland. Thereafter the officials at home would no longer be entirely dependent upon the information received from prejudiced persons attached to one or the other of the rival groups, for the commodore of the convoy was expected to furnish a great deal of formal detail regarding the observance of the regulations, the condition of the fishery and settlement, and the activities of competitors and enemies. Such a procedure had been recommended during the investiga-

[16] Order in council, Mar. 28, 1677, *Acts, Priv. Coun., Col.*, 1613-1680, §1128.

[17] Besides the order in council cited above, see especially: general orders of the Lords of Trade to the fishing admirals, Mar. 30, 1677, C.O. 1: 39, no. 56; and a particular order for St. John's harbor, same date, C.O. 1: 39, no. 57.

tion of 1671 and had subsequently been adopted at the instance of the Lords of Trade in 1675, but Sir John Berry had now demonstrated the practicability of the scheme. In 1677, the first formal investigation of this sort was authorized by the Privy Council. The Lords of Trade prepared twenty-six heads of inquiry which were transmitted to the commodore through the Admiralty. The inquiries related to all phases of activity at Newfoundland: the condition of the inhabitants, their number and occupations; the observances of the charter regulations; the difficulties between the planters and the fishermen; and the activities of the French in their sphere of control. The heads for 1677 are the first official queries issued by the royal authorities for the guidance of the commodore in his annual investigation of the fishery.[18] Thereafter the practice of issuing these questionnaires became regular, although from time to time the questions were altered in form, and others added or substituted. Henceforward the annual "answers to heads of inquiry" form important sources of information relative to the fishery and settlement, while the questions themselves are useful guides to the changes in royal policy regarding Newfoundland. In the later years of the seventeenth century and in the eighteenth century the investigations conducted by the commodores were often perfunctory and their answers stilted and incomplete, but they are none the less valuable because of their formality and brevity.

The first investigation under formal orders from the Privy Council was undertaken by Sir William Poole, R.N., in the summer of 1677. His report was based upon

[18] The action taken by the Lords of Trade in preparing the heads of inquiry for the commander of the convoy is indicated in the journal, May 18, 1677, *Cal. State Paps., Col.*, 1677-1680, §260. Their report transmitting the heads of inquiry to the Privy Council is contained in the committee report and order in council, May 18, 1677, C.O. 1: 40, no. 84; *Acts, Priv. Coun., Col.*, 1613-1680, §1136. The heads of inquiry, C.O. 1: 40, no. 84 (i).

his own observations and upon conversations with the masters of fishing ships, "sackers," and planters, whose grievances he had considered. He discovered that St. John's, the most frequented and populous harbor, was the center of discord, and that more trouble was experienced there than in any of the other fishing communities. He confirmed Berry's earlier opinion that the charter regulations could not be enforced because of the conflicting statements of the rival groups as to the cause of the disorders. Like his predecessor, Poole felt that the western adventurers were dishonest in laying the entire blame for the decline of their fishery upon the planters. He felt that they exaggerated the villainies of the inhabitants and underestimated the detrimental effect of several bad fishing seasons upon the entire industry. He pointed out that were bad seasons to recur frequently the inhabitants would abandon the island voluntarily.[19] In their blind hatred of their competitors the West Country fishermen frequently lost sight of their true position, and failed to appreciate also that the planters were faced with ruin quite as often as they were themselves.

After Poole's return to England the planters and western adventurers were ready for another attempt to settle definitely the matter of the regulation. The Lords of Trade resumed their investigation in December 1677, when both parties to the perennial dispute brought forward the usual charges, proposals, and remedies. The agents of the planters again made an attack upon the validity of the charter of 1676, which they alleged had been obtained surreptitiously and which they claimed violated the proprietary patents under which their ancestors had settled in the island. The provisions of the new charter, they claimed, were absolutely destructive of their in-

[19] Sir William Poole to the Lords of Trade, Sept. 10, 1677, *Cal. State Paps., Col.*, 1677-1680, §405.

COLLAPSE OF CHARTER REGULATION 161

terests, and they asked that the clauses concerning the restriction or removal of the inhabitants be permanently rescinded. They renewed the old plea for a civil government and military protection, offering to support a governor and a minister of the Church of England at their own expense.[20] The adventurers did not attempt to defend the legality of their charter and no longer insisted on the complete removal of the people, but they did seek to check the growth of the plantation by demanding the strict enforcement of the clause forbidding the transportation to Newfoundland of persons other than members of ships' companies. The efforts of the planters to invalidate the Western Charter came to naught, for the Lords of Trade, on December 18, 1677, decided to report to the Privy Council that the Newfoundland fishing trade should be regulated in accordance with the charter of 1676. The committee took the stand that invalidation was unnecessary because the recall of the restrictive orders gave the planters opportunity to continue in the fishery.[21]

In spite of their great influence the western adventurers were none too well favored by conditions in the trade and fishery between 1675 and 1680. During these five years there was a marked decline in the number of fishing ships which they sent out, and a corresponding increase in the quantity of fish produced by the inhabitants. Sir John Berry reported in 1675, that the fishing season had been only indifferently good, the catch averaging about 200 quintals per boat. The total production was 172,000 quintals valued at £103,000, at the rate of 12s. per quintal. There were about 175 ships fishing between Cape Race

[20] Petition of the inhabitants of Newfoundland to the king [Dec. 19, 1677], C.O. 1: 40, no. 128.
[21] Journal of the Lords of Trade, Dec. 18, 1677, *Cal. State Paps., Col.*, 1677-1680, p. 193; order in council, Jan. 16, 1678, *Acts, Priv. Coun., Col.*, 1613-1680, §1195.

and Cape Bonavista, but the inhabitants had produced about £47,000, or about 27 per cent of the total catch. Besides the fish, a proportionate amount of train oil had been refined, and the total production of the fishery amounted to over £163,000. In the summer that Berry was at Newfoundland about forty sack ships called there to buy fish.[22] In the following season of 1676, there was a decided reduction in the number of fishing ships, only 125 appearing, and a corresponding falling off in the number of sack ships. The planters produced about the same proportion of the total catch, in this case about 30 per cent.[23] A comparison made in 1677 with the fishery of 1615, showed that the West Country fishermen had made no appreciable progress in sixty-two years. The figures for 1677 compare very unfavorably with those of the earlier year, the total catch being worth approximately £144,000, as compared with £133,000 in 1615, an improvement of only £11,000. During this long period the number of fishing ships had declined from 250 to 110. Had there been an increase in the size of the vessels the difference would not have been significant, but there had also been a proportionate decline in tonnage and a reduction in the number of men employed. Besides the decline in production, shipping, and labor, there had been an increase in the price of dried cod from 8$s.$ to 12$s.$ per quintal, so that the actual difference between the quantities produced in 1615 and 1677 was less than the above figures indicate. During the very bad season of 1677, the 1,900 inhabitants produced

[22] Berry to Williamson, July 24, 1675, C.O. 1: 34, no. 118; the same to the same, Sept. 12, 20, 1675, C.O. 1: 35, nos. 16, 16 (i, ii), 21.

[23] Abstract of papers concerning Newfoundland: Capt. Russell's account, Dec. 6, 1676, C.O. 1: 38, no. 91; his account of ships making fishing voyages, C.O. 1: 38, no. 81; Capt. Wyborne's account of ships fishing, 1676, C.O. 1: 38, nos. 88, 89; a total account of the inhabitants, etc., 1676, C.O. 1: 38, no. 90. An account of the condition of the planters, received from John Downing, their London agent, Dec. 14, 1676, Egerton MSS., 2395, fol. 564.

COLLAPSE OF CHARTER REGULATION 163

about 20 per cent of the annual catch. In fact the season was so extremely poor that the planters did not make enough to pay wages or meet the expense of provisions.[24] During the ensuing years the situation failed to improve. No figures are available for 1678, but in 1679 the season was nearly as bad as in 1677; only about 159,000 quintals of fish were made, which were carried away by about 140 sack ships.[25] In 1680 over 100,000 quintals were shipped in seventy-two sack ships to Spain and Portugal. By this time the inhabitants were keeping about a quarter of the fishing boats, and had become a decided factor in the industry. In competing with the English fishermen they were handicapped because their expenses were greater than those of the West Country ships. The planters had to pay more for materials and provisions, much of which had to be imported; and they had to offer higher wages in order to attract and retain labor in Newfoundland. On the other hand, these expenses kept them from underselling the English fishermen, so that the West Country industry did not have the cut-throat competition which it feared. However, the planters managed to keep on increasing the productivity of their fishery, while that of their rivals continued to decline. The falling off in the number of ships went on, there being only ninety-seven there in 1680, when the total catch was worth £126,000, including train oil. In that year the inhabitants produced a

[24] Account of the colony and fishery [1677], Additional MSS., 13,972, fols. 15-31 v. See especially fols. 26-28; other copies: C.O. 1: 67, no. 37; Sloane MSS., 2902, fol. 196. See also Sir William Poole to the Lords of Trade, Sept. 10, 1677, C.O. 1: 41, no. 62.

[25] Estimate of sack ships in St. John's harbor, etc., 1679, received Feb. 15, 1680, C.O. 1: 43, no. 112; Capt. Charles Talbot's answers to inquiries, clause 9, inclosed with his letter to Southwell, Sept. 15, 1679, C.O. 1: 43, no. 121; a list of ships under convoy of H.M. frigates *Assistance* and *Assurance*, under command of Sir Robert Robinson, Oct. 4, 1680, C.O. 1: 46, no. 17.

third; and by this time, too, about 60 per cent of the fish was carried to market in sack ships, the remainder still being transported in fishing ships, though there was an increasing tendency of the West Country people to discontinue the use of fishing vessels for market voyages. Captain Sir Robert Robinson, R.N., who was commodore in 1680, reported that there had been about 200 ships in the English zone, besides about 100 more foreign and colonial vessels which came to trade in the settlements. Robinson reported the annual catch as quite poor, averaging only 170 to 180 quintals per fishing boat, as compared with more prosperous years when the fish ran from 250 to 300 quintals per boat.[26]

Ever since 1660 there had been numerous years when the fishermen had reported poor catches. These poor years may have been caused by changes in the salinity and temperature of the currents as is known today, but it is also probable that the region occupied by the English was becoming too congested. There was insufficient room for both West Countrymen and planters in the narrow zone between Cape Race and Cape Bonavista, with the result that overcrowding led to an intensification of the bitterness between the two groups. Congestion was the real reason why the western adventurers demanded the removal of the inhabitants and the prohibition of immigration into Newfoundland. Expansion would have relieved the tension, but this was impossible as long as the French controlled the southern coast west of Cape Race and fished to the northward above Cape Bonavista. The failure of the fishery to improve after the adoption of the new regulations in 1675 and 1676 shows that the funda-

[26] A list of planters and inhabitants, an account of ships fishing there, and a list of sack ships at Newfoundland, 1680, C.O. 1: 46, no. 79; Robinson to Blathwayt, Sept. 16, 1680, C.O. 1: 46, no. 8; Robinson's answers to inquiries, 1680, clause 11, C.O. 1: 46, no. 8 (x).

COLLAPSE OF CHARTER REGULATION

mental causes for the difficulties in the fishery had not been discovered. The persistent reports of poor catches which came in year after year indicate that the difficulties were occasioned by the uncertainties of the natural environment of the fish, but congestion and the absence of any attempts at conservation contributed also to unsatisfactory fishing voyages. The reports of the western adventurers, planters, and commodores are filled with statements regarding the destruction of cod fry, the filling in of the harbors with ballast and refuse, as well as the destruction of the scanty forest resources of this part of Newfoundland. Overcrowding, although actually responsible for the bad conditions, was seldom acknowledged by either the Newfoundlanders or the West Countrymen as the real reason for the decline of the industry. The fishing trade itself required an adventurous spirit, the daily dangers were considerable, and, therefore, neither the planters nor the English fishermen had any desire to become pioneers and force the French out of the island. The absence of an aggressive, pioneer spirit is one of the most noteworthy features of seventeenth century Newfoundland, because it is in such striking contrast to the persistent adventurousness which was characteristic of both the English and French colonists on the American mainland during this period.

The investigation begun in 1677 by the Lords of Trade dragged slowly through these years of depression until 1681. The questions before them were still the enforcement of the charter rules requiring the return of fishermen and sailors to England at the end of the voyage, and the proposal to remove the restrictions upon the planters. William Downing, brother of John Downing, and Thomas Oxford represented the planters. They advocated the establishment of a more stable government than was provided by the Western Charter of 1676. Depositions from fishermen were taken, and from these it was made clear that the presence of the planters was necessary to the success of the West Country trade. The inhabitants still complained of violence committed by the fishermen, and insisted that the restrictions upon their own activities and the growth of the plantation should be permanently rescinded. William Downing emphatically denied that the planters wished to prevent the fishing ships from visiting Newfoundland, but that they did want equality of treatment and an equal share in the control of the fishery in each of the harbors. He asserted that the Newfoundlanders were willing to furnish a quota of seamen for the Royal Navy, and were prepared to coöperate with the West Countrymen in the proper regulation of the fishery. Above all, the islanders sought the appointment of a governor to supervise affairs in the fishery and settlements. The western adventurers, represented at the hearings of the Lords of Trade by Messrs. Scutt and Parrett, claimed that the presence of the planters was detrimental not only to the fishery but also to English commerce in general because they encouraged the New Englanders. The agents of the adventurers also stoutly maintained the validity of the charter of 1676 without reservation. At one time during the winter of 1680-1681, the crown appears to have been on the point of establishing civil government. William Hinton claimed that the king had promised to make him governor, but Sir Robert Robinson and a Roman Catholic gentleman named Coney were also possible candidates for the position. No definite action was ever taken, and the investigation which had begun in 1677 continued intermittently until ended by the Revolution of 1689. The restrictions upon the inhabitants were removed by recalling the orders of 1675, but the objectionable clauses were never expunged from the charter of 1676. During the last years of Charles II and dur-

COLLAPSE OF CHARTER REGULATION 167

ing the brief reign of James II the Lords of Trade considered Newfoundland only perfunctorily, chiefly when the heads of inquiry were being prepared for the commodore or occasionally when pressure was brought to bear upon the committee by Downing, Oxford, and Hinton.[27]

During the period from 1677 to the flight of King James II in 1688, there was a slight but unmistakable tendency on the part of the authorities to change their attitude toward the problems of Newfoundland. The crown was less ready than formerly to accept the pretensions of the western adventurers at their face value, and appears to have been more willing to pay attention to the proposals advanced by the Newfoundlanders. Although both the Lords of Trade and the Privy Council remained steadfast in their insistence that the charter of 1676 be enforced, they readily agreed to a virtual suspension of the oppressive clauses by recalling the orders of 1675 which had been designed to enforce them. During this period the western adventurers were placed more on the defensive than before. The attack of John Downing upon the validity of the charter of 1676 succeeded in giving the initiative to the planters, who from this time on became much more aggressive. The reason why the cause of the Newfoundlanders was placed in a more favorable light from 1677 onward was that the persons representing their interests were men of greater ability than those who had been their earlier advocates. Hinton, the Downings, and Oxford had all lived in Newfoundland and knew

[27] Depositions from persons favorable to the retention of the inhabitants were received in 1679, C.O. 1: 42, nos. 20, 21, 22. Petitions and proposals were also received from representatives of the planters, such as William and John Downing, Thomas Oxford, and William Hinton, C.O. 1: 43, nos. 16, 40, 41, 51, 83; C.O. 1: 44, nos. 17, 18, 23, 27, 46; C.O. 1: 47, no. 15. Reports and suggestions from the commodores in 1679 and 1680: Capt. Charles Talbot, C.O. 1: 43, nos. 121, 121 (i); Sir Robert Robinson, C.O. 1: 44, nos. 46, 50; C.O. 1: 46, nos. 8, 8 (i-x).

intimately the condition of the planters. Previously, the Newfoundland cause had been represented by London merchants, who were interested in developing their own business with the planters but who knew little about actual conditions in the island. Moreover, the inspiration given by Sir John Berry in 1675 must have encouraged the Newfoundlanders to assume the defense of their own interests. At any rate, from that time forward there was a decided change in their attitude from one of resignation to one of hope.

The change in the point of view of the crown was not due altogether to the forcefulness of the Newfoundland agents; outside influences had their effect. During this period the crown was actively endeavoring to extend the royal prerogative over the proprietary and charter colonies in America. Hence the advocates of civil government for Newfoundland found themselves supported by the advocates of royal supremacy in the plantations, and they were consequently in a more strategic position to attain their objective than at any time since the restoration of Charles II. Although the royal officials toyed with the idea of erecting a government in Newfoundland, the influence of the West Country was still sufficiently strong to prevent their making a definite move in that direction.

After 1680 there was no improvement. The seasonal catches were only indifferently good and the West Country industry continued to suffer, some of the adventurers anticipating its total disappearance. Many shipowners laid up their vessels, and others contemplated doing the same. By this time, too, the West Country merchants were willing to admit that the confinement of the English fishery to the coast between Cape Race and Cape Bonavista was an important reason for their ill success. The situation was so bad that they considered themselves fortunate if they made one good voyage in three years. The

COLLAPSE OF CHARTER REGULATION 169

crowded condition could only be relieved either by employing only half the number of fishing boats or else by sending half the fleet to fish on the southern coast where the French were already established, but where the English fishermen would enjoy the same climatic and geographic advantages as did their Gallic rivals. The industry was also handicapped by the war between England and the Barbary pirates which hindered deliveries to southwestern Europe and the Mediterranean. Though prices had risen, the market for fish was poor, and the West Country business men were obtaining less profit than formerly. The large number of sack ships calling at Newfoundland for ladings was partly responsible for the unfavorable marketing conditions. They usually arrived at about the same time, and though the resultant sharp competitive bidding often proved advantageous to the West Countrymen, its good effects were offset by the practice of the local factors in holding over fish from the previous season on the ground that it would cause the merchantmen to call earlier for their ladings. It was supposed that this method of disposing of the fish would enable the English to compete on more even terms with the French in the early markets of southern Europe, but it actually had the effect of glutting the local market. The factors and planters profited by the arrangement, but the western adventurers, who did not finish making their fish until the latter part of August, found it disadvantageous. In 1684, there were only forty-three fishing vessels engaged in the industry between Cape Race and Cape Bonavista. These ships employed less than 2,000 men and fewer than 300 boats. The byboatkeepers alone had 350 men engaged in fishing, and the less than 2,000 inhabitants made nearly as much fish as the adventurers.[28] The

---

[28] Conditions in the fishery are described in the answers to inquiries submitted by the commodores from 1682 to 1684 inclusive. Capt. Daniel Jones

end of the century found the situation little improved, although the fishery was generally more prosperous. The western adventurers were no longer dominant, for by 1698 the inhabitants were producing about the same annual catch as their competitors, but they still clung tenaciously to the control of administration.[29] Though the crown considered the possibility of erecting a civil government, it is evident that the authorities were loath to arouse the ire of the West of England, which was influential in parliament. Besides the project of sending out a governor which was considered in 1680-1681, another plan of the same nature was considered after the accession of William and Mary, in May 1689, when the king in council ordered that a governor be sent to Newfoundland, together with materials for the construction of a fort at St. John's, and guns and stores for its defense and support. Though the project was abandoned, these proposed steps were prophetic of later developments.[30]

The reluctance of the crown to approve the proposed changes in the management of Newfoundland affairs was

to [Blathwayt], Sept. 12, 1682, C.O. 1: 49, no. 51; Capt. Talbot to the king, read Feb. 14, 1683, C.O. 1: 57, no. 29; Capt. Francis Wheler to Blathwayt, answers to inquiries, especially clauses 13, 16, Oct. 27, 1684, C.O. 1: 55, no. 56.

[29] Board of Trade representation, Mar. 30, 1699, *Cal. State Paps., Col.*, 1699, §217; a state of the trade to Newfoundland, 1706, Egerton MSS., 921, fols. 3-8. There is considerable difference between the above and the report of Capt. Charles Norris for 1698, *Cal. State Paps., Col.*, 1697-1698, §§852, 852 (i, ii). Compare with Sir Stafford Fairborne's account, 1700, *ibid.*, 1700, §744 (i); and also with Commodore Graydon's answers to inquiries, Mar. 13, 1702, *ibid.*, 1701, §879 (xiii).

[30] In May 1689, the Lords of Trade recommended that a governor be sent to Newfoundland for the duration of the anticipated war with France, and also proposed that forts be erected at St. John's, *Cal. State Paps., Col.*, 1689-1692, §124; order in council, May 18, 1689, *ibid.*, 1689-1692, §132; but the abstract of the order in *Acts, Priv. Coun., Col.*, 1680-1720, §289, makes no mention of a governor, and this proposal was probably stricken out in the final draft.

COLLAPSE OF CHARTER REGULATION

not entirely due to fear of the wrath of the West of England. It was also dictated by anxiety lest the suggested reorganization should injure or even destroy the nursery for seamen. Since 1615 the training of sailors had been recognized as an important function of the fishery.[31] The voyage to the fishing grounds was considered an admirable apprenticeship for prospective mariners. After a cruise or two green landsmen were transformed into experienced and able seamen, ready for employment in other branches of the mercantile marine, and available for impressment into the Royal Navy. Because the navy found the Newfoundland trade a bountiful source of recruits and because the West Country shipowners and operators had become increasingly dependent upon the navy during the wars with France, Spain, and Holland, for protection for their shipping, there had developed an alliance between the crown and the western adventurers which sought to protect and extend the training facilities offered by the fishery. Therefore, both the adventurers and the royal officers were keenly aware of the necessity of maintaining the fishery in an undisturbed condition, and they constantly sought to eliminate any factors which threatened to destroy its usefulness as a training school.

The problem of the participation of the byboatkeepers and planters in the fishery would probably have been settled much more easily and in a manner much more favorable to these groups had the government not always been a party to the disputes. The importance of the training-school idea tended to prejudice the crown in favor of the maintenance of an industry conducted in vessels having their home ports in England. Provisions in the charter of 1676 designed to prevent men from remaining permanently in Newfoundland were included not only to check the growth of the plantation but also to insure the

[31] *The Trades Increase* (London, 1615), pp. 3, 13, 26.

return of trained seamen to England. Similarly, the clause requiring that one-fifth of the complement of each vessel going to Newfoundland should be composed of green men was also intended to encourage the same. The problem of the constant emigration of trained fishermen and sailors from Newfoundland to New England would never have become an important issue had it not been detrimental to the training function of the fishery. The political factors involved in the question of proper regulation for Newfoundland were quite as important as the economic. The persistent support which the crown accorded the western adventurers throughout the late seventeenth century cannot be explained in any other way than that the government was tremendously concerned with building up the navy during these years and, therefore, felt that the maintenance of the West Country industry was of great advantage to the nation.

Had England been the only country which recognized the possibilities of Newfoundland as a training school for seamen, this feature would have undoubtedly received less attention than was actually the case. France, however, appreciated this use of the fishery quite as much as did England. As the two nations gradually drew face to face for the long struggle for political and economic supremacy, it became apparent to Englishmen that the reduction of France's fisheries at Newfoundland and in the neighboring Acadian and Laurentian regions was quite as essential from the point of view of naval supremacy as it was from the standpoint of commercial dominance. Consequently quite as many valued the fishery as a contributor to the maintenance of the balance of power, as did those who thought of it only in terms of national economic supremacy.[32]

[32] Sir William Petty, *Political Arithmetic* (London, 1690), reprinted in Charles H. Hull, ed., *The Economic Writings of Sir William Petty*, I, 232-

COLLAPSE OF CHARTER REGULATION 173

This function of the fishery had been appreciated from the earliest times, but it was not until after the return of Charles II that any specific provisions were included in the regulations looking toward the strengthening of the training school. Like its predecessor of 1634, the charter of 1661 declared that the Newfoundland fishery was valuable because "the navigation and mariners of this Realm had been much increased. . . ." However, it also contained a more specific provision in the twelfth clause which prohibited the transportation of any persons to Newfoundland other than *bona fide* members of the crews of fishing vessels or persons who were resident in the island.[33] This important provision, as has already been explained, was intended to prevent the growth of the fishery carried on by the byboatkeepers and to put a stop to further emigration from England to Newfoundland, not only because the byboatmen and inhabitants competed with the English fishermen, but also because they were diverted from service in the merchant marine and therefore unavailable for impressment in the navy. This clause was not observed, and the Privy Council was obliged to call upon the mayors and customs officers of the outports to enforce it in 1663 on the ground that failure to do so would eventually destroy the entire trade.[34] The byboatkeepers and their employees went to Newfoundland as passengers, conducted their fishery independently, and returned to England as passengers, unless they chose to

---

313. See especially p. 281. "J.B.," *An Account of the French Usurpation upon the Trade of England* (London, 1679), p. 14; Sir Henry Pollexfen, *A Discourse of Trade and Coyn* (London, 1697), pp. 90-91.

[33] The Western Charter, Jan. 26, 1661, C.O. 1, 15, no. 3. See especially the preamble and clause 12.

[34] Circular letter from the Privy Council to the mayors of Southampton, Weymouth, Lyme, Dartmouth, Plymouth, Fowey, and Barnstaple, Dec. 4, 1663, *Acts, Priv. Coun., Col.*, 1613-1680, §612; also the reinforcing order, Feb. 25, 1670, *ibid.*, 1613-1680, §891.

174 BRITISH FISHERY AT NEWFOUNDLAND

remain in the island permanently. In 1665, during the second war with Holland, it was estimated that a thousand men remained in Newfoundland at the end of the season in order to avoid impressment.[35] Thus, from the point of view of those interested in building up the Royal Navy, the great defect of a fishery conducted either by planters or byboatmen was its tendency to divert men from employment in general shipping and its failure to train them beyond the status of mere fishermen, leaving them ignorant of the higher branches of seamanship and useless as recruits for the carrying trade or the navy.

The western adventurers were always careful to call the attention of the crown to these weaknesses. One of the principal arguments advanced against the encouragement of a plantation and the erection of local government in Newfoundland was that the increase of mariners and shipping would cease.[36] The proposal to remove the inhabitants undoubtedly designed in part as a means of improving the nursery for seamen, for were there no residents in the island, that portion of the fishery formerly in their hands would be retrieved by the adventurers and result in an increase in the number of trained sailors, not only because more fishing ships would go to Newfoundland but also because the deserting sailors could not seek protection from the Newfoundlanders. Thus, a greater number of men would be returned to England for service in the navy.[37] On the other hand, those who advocated royal encouragement for the planters pointed out that the

[35] John Collins, *A Plea for the Bringing in of Irish Cattel, and Keeping out of Fish Caught by Foreigners* (London, 1680), pp. 20-21.

[36] Order in council, Aug. 28, 1667, *Acts, Priv. Coun., Col.*, 1613-1680, §716; petition of Plymouth merchants, Aug. 28, 1667, *Cal. State Paps., Col.*, 1661-1680, §1561; order in council referring a petition from the Bristol merchants to a committee, Dec. 4, 1667, *Acts, Priv. Coun., Col.*, 1613-1680, §735.

[37] Reply of the merchants, etc., of the West of England to Capt. Robinson's allegations [1668], C.O. 1: 22, no. 71.

COLLAPSE OF CHARTER REGULATION 175

removal of the inhabitants would cause the French to gain complete control of Newfoundland, with the result that the English navy would not only suffer the loss of potential sailors, but would also see the naval strength of its rival increased.[38] Both groups of contestants for the control of Newfoundland were interested in the training of seamen, but the views of the western adventurers were the ones that carried most weight in official circles.

During the investigation of 1671, the Council for Foreign Plantations took into consideration the best means of increasing the number of seamen and insuring a regular supply of trained men for service under the flag. Their proposed rules were intended to encourage the training of seamen or to make the trained men available for impressment. With the exception of the rule prohibiting the transportation of civilians on warships, all of their suggestions were approved by order in council on March 10, 1671, and were intended as clauses additional to the new Western Charter.[39] The recommendation that the commanders of naval convoys be forbidden to carry civilians was included in directions given on the same day to the Lord High Admiral as one of the articles of instruction for the commodore.[40] In spite of their rigorousness and their obvious intent they were not executed until after the reconsideration of the fishery regulations in 1675.

During the investigations held in 1675 there was a decided difference of opinion as to the commercial phases of the fishery, but both the London merchants and the western adventurers agreed that the training school for seamen must be preserved and encouraged. Those who

[38] Capt. Robinson's proposals, 1668, C.O. 1: 22, no. 70.
[39] Order in council, Mar. 10, 1671, *Acts, Priv. Coun., Col.*, 1613-1680, §915.
[40] Clause 15 of the report of the Council for Plantations was incorporated as clause 1 in the order in council, Mar. 10, 1671, *ibid.*, 1613-1680, §916.

sought to establish civil government pointed out that in its most prosperous days the fishery had trained about 10,000 seamen yearly. They stated that the training of sailors constituted one of the major reasons for England's interest in Newfoundland, but that under the direction of the western adventurers it was a question whether the maximum benefit accrued to the nation, owing to the objection of the sailors to the arduous toil involved in their fishing activities. The Londoners prophesied that the fishery would continue to decline not only until the number of seamen would fail to increase, but that English shipping would fall off as well.[41] Though the critics of the western adventurers talked a great deal of the inefficiency of charter regulation by the West of England, they offered no practical plan to improve the nursery for seamen. On the other hand the West Country interests took the stand that it was their control which encouraged the training of sailors. They pointed out that one-fifth of their employees were green men who became able seamen in two or three voyages; that formerly the Newfoundland trade had developed more sailors than all the other English trades put together, with the single exception of the "river of London," and that the operation of the fishery by byboatmen and planters would result in a loss of men for the navy.[42] As we have seen, the counsels

[41] Reasons for the settlement of Newfoundland and the trade thereof under government [1668?], C.O. 1: 22, no. 69; C.O. 1: 67, no. 16 (i); some modest observations and queries by "An Englishman," Mar. 25, 1675, C.O. 1: 34, no. 32; minutes of the Committee of Trade and Plantations, Mar. 25, 1675, *Cal. State Paps., Col.,* 1675-1676, p. 191; James Houblon to [Southwell], Mar. 30, 1675, C.O. 1: 34, no. 27.

[42] Remonstrance of the merchants, etc., of Plymouth, Dartmouth, etc., to the king [1675], C.O. 1: 67, no. 26; petition of the gentry, merchants, etc., of Exeter, Dartmouth, etc., to the king, read Mar. 25, 1675, C.O. 1: 67, no. 25; papers presented by Parrett to Williamson, read Mar. 25, 30, 1675, C.O. 1: 67, nos. 27, 29; Richard Hooper and Thomas Yearing, mayors of Barnstaple and Bideford, to the Lords of Trade, Mar. 30, 1675, C.O. 1: 34, nos. 38, 39.

COLLAPSE OF CHARTER REGULATION 177

of the West Country people prevailed at court during the investigation of 1675, and the government set about to put into effect the new rules which had been approved in 1671. These were incorporated with only slight changes in the new charter of 1676.

But the problem of encouraging the training of sailors was not solved by the inclusion of these new prohibitions and requirements in the charter. It is very clear that a considerable number of the English fishermen were sufficiently unpatriotic not to appreciate the naval function of the fishery, and were unwilling to sacrifice their own personal gains in order to benefit the nation. Some of the more conscientious masters of fishing ships tried to hold their men by refusing to pay them until they had returned to England, but others encouraged their men to return home as passengers in order to profit from the passage money. Many fishermen preferred to save money by remaining in Newfoundland during the winter, making the return voyage to England once every two years, while others, as the result of drink or failure to obtain their wages, were left behind. These men formed a large but variable portion of the population, their number depending somewhat upon the prosperity of the fishery. During the winter of 1683-1684, following a fairly good season, only 120 men remained, but it was expected that following the poor season in the summer of 1684, more would winter in the island. As "long as their [sic] comes no Women," these men formed a floating population, which contributed nothing to the improvement of English shipping.[43]

The impermanence of a large part of the population was a problem which continually engaged the attention of

[43] Capt. Wheler's answers to inquiries, received Oct. 27, 1684, C.O. 1: 55, no. 56, especially clause 13; observations by Capt. Wheler upon the articles of the Western Charter, received Oct. 27, 1684, C.O. 1: 55, no. 56 (i), especially clause 7.

178 BRITISH FISHERY AT NEWFOUNDLAND

both the crown and the English merchants. Had the fishermen and sailors remained in Newfoundland the matter of obtaining their services would have been much simpler than was actually the case.[44] Unfortunately, from the point of view of those interested in obtaining skilled men for the navy and mercantile marine, there was a steady emigration to New England. In 1684, Captain Daniel Jones, commander of the convoy, stated that none violated the rules of the western charter as frequently as the New England traders who visited Newfoundland annually, and "spirited away" Englishmen "to the utter ruin of both merchant adventurer and planters, and the decay of the fishing." Jones reported that one New England vessel came to St. John's with eleven men and tried to sail away with twenty more. The commodore made him put all his passengers ashore and took bonds from the other New England masters not to carry men away to the continental colonies.[45] Toward the close of the century numerous reports were received by the authorities concerning the extent of this traffic. Captain Charles Hawkins, commodore in 1691, stated that the yearly exodus amounted to between 100 and 150 seamen and fishermen. The emigration of these skilled fishermen and sailors was regarded as very harmful to the adventurers and planters

[44] Capt. [William] Davies to M. Wren, Sept. 9, 1671, *Cal. State Paps., Col.*, 1669-1674, §616; Capt. Davies' reasons for the decay of the trade [1672], C.O. 1: 29, no. 78; Berry to Williamson, July 24, 1675, C.O. 1: 34, no. 118; Williamson's notes of the evidence of Gould and Parrett, Feb. 27, 1675, C.O. 1: 34, no. 16; a brief narrative concerning Newfoundland, by John Downing [Nov. 4, 1676], Egerton MSS., 2395, fols. 560-563; C.O. 1: 38, no. 70.

[45] Robinson to Blathwayt, Sept. 16, 1680, C.O. 1: 46, no. 8; Capt. Daniel Jones to the same, Sept. 12, 1682, C.O. 1: 49, no. 51. Jones inclosed bonds which he had exacted from John Tawley [Sayley?], George Snell, Thomas Harvey, and William Pepperel not to take Englishmen from Newfoundland, C.O. 1: 49, no. 51 (i). This is the first recorded instance of this practice which became regular, especially after 1700.

COLLAPSE OF CHARTER REGULATION 179

who could not meet the higher wages offered in the colonies.[46] The home authorities were greatly concerned lest the traffic eventually destroy the Newfoundland fishery and consequently harm the general merchant marine and the Royal Navy. Consequently, the crown and the western adventurers sought to destroy the influence of New England in the northern fishery.

Like all other English shipping trades, that of Newfoundland was subject to embargo from time to time, in order that the press gangs might procure men for service in the fleet. The embargoes were most frequent during the wars with France at the end of the seventeenth century, but some were laid upon the shipping during the second and third Dutch wars. The Newfoundland trade was usually fortunate in having the restrictions removed in a relatively short time, partly because of the seasonal nature of the fishery and partly because the government appreciated its function as a training school. The embargoes were often lifted as soon as the navy had obtained its quota of new men from the fishing fleet, though sometimes a general press caused the western adventurers to undertake the summer's fishing with a larger proportion of inexperienced men than was usual in times of peace. In general, during the seventeenth century at least, the Newfoundland trade was singularly free from embargoes, or at least freed from them much more expeditiously than was the case in other shipping trades. Impressment was unpopular among fishermen and sailors and unquestionably was the cause of a great deal of the desertion at Newfoundland and subsequent emigration to New England. On a number of occasions embargoes

[46] Capt. Charles Hawkins' answers to inquiries, Dec. 4, 1691, *Cal. State Paps., Col.*, 1699, §1287 (i); Capt. Leake's answers, Sept. 17, 1699, *ibid.*, 1699, p. 440; Capt. Fairborne to [William Popple], Sept. 12, 1700, inclosing answers to inquiries, *ibid.*, 1700, §774 (i), especially clause 27.

caused the fishing fleet to remain at home, but it was not until the War of the Spanish Succession that the fishing ships were impeded by embargoes to any considerable extent.[47]

The western adventurers were not entirely disinterested in their concern for the nursery for seamen. They were anxious to obtain adequate naval protection for the fishing fleet during the voyage to Newfoundland and return in times of war, and desired protection against fish pirates while on the coasts of the island, as well as for the market fleet bound for Spain, Portugal, and the Mediterranean ports. In time of war there was danger from Dutch and French warships and privateers, and even in times of peace the Barbary rovers were a constant menace. To obtain naval protection the western adventurers had to offer something to the crown in return, and we find them, therefore, willing to permit the introduction of regulations for the encouragement of the nursery for seamen, and after 1675 consenting to naval supervision over the fishing fleet.

By the end of the seventeenth century the ineffectiveness of charter regulation had been clearly demonstrated. The rules which the Lords of Trade had devised in 1675 had proved unsatisfactory from the start, particularly those clauses, designed to give the West Country mastery of the industry, which had been suspended almost immediately after they had been approved by the crown. The Newfoundlanders had shown a surprising amount of persistence in their attack upon the validity of the restrictions contained in the charter of 1676, and had kept up their agitation for equality of treatment. Moreover, from 1677 onward the English government had wavered in its

[47] There are numerous references to embargoes and passes for ships from 1689-1700 in *Acts, Priv. Coun., Col.*, 1680-1720, pp. 158-174, 219-226, 279-282, *passim*.

COLLAPSE OF CHARTER REGULATION

hitherto steadfast support of the western adventurers, and on at least two occasions between 1680 and 1699 had been on the point of erecting a civil government in Newfoundland, and recognizing the rights of the planters. Beset by these difficulties, and the failure of their industry to show any marked improvement, the West-of-England operators lost confidence in the charter regulation and besought the help of parliament in solving the problems of their overseas enterprise. The period of charter regulation closed with the passage of the Newfoundland Act in 1699, terminating a phase of the history of the Newfoundland fishery which had begun when John Guy brought his first settlers to the shores of the island, but leaving the planters still resident, in possession of their homes and fishing conveniences near the shore, and playing a more important part in the industry than had been the case earlier in the century.

CHAPTER VI

THE MENACE OF FRANCE AND NEW ENGLAND, 1660-1699

ALTHOUGH the French and the English had fished near each other at Newfoundland for many years, the latter had not considered France a competitor until about the time that Sir David Kirke tried to force an "acknowledgment" of five per cent from foreign shipping at Newfoundland. It was not until the days of Louis XIV and Colbert that the English began to realize that their ancient rival was making a serious attempt to gain control of the entire North American fisheries, and was seeking to oust them from Newfoundland. In 1662 a French settlement was established at Placentia on the southern coast, which was peopled by about sixty families, fortified, and garrisoned by troops from France. Subsidized by the crown, the French colonists set about making dry fish, a branch of the industry which they had heretofore neglected at Newfoundland, although they had prosecuted it upon the coasts of Acadia and in the St. Lawrence for some time. The first complaints of their activities reached England in 1666, when it was reported that besides establishing the settlement and fortifications, the French had ousted an Englishman and declined to pay the "acknowledgment." The news of this event was received in England with mixed feelings. The western adventurers refused to become alarmed but the London merchants, supported by the Newfoundland planters made a great deal of it. The adventurers pointed out that the French had in no way encroached upon the English fishery. They maintained that the fort at Placentia was not intended as a base of operations against their own fishery, but had been established to protect the French colonists against the English planters who annoyed them.[1] It is impossible to tell just how well they understood the actual situation. The western adventurers were notoriously lacking in the nationalistic feeling which inspired the London merchants, and it is probable that they were more concerned with the disorganized state of their own industry than with the activities of foreigners. One wonders too, whether the West Country group did not capitalize the pro-French leanings of the court of Charles II during these years.

Meanwhile the French were rapidly expanding their North American fisheries. They already had established themselves over a large area in the northeastern part of the continent. Should they gain control of Newfoundland, they would enjoy a practical monopoly of the fisheries, as their remaining competitors would be the people of New England. The selection of the southern coast gave France a decided advantage, as the climate there was superior to that of the English zone, and the topography of the shore was such that stages could be dispensed with and the fish cured on the beaches. These geographic advantages, coupled with the fact that the French government had not only subsidized the settlement at Placentia, but provided fortifications, a garrison, and patrol ships to protect the fishery, as well as a civil administration, gave them an

[1] Address to the king showing the present condition of Newfoundland [Jan. 8], 1668, C.O. 1:22, no. 5; reply of the merchants, etc., to Capt. Robinson's allegations, 1668, C.O. 1:22, no. 71. Cf. "Some remarks on the Decayes of that so Useful Trade the English had before the Year 1661, to Newfoundland, and Nova Scotia, &c., in the Northern parts of America," n "A book in Folio consisting of divers Tracts bound up together," no date, Harleian MSS., 1223, no. 1, fols. 1-6.

opportunity to make earlier deliveries than did the English.

By 1676 it was apparent that French competition was to be taken seriously. During the seven years previous the English felt that they had lost while their foreign competitors had gained. By this time St. Malo alone was sending 120 vessels to the fishery north of Cape Bonavista, and altogether there were employed on the coast of the island or on the banks about 600 French fishing ships which hailed from the ports lying between Calais and St. Jean de Luz. Such figures as these caused a considerable amount of apprehension in English mercantile circles, and during the investigation of 1675-1676, attempts were made by the Lords of Trade to obtain accurate data in regard to all French activities in Newfoundland.[2]

Information gathered at this time made the English fully aware of the French competition and its implications. In general, the French industry appeared to be on a sounder financial basis than the English, for the leading merchants of St. Malo, Dieppe, Rouen, Nantes, La Rochelle, Bordeaux, and Bayonne, had not only a good deal of capital invested in the fishery, but also sufficient resources to prevent them from mortgaging their ships in order to undertake voyages to Newfoundland. Furthermore, their production costs were less because they could fit out, provision, and man their ships more cheaply than could the English. Besides these commercial advantages, the French industry was encouraged by the government of Louis XIV. Not only was the domestic trade in dried

---

[2] Capt. Robinson's proposals, 1668, C.O. 1: 22, no. 70; Capt. Davies' reasons for the decay of the trade, 1672, C.O. 1: 29, no. 78; reasons for the settlement of Newfoundland and the trade thereof under government [1668] C.O. 1: 22, no. 69; Williamson's notes on the evidence of Gould and Parrett, Mar. 1, 1675, C.O. 1: 34, no. 17; George Pley's reasons for a settlement in Newfoundland, Mar. 17, 1675, C.O. 1: 34, no. 25; Nath. Osborne to Williamson, same date, C.O. 1: 32, no. 23.

fish protected by a tariff, but the crown subsidized the settlement at Placentia and offered similar assistance to the merchants for the construction of vessels to be used in the industry. Furthermore, the French had been granted admission to the Spanish market by the Treaty of the Pyrenees, and were also favorably situated for selling their product to better advantage in Portugal and Italy than were the English, because they employed improved methods of curing the fish. At Newfoundland the French had encouraged English planters to accept their protection, had sent ships to the English settlements to buy fish before the arrival of their own sack ships, and were operating elsewhere under the guise of Jerseymen.[3] Despite all this evidence there still remained considerable difference of opinion in England as to the real danger to be expected from France. The conservative West Countrymen were dubious in their attitude toward the French menace until after 1675, and even those who were most emphatic in describing its dangers, appear to have been very indefinite in their knowledge of the real situation.

Uncertain facts and extravagant claims made the Lords of Trade discount a great deal of the information received. In their report of April 15, 1675, they admitted that France had recently applied herself industriously to the fishery because of the governmental aid granted to her merchants and recognized that England was having to meet vigorous competition abroad, but they accounted for France's real success on the ground that her fishery was carried on by ships operated from her home ports. The committee report minimized the danger of the French fort at Placentia, which was intended to protect the fur

[3] Gould to Southwell, Mar. 18, 1675, C.O. 1: 67, no. 22; Houblon to the same, Mar. 20, 1675, C.O. 1: 34, no. 27; reasons for the settlement of Newfoundland, etc. [1668?], C.O. 1:22, no. 69; Thomas Farr to the Lords of Trade, Mar. 21, 1675, C.O. 1: 34, no. 28.

trade rather than the fishery, and maintained that the English still had an advantage in the fishing trade. "They do catch it as cheap, cure it as well, come as early to market, can there sell as dear, and afford it as cheap as any the French can do." Nevertheless, the Lords of Trade were sufficiently concerned with French activities to recommend in the same report that the convoy commander be instructed to find out all he could about the trade and fishery of England's rival.[4]

As in the case of the disputes between the West Countrymen and the planters, it was Sir John Berry who gave the authorities their first independent account of the French at Newfoundland, although he found it difficult to obtain accurate information. However, he did learn in 1675 that they had two warships of thirty and forty guns to protect their south-coast fishery, and that these vessels acted as convoy for a fleet of forty or fifty ships, which had its rendezvous at Trepassey. Besides this fleet, there were about twenty large vessels from St. Malo which fished north of Cape Bonavista without naval protection. Berry felt that the removal of the planters from the English fishing zone would force them to seek French protection, which had already been offered. He also asserted that the fort at Placentia was intended to protect the fishery, and not the fur trade as the western adventurers maintained.[5] Corroboration of Berry's statements was

[4] Paper from Parrett to Williamson relative to French fishing method [1675], C.O. 1: 67, no. 28; minutes of the Committee of Trade and Plantations, Mar. 30, Apr. 1, 1675, *Cal. State Paps., Col.*, 1675-1676, pp. 197-198, 201, 204-205; George Pley's reasons for a settlement, etc., Mar. 17, 1675 C.O. 1: 34, no. 25; report of the Lords of Trade, Apr. 15, 1675, C.O. 1: 67, no. 30 (i); order in council approving report, May 5, 1675, *Acts, Priv. Coun. Col.*, 1613-1680, pp. 621-626.

[5] Berry to Williamson, July 24, 1675, C.O. 1: 34, no. 118; the same to the same, Sept. 12, 1675, C.O. 1: 35, no. 16; Berry's observations, 1675, C.O. 1: 35, no. 82.

MENACE OF FRANCE AND NEW ENGLAND 187

received in 1676 from the convoy commander who reported ninety French families at Placentia, with more arriving annually, so that by 1677, there were 250 there. The prospective settlers received two years' supply of bread and salt from the government, which also paid their passages to Newfoundland. Generally, the French made good fishing voyages. In 1676 they took between 400 and 500 quintals per boat, although the difference between their catch and that of the English was offset by the fact that their boats were longer and better built than those used by the western adventurers and planters. The French boats, probably small decked vessels, were manned by six fishermen, and were left behind at the end of the season. The French ships arrived at Newfoundland as early as February or March, stopped fishing at the end of July, and left for market in August. The ships were of considerable size, employing from twelve to fourteen boats each. In 1676, there were forty-nine of them reported on the south coast besides fifty more north of Cape Bonavista. Some of the French vessels were so large and so well armed that they were able to dispense with the protection of the convoy.[6]

By 1677 the Lords of Trade were sufficiently impressed to contemplate sending three commissioners with the convoy to investigate the French settlement and fishery. Although instructions for the commissioners were actually considered in committee, no definite action was taken to appoint them, and it was finally decided to have the commodore learn what he could. When the heads of inquiry

[6] Account of the French trade, etc. [1675], C.O. 1: 38, no. 82; John Downing's brief narrative, received Nov. 24, 1676, Egerton MSS., 2395, fols. 560-563; C.O. 1: 38, no. 10; account of the colony and fishery [1677], Additional MSS., 13,972, fols. 15-31 v; C.O. 1: 67, no. 37; C.O. 1: 38, nos. 3, 74; abstract of papers, Dec. 6, 1676, C.O. 1: 38, no. 91; Capt. Wyborne's account of French ships, 1676, C.O. 1: 38, no. 84; Journal of the Lords of Trade, Mar. 3, 4, 1677, *Cal. State Paps., Col.*, 1677-1680, §215.

were drafted a considerable number of the queries were devoted to this aspect of Newfoundland affairs. In general, the tenor of these clauses was similar to that of the questions concerning the English settlement and fishery, and they were intended to furnish a basis for comparison of the two fisheries.[7]

The situation became gradually worse after 1677. The planters complained that the western adventurers refused to recognize the true situation, minimizing the French danger. Meanwhile, the French continued to improve their fishery until they were either able to undersell or to meet the prices of their rivals. However, during the period from 1680 to the outbreak of the war with France in May, 1689, the Lords of Trade continued to manifest remarkable indifference. In 1690 the merchants of London trading to New York and New England expressed the opinion that their trade was now the most important nursery for seamen since the French had beaten the English out of the Newfoundland trade.[8] The narrow regionalism of the western adventurers, the disorganized state of the fishing industry, the controversies between the West Countrymen and the planters, and the evident lack of influence of the London mercantilists in official circles all contributed to the lack of interest in the danger of French competition. Moreover, the disturbed political

[7] Order in council, May 18, 1677, C.O. 1: 40, no. 84; *Acts, Priv. Coun Col.*, 1613-1680, §1136; heads of inquiry, 1677, especially clauses 15-27, C.O. 1: 40, no. 84 (i).

[8] Besides the references indicated in note 6, see proposals of William Downing and Thomas Oxford, Apr. 29, 1679, C.O. 1: 43, no. 51; Capt. Talbot's answers to inquiries, Sept. 15, 1679, C.O. 1: 43, no. 121 (i); William Hinton's observations, etc. [Feb. 26, 1680], C.O. 1: 44, no. 32; Robinson to Blathwayt, Sept. 16, 1680, C.O. 1: 46, no. 8 (x); petition of the merchants of London trading to New York and New England, May 14, 1690, Historical MSS. Commission, *Calendar of House of Lords Manuscripts, 1690-1691* §271; the agents of Massachusetts Bay to the Lords of Trade [Apr. 2 1691], *Cal. State Paps., Col.*, 1689-1692, §1418.

situation in England during the reign of James II and until William and Mary were safely installed on the throne accounts for the apathy in official circles. It was not until the wars with France in the closing years of the seventeenth century and in the early part of the eighteenth had aroused national feeling in England to a high pitch, that any propaganda against French rivalry in Newfoundland or elsewhere could receive popular approval and arouse official concern. Lack of interest in French competition during the period from 1660 to 1692 is another proof that the English crown had not accepted this cardinal principle of the mercantilist doctrine, and that the governments of that period had no fixed commercial policy.

While the western adventurers and the crown steadfastly refused to face the facts with respect to France, they were tremendously concerned with the relationship of Newfoundland to the other parts of the English commercial and colonial world. The merchants had recognized the value of the island as a good market for the sale of English products and manufactures. The fishing fleet was fitted out, provisioned, and equipped in England. Moreover the planters, never self-sustaining, were dependent upon the mother country or other sources for provisions, clothing, and equipment. Consequently a cardinal principle of the western adventurers had been the retention of this profitable trade in their own hands, and the elimination of possible competitors.

Previous to 1660 the English merchants had exported most of the commodities needed in the fishery, and as long as they continued to dominate the industry they could control the outgoing trade as well. The slow but steady increase in the permanent population of the island, the introduction of byboatkeepers, and the rise in prices of English commodities used at Newfoundland which took place after 1660, obliged the western adventurers to face competition from Ireland, Scotland, France, and the northern colonies in America. When it was first proposed in 1668 to remove the inhabitants, one of the reasons put forth in support of the plan was that thereafter provisions and other necessaries would be carried in fishing ships from England rather than be purchased in New England. The rise of New England's commerce caused the merchants of the West Country to fear that if the fishery were controlled by the Newfoundlanders they would procure all their supplies in New England.[9]

The investigation of 1675 brought out this West Country fear of the continental colonists, for it showed that their export trade to Newfoundland had declined as much as their fishery. For example, about 1655 the western adventurers had invested £120,000 annually in this branch of commerce, but by 1675 they were being supplanted by Ireland, New England, and France, a situation which caused a reduction by two-thirds of the number of ships in the trade. During the investigation the western adventurers voiced their fear that were a civil government established the Newfoundlanders would sell their fish to the Irish and New Englanders in exchange for provisions, thus depriving English merchants not only of their export trade to the island, but also of the carrying trade in fish, which they considered an "unspeakable loss" to England. Some even went so far as to request that a frigate be sent to Newfoundland with orders to seize any New England ships found trading there. On the other hand, John Gould, a London merchant favorable to the

[9] The reply of the merchants, etc., to Robinson's allegations [1668], C.O. 1: 22, no. 71. For a summary of seventeenth and eighteenth century New England trade see Ralph G. Lounsbury, "Yankee Trade at Newfoundland," *The New England Quarterly*, III (1930), 607-626.

MENACE OF FRANCE AND NEW ENGLAND 191

planters, pointed out that "in truth victuals do come from New England . . . but still that is from England that is from one of its colonies and the trade of which circles to England." This liberal view is in striking contrast to the opinions held by most of Gould's contemporaries, who were distinctly hostile to the New Englanders and the Irish, whose trade they considered subversive of the best national interests. The ideas of the majority prevailed with the Lords of Trade, and as we have already seen, caused them to announce to the Privy Council that a Newfoundland under government would be drawn within the sphere of New England influence[10] and should not be encouraged. During these years, it is safe to conclude, the western adventurers and the crown were much more fearful of the pernicious influence of New England than they were alarmed at French activities at Placentia and elsewhere in Newfoundland.

In spite of their fear of New England no one seemed to have a very clear idea of the exact nature of this trade. Englishmen generally believed that the colonial traders were responsible for supplying the planters with French brandy and Madeira wines, as well as provisions, though Sir John Berry had reported in 1675 that these spirits and wines were brought to Newfoundland by the English salt ships, and that he had found no evidence pointing to such a traffic on the part of the northern colonists. All that he was able to discover was that the New Englanders

---

[10] Williamson's notes of the evidence of Gould and Parrett, Feb. 27, 1675, C.O. 1: 34, no. 16; Parrett's paper presented to Williamson, received [Mar. 25, 1675], C.O. 1: 67, no. 27; another paper from Parrett, read Mar. 30, 1675, C.O. 1: 67, no. 29; remonstrance of the merchants, etc., of Plymouth, Dartmouth, etc., to the king [Mar. 25, 1675], C.O. 1: 67, no. 26; minutes of the Committee of Trade and Plantations, Apr. 8, 1675, *Cal. State Paps., Col.*, 1675-1676, §517; report of the Lords of Trade, Apr. 15, 1675, C.O. 1: 67, no. 30 (i); order in council approving the report, May 5, 1675, *Acts, Priv. Coun., Col.*, 1613-1680, pp. 621-626.

bartered their commodities to the inhabitants in exchange for fish or English goods. Apparently, there were not a great many colonial vessels engaged in the trade at this time. John Downing stated that there were usually from five to ten, whose burdens ranged from sixteen to fifty tons. These vessels brought ladings of bread, peas, flour, beef, pork, butter, tar, boards, brown sugar, and molasses, and received in exchange brandy, clothing, fishing gear, hard money in the form of Spanish pieces of eight, if they were obtainable, French and Spanish wines, as well as "some red stinking fish" otherwise known as "refuse" fish which they took to Barbados to serve as food for the negro slaves.[11]

In 1677, the commodore was instructed to investigate the origin of the commodities consumed by the byboat-keepers and inhabitants. He was directed to ascertain also how far the wine, brandy, and rum imported from New England was responsible for the debauchery of the fishermen.[12] Thereafter a great deal of valuable information concerning the activities of the New Englanders and other traders was received annually from the commodores. In this way the home government obtained sufficient information concerning illegal trade to enable it to take steps to check that trade in the continental colonies.

By 1679 the New Englanders had developed a three-cornered trade at Newfoundland. They imported tobacco, sugar, molasses, rum, meat, bread, flour, tar, and boards,

[11] Report of the Lords of Trade, Apr. 15, 1675, C.O. 1:67, no. 30 (i); Berry to Williamson, July 24, 1675, C.O. 1:34, no. 118; the same to the same, Sept. 12, 1675, C.O. 1:35, no. 16; the same to Southwell, same date, C.O. 1:35, no. 17; Berry's observations, 1675, C.O. 1:35, no. 82; John Downing's brief narrative [Nov. 24, 1676], C.O. 1:38, no. 70; an account of the French trade, Sept. 1676, C.O. 1:38, no. 82 (included in Berry's observations).

[12] Order in council, May 18, 1677, C.O. 1:40, no. 84; *Acts, Priv. Coun., Col.*, 1613-1680, §1136; heads of inquiries, 1677, C.O. 1:40, no. 84 (i).

which they traded to the inhabitants for fish. They also supplied the Newfoundlanders with nets and fishing tackle. The fish which they obtained from the inhabitants was then exchanged with the sack ships for wine, brandy, piece goods, and other articles, which they carried home with them. During the late seventeenth century the rum traffic does not seem to have been as extensive and devastating in its effects as was the case later, the prevalent debauchery in Newfoundland being occasioned by the consumption of foreign liquors. According to Commodore Charles Hawkins, the colonial traders brought about 300 tons of provisions from New England and the West Indies in 1691. Besides accepting return cargoes of refuse fish for delivery in Barbados, the colonial shipmasters by this time were accepting ladings of merchantable fish for Spain and Portugal, and thus participating in the carrying trade in competition with the English merchantmen.[13]

In the last decade of the seventeenth century the New England-Newfoundland trade became even more extensive. In 1696, the merchants of Plymouth complained that the New Englanders, Irish, and others were dealing widely with the Newfoundlanders. By this time too, the New England traders were selling their cargoes of rum, molasses, tobacco, and some provisions, either for hard money or bills of exchange, both of which were at a premium in New England. The bills of exchange were especially attractive as they brought a profit of twenty-five per cent on resale in the colonies. Previous to 1675 provisions had been the important item in New England's

---

[13] Capt. Talbot to Southwell, Sept. 15, 1679, C.O. 1:43, no. 121; his answers to inquiries, 1679, C.O. 1:43, no. 121 (i); Robinson to Blathwayt, Sept. 16, 1680, C.O. 1:46, no. 8; Capt. Hawkins to the Lords of Trade, Dec. 4, 1691, *Cal. State Paps., Col.*, 1699, §1287; his answers to inquiries, 1691, *ibid.*, 1699, §1287 (i); Capt. Crawley's answers to inquiries, 1692, *ibid.*, 1699, §1293 (i).

commerce with Newfoundland. Thereafter tobacco, rum, and molasses became leading articles of import. In 1698, Captain Charles Norris, the commodore, reported that the shipmasters were constantly complaining of the debauchery of the seamen which resulted from the large quantities of rum which were brought from New England and the West Indies, and Commodore Leake made a similar report in the following year.[14] In the absence of records of imports and exports it is difficult to estimate the real extent of the colonial trade. Information furnished by the merchants and shipmasters who opposed it naturally exaggerated its volume and ill effect, and the reticence of the smugglers made it difficult for the commodores to ascertain its exact nature and volume.

The English merchants were opposed to the colonial traders on four grounds: (1) that they competed with the provision trade from the mother country; (2) that they competed in the carrying trade to foreign countries; (3) that they enabled the colonists to use Newfoundland as an entrepôt for the illegal transshipment of enumerated plantation commodities to foreign countries; and (4) that they served as a center for the receipt of illegal foreign goods for carriage to the continental and West Indian colonies. To these may be added the fact that the presence of the colonial vessels gave the English fishermen and seamen opportunities to escape to other parts of the colonial world. From 1660 to 1675 the colonial trade in provisions seems to have caused great concern in England, but afterward the clandestine trade carried on through Newfoundland in both directions came to be regarded as more important. This was serious because it not only affected

[14] Address of the merchants of Plymouth to the Board of Trade, Dec. 11, 1696, *Cal. State Paps., Col.*, 1696-1697, §492; Capt. Charles Norris' answers to inquiries, 1698, *ibid.*, 1697-1698, §990 (i); Capt. Leake's answers to inquiries, Sept. 17, 1699, *ibid.*, 1699, §793.

MENACE OF FRANCE AND NEW ENGLAND 195

the fortunes of the West Country but also harmed the entire English commercial system.

Most of the clandestine or illegal trade to and from the colonies was carried on in four different ways: by English ships which came from ports of continental Europe, ostensibly laden with salt for the fishery; by ships claiming English registry, although operated in the interests of foreigners; by plantation vessels; and lastly, by French ships or vessels of other foreign registry. The Navigation Act of 1660 closed the fishery to foreign ships if their cargoes were bound for England, but no restriction was placed upon vessels bound to foreign markets with fish, which was the most important part of the Newfoundland carrying trade. Moreover, the Navigation Act of 1663 did not alter this freedom in any way. Indeed, it specifically provided for the free importation of salt for the New England and Newfoundland fisheries, a provision which was responsible for a good deal of the illegal importation of foreign goods carried on by the salt ships. It is doubtful whether the restrictions placed on colonial commerce by the Navigation Act of 1663 were intended to apply to Newfoundland, although shortly after the passage of the law attempts were made to enforce it there as well as the act of 1660.[15] John Rayner, who acted as deputy governor for Lord Baltimore in Avalon, considered himself authorized to enforce the act of 1660, and seized a Dutch built ship for trading at Newfoundland. Captain Sir Robert Robinson, commodore of the convoy in 1668, seized some goods from a small French vessel which he captured, and also the cargo landed from two others which escaped.[16] In neither case was any attention

---

[15] "An Act for the Encouraging and increasing of Shipping and Navigation," 12 Charles II, c. 18, clause 5. This act was made permanent by 13 Charles II, c. 14, and 15 Charles II, c. 7, clause 5.

[16] Orders in council, Nov. 12, 14, 1662, *Acts, Priv. Coun., Col.*, 1613-1680, pp. 339-340, 341; another order, Aug. 28, 1669, *ibid.*, 1613-1680, p. 530.

196   BRITISH FISHERY AT NEWFOUNDLAND

paid by the crown to these seizures, from which we may assume that neither the act of 1660 nor that of 1663 were intended to apply to Newfoundland.

Foreign vessels were not the only offenders. In 1675, the merchants complained that foreign goods were furnished the planters by ships flying the English flag. Sir John Berry submitted a list of thirty-four English ships, hailing from London and the West Country, which brought brandy and wines. Sometimes these goods may have been cleared through the customs in England, but the salt ships, which called at the ports of France, Spain, Portugal, and the Cape Verde Islands before proceeding to Newfoundland, had ample opportunity to acquire foreign goods after leaving home, and were considered among the worst offenders. One of the reasons advanced for establishing civil government was that it would be able to prevent the entry of uncustomed goods into Newfoundland. In 1680, Captain Sir Robert Robinson reported that there were as many as a hundred vessels annually engaged in illegal trade, and that they were so numerous that it was impossible for the warships to keep track of them.[17]

For many years no action was taken by the government to prevent this illegal trade. It was not until 1692 that the Lords of Trade directed the commodore to put a stop to trading in foreign bottoms, and requested the Admiralty to instruct the convoy commander to seize all foreign vessels found in the English zone. The committee ruled that this trade was contrary to law, but nothing definite

[17] Paper presented by Parrett to Williamson [Mar. 25, 1675], C.O. 1: 67, no. 27; Berry to Southwell, Sept. 12, 1675, C.O. 1: 35, no. 17; the same to Williamson, same date, C.O. 1: 35, no. 16; John Downing's brief narrative, etc. [Nov. 24, 1676], C.O. 1: 38, no. 70; William Hinton's observations [Feb. 26, 1680], C.O. 1: 44, no. 32; Robinson to Blathwayt, Sept. 16, 1680, C.O. 1: 46, no. 8; Talbot's answers to inquiries, 1679, C.O. 1: 43, no. 121 (i).

was done to execute these orders.[18] It was hoped that the new Board of Trade would take some action against foreign ships in the fishery. By 1698 many foreign vessels were going directly to Newfoundland from Spain and Portugal to buy fish from the English and to dispose of their cargoes of foreign goods there. The foreign shipmasters enticed English seamen to desert in order to secure their services. In the last years of the century a number of cases of Spanish, Portuguese, and Italian enterprises of this nature were reported to the Board of Trade, which urged the Admiralty to oust the foreigners because they could offer better prices for the fish on account of their more intimate commercial contacts in their own countries. In 1698, Commodore Charles Norris reported that about a quarter of the ships engaged in the carrying trade were not of English registry. But in spite of all this nothing definite was done to stop it.[19]

Another important problem was the smuggling carried on from the island. Newfoundland could absorb neither the large quantities of goods of European growth or manufacture, nor the equally large amounts of enumerated plantation commodities such as tobacco, sugar, rum, and molasses which were imported. Indeed, the island became the center of an extensive reëxport business. The absence of any permanent officials charged with its pre-

[18] Journal of the Lords of Trade, June 27, 29, 1692, *Cal. State Paps., Col.*, 1689-1692, pp. 661, 662; Capt. Crawley's answers to inquiries, 1692; *ibid.*, 1699, §1293.

[19] Paper presented by Richard Enstead to the Board of Trade, Nov. 27, 1696, *ibid.*, 1696-1697, §428; John Pym to Simon Cole, Mar. 19, 1698, *ibid.*, 1697-1698, §306; Board of Trade to the king, Mar. 31, 1698, *ibid.*, §339; Copy of a letter from Leghorn regarding permission, granted by the English consul, Mar. 31, 1698, to a Jewish merchant to send a French-built ship to buy fish at Newfoundland, *ibid.*, §341; Board of Trade request to the Admiralty, Apr. 21, 1698, *ibid.*, §389; Board of Trade journal, May 2, 1698, *ibid.*, p. 187; Capt. Charles Norris to the Board of Trade, Aug. 31, 1698, *ibid.*, §787; his answers to inquiries, Nov. 13, 1698, *ibid.*, pp. 552-555.

vention, naturally encouraged it. The commodore had neither the time nor the authority to stop it. Moreover, had he been empowered to prevent smuggling there were too many fishing ships, merchantmen, and trading sloops there during the summer to permit very close scrutiny, particularly as the commodore was increasingly occupied in settling disputes between the fishermen and planters and otherwise attempting to enforce the regulations. Smuggling had become very general by the end of the seventeenth century, although it is impossible to determine its real extent not only because it was carried on in conjunction with legitimate commercial activities, but also because the reticence of those engaged in it prevented the commodores from obtaining definite information. Nevertheless, it is certain that a great many persons in the British mercantile marine, Newfoundland, and the colonies were participating actively, including high officials and important merchants in the plantations.

The most notable attempt made during the seventeenth century to check the clandestine trade was undertaken in the continental colonies. Edward Randolph, surveyor general of the customs first in New England and then in the plantations generally from 1680 to 1701, was responsible for the official drive against smuggling which took place during those years. He discovered that a great many colonial merchants carried on illegal trade with foreign countries under the guise of dealing with Newfoundland, and cited a number of instances in his reports to the commissioners of customs. In 1680 the master of a Charlestown bark pretended to be bound for Newfoundland, although Randolph could not see the use to which his logwood cargo could be put there. Many ships which were laden with tobacco gave bonds at the Naval Office in Boston that they were bound for Newfoundland, but they proceeded directly to Scotland, Canada, or other foreign countries.[20] This last practice makes it difficult to tell just how much traffic there actually was between the continental ports and Newfoundland. Besides attempting to prevent the export of enumerated commodities under pretext of trading with Newfoundland, Randolph and the other customs officials seized and prosecuted many ships and masters for importing prohibited goods from Newfoundland. Sir Edmund Andros was instructed in January 1687 not to permit this traffic. He was told that the customs commissioners had found large quantities of European goods imported into the colonies, particularly New England, under pretext of trading with Newfoundland for fish, and was ordered to inform the people of New England that Newfoundland was not a plantation similar to the others, and did not enjoy the privileges accorded the colonies by the navigation acts. Therefore, all illegal goods that came from there would be seized. New York, too, was particularly remiss, and Governor Dongan complained in 1687 that one of the obstructions to legitimate trade at Newfoundland was the absence of a royal officer to enforce the acts of trade. Like New England, New York seems to have developed its clandestine activities in connection with the legal traffic in provisions.[21] Sometimes miracles occurred in the smuggling trade from Newfoundland, as when a brigantine was seized in Maine for bringing in brandy and, "the master sayeth he took it up floating in

[20] Edward Randolph to the Commissioners of the Customs, June 7, 1680, R. N. Toppan, *Edward Randolph* (Prince Society), III, 71-72. Cited hereafter as Toppan, *Randolph*. Abstract of letter from the same to the same, May 16, 1682, *ibid.*, p. 166.

[21] Account of charges and disbursements made by Randolph, 1679-1684, *ibid.*, III, 350; Randolph to the Governor of New York, Sept. 20, 1684 [1686], *ibid.*, IV, 125; instructions from the Commissioners of the Customs to Gov. Edmund Andros, Jan. 12, 1687, *Cal. State Paps., Col.*, 1685-1688, §1097; Gov. Dongan to the Lords of Trade, Mar., 1687, *ibid.*, p. 330.

the Sea and might have taken up a great deal more."[22] The opposition of the Boston merchants to the attempts of Governor Andros and Edward Randolph to suppress smuggling was largely responsible for the ejection of these officials and their incarceration in the common jail during the uprising of 1689. With Andros and Randolph out of the way, vessels arrived openly from Holland, Scotland, Newfoundland, and elsewhere laden with wine, oil, and brandy, and the customs officers who attempted to seize the goods were threatened with violence. Both the governor and surveyor general believed that the Boston revolt had been engineered by the former magistrates under the charter government, who desired to free Massachusetts from English control in order that they might disregard the navigation acts. These charges were denied by the Boston merchants, who affected an air of injured innocence when accused. From his cell Randolph continued to keep his eye on occurrences in Boston, and reported to the home government that since the uprising all the New England harbors had become free ports, whence enumerated commodities were carried without hindrance to Newfoundland and other illegal places. After Massachusetts Bay obtained its new charter and came under a modified royal control, the situation improved, but the clandestine exportation of enumerated commodities continued for many years.

New England and New York were not the only colonies which received and shipped illegally through Newfoundland. In Maryland in 1694, Randolph brought suit against an English shipmaster who had brought a large consignment of European goods from Newfoundland, but owing either to the ignorance or connivance of the customs col-

[22] Thomas Treffry to Gov. [Andros?], Oct. 9, 1688, *Documentary History of Maine* (*Collections* of the Maine Historical Society, Second Series), VI, 439.

lectors he was unable to secure a conviction. Governor Nicholson also reported that Maryland was too frequently furnished with European goods by way of Newfoundland in exchange for tobacco and other enumerated commodities which were transshipped at the island to vessels bound for European ports.[23] There was also a lively traffic in tobacco from Virginia to Newfoundland. In 1685 Captain Allen, of H.M.S. *Quaker,* reported that the Virginians were very angry at his remaining on their coast, but when he went away for a time they hired small vessels and shipped tobacco to New York and Newfoundland and thence to Holland. On their return voyages these vessels brought in French brandy. The clandestine tobacco trade was very general throughout the colonies. Ports as far north as Boston engaged in it. Randolph reported numerous examples in Pennsylvania, Maryland, and Virginia in 1695.[24] Three years later Robert Quary told the Board of Trade that the collectors in Pennsylvania, Maryland, and Virginia countenanced the clearing of tobacco-laden vessels for Newfoundland, and that the trade in Delaware Bay was extensive. By 1699 the traffic from Pennsylvania to Newfoundland and thence to Scotland was considerable.[25] Probably the colonial officials

[23] There are numerous references to Randolph's efforts to suppress illegal trade by way of Newfoundland. See especially C.O. 5: 855, no. 43, fol. 108; *Cal. State Paps., Col.*, 1689-1692, §§468, 468 (i); Toppan, *Randolph*, IV, 279, 283, 300, 303; V, 23, 40, 41, 42, 49, 50, 57-58; VII, 459; *Andros Tracts* (Prince Society), II, 209-210.

[24] Secretary to the Treasury to William Popple, Oct. 14, 1698, inclosing extract from a letter from Gov. Nicholson of Maryland, *Cal. State Paps., Col.*, 1697-1698, §§894, 894 (i); Pepys to Blathwayt, Nov. 11, 1687, inclosing letters from Captains Allen and Crofts, *ibid.*, 1685-1688, §§1507, 1507 (i); Lord Howard of Effingham's instructions to Capt. Thomas Perry, Dec. 31, 1688, C.O. 1: 65, no. 92.

[25] Toppan, *Randolph*, V, 40, 41, 42, 118; Robert Quary to the Board of Trade, Aug. 26, 1698, *Cal. State Paps., Col.*, 1697-1698, §772; Board of Trade representation to the Lords Justices, Aug. 4, 1699, *ibid.*, 1699, p. 382.

only brought to light a very small proportion of the total clandestine trade, but it was more easily controlled in the plantations than was possible at Newfoundland. The commissioners of the customs were particularly concerned over this tobacco trade but it was too well established to be easily suppressed, antedating the passage of the acts of trade by a good many years. The commissioners felt that its continuance was due to a misunderstanding as to the status of Newfoundland, which was not a true plantation.

Illegal trade carried on through Newfoundland was regarded as a threat to England's commercial position, but in spite of continual complaint from the merchants and officers responsible for the enforcement of the acts of trade, the traffic continued. The chief difficulty arose from the uncertain position occupied by Newfoundland in the commercial system. No one was at all certain whether the acts of trade applied there, and even the commissioners of the customs classified it as a foreign country in their export and import returns.[26] Even had there been any certainty as to the intent of the commercial laws, enforcement would have been difficult because there were no resident officials to put a stop to it. The unwillingness of the western adventurers to surrender control of the fishery to anyone constituted an effective barrier to the proper enforcement of the acts of trade. During the seventeenth century, Newfoundland was a great hole in the wall of national self-sufficiency that the mercantilists sought to erect around the mother country and her colonies. The willingness of the crown to support the West Country merchants in their stand for a loosely regulated fishery,

---

[26] Custom House Papers, Accounts, Ledgers of Imports and Exports, 1697-1780, *passim*. In these ledgers Newfoundland is always classified among foreign countries, while the continental and West Indian colonies are classified as plantations.

MENACE OF FRANCE AND NEW ENGLAND

is convincing proof that the government had no consistent commercial policy with respect to its American possessions. Although dislike for New England was pronounced in English mercantilist and official circles, the menace of colonial competition was probably never as great as the danger to be anticipated from France, but it was much less obvious and, therefore, more difficult to appreciate until after 1699.

CHAPTER VII

LAW AND WAR, 1699-1713

THE passage of the Newfoundland Act in 1699 closed a period of sixty-five years of semi-private control of the fishery under royal charters, and inaugurated an era of restricted private management in accordance with the dictates of parliament. In spite of the extremely narrow views of the western adventurers, certain more or less definite theories had been evolved between 1660 and 1699 relative to the function of the fishery and trade of Newfoundland in the national commercial system. Some of these ideas were to be applied with varying success in the ensuing years. It was recognized early in the century that the Newfoundland fishery was of commercial benefit to the nation.[1] This was followed by the realization that fishing was the only industry worth following in the region, and that the resultant trade was an important branch of England's foreign commerce because it contributed indirectly to the national well-being.[2] Unlike the plantation trade, which occupied a central position in the national economy and contributed directly to the national wealth, the Newfoundland trade performed a function which was largely supplemental. This difference between the two branches of Anglo-American commerce

[1] Thomas Jenner, *Londons Blame, if not its Shame: Manifested by the Great Neglect of the Fishery* (London, 1651), *passim*. Thomas Mun, *Englands Treasure by Forraign Trade* (London, 1664), *passim*. William Petyt, *Britannia Languens, or a Discourse of Trade* (London, 1680), p. 31.

[2] Remonstrance of the merchants, etc., of Plymouth, Dartmouth, etc., to the king, read Mar. 25, 1675, C.O. 1: 67, no. 26.

LAW AND WAR, 1699-1713

was appreciated, and it was understood that the respective values of the two were to be judged by quite different standards. Nowhere is the difference in the status of the plantation trade and that of the fishery more apparent than in connection with the suppression of smuggling. The attempts to suppress the traffic at Newfoundland were weak and futile because it was never intended that Newfoundland should be brought under the regulations applicable to the continental and West Indian colonies.[3] Newfoundland's position outside the English colonial system was determined by the absence of a domestic market for fish, which obliged the industry to seek foreign outlets at all times.[4] The emphasis upon this one industry was also the result of the severity of the climate and the infertility of the soil which precluded the development of any other significant economic interest.[5] These factors have been referred to previously, but their importance in determining the *rôle* played by Newfoundland in the seventeenth and eighteenth centuries cannot be overemphasized.

People of that day appreciated the economic benefits which the nation derived from the trade. They recognized that the exportation of Newfoundland fish to foreign markets was valuable for several reasons: the merchants profited directly from the fishing industry and the export trade, the nation profited from the direct returns from foreign countries, and individuals profited because English labor was employed and English commodities con-

[3] Order in council, May 25, 1675, and the report of the Lords of Trade, Apr. 15, 1675, *Acts, Priv. Coun., Col.*, 1613-1680, p. 623.

[4] During the second Dutch war the government considered importing Newfoundland fish into England, order in council, Jan. 4, 1665, *ibid.*, 1613-1680, pp. 388-389. This is the only example after 1660 of such an intention that the author has encountered.

[5] Sir Josiah Child, *A New Discourse of Trade* (3d ed., London, 1718), p. 205.

sumed in the fishery.⁶ They knew also that the crown gained in revenue from goods imported into England as a result of the foreign trade in fish.⁷ Furthermore, those interested in the fishery had come to appreciate it as a training school for sailors. As the political position of England in world affairs became increasingly dependent upon naval strength this function was emphasized more and more and greater pressure was brought to bear upon those engaged in the industry to improve its efficiency in the interests of national safety.⁸ All these economic and political reasons were enunciated previous to 1699; but the following years witnessed more serious attempts to apply them than had been the case before. During the ensuing years no new ideas relative to the function of the Newfoundland trade were evolved, although changes in emphasis took place from time to time as new factors modified to some extent the original conceptions.

By 1699 the planters were firmly established albeit not prosperous. All continued to live as close to the water as possible for the crown had made no attempt to inforce the restrictive clauses of the charter of 1676 requiring their emigration or withdrawal six miles inland. The kitchen gardens and barnyards of the planters were adjacent to the spaces devoted to salting and drying fish. Wandering livestock knocked over piles of cod and trampled them much to the annoyance of the West Countrymen. Al-

⁶ Remonstrance of the merchants, etc., of Plymouth, Dartmouth, etc., read Mar. 25, 1675, C.O. 1: 67, no. 26.

⁷ During the years when the royal finances of Charles II were in poor condition, the crown was interested in Newfoundland because of the customs revenue derived indirectly from its trade with foreign countries. There are numerous references: *Acts, Priv. Coun., Col.*, 1613-1680, §§716, 735; C.O. 1: 22, no. 69; C.O. 1: 34, no. 32; C.O. 1: 44, no. 18; C.O. 1: 51, no. 29; C.O. 1: 67, nos. 16 (i), 25, 27. Later when the financial stringency became less acute references to this function of the fishery receive less emphasis and are less frequent.

⁸ Sir Josiah Child, *A New Discourse of Trade*, pp. 205-212.

though no one had a legal title to his riparian land continuous occupancy for several fishing seasons gave the planters and byboatmen a vested interest which amounted to a prior claim in case of dispute with the fishing admirals. Holdings were not formally registered, although from time to time the commodores drew up lists of the occupiers in the more important harbors. Later these lists served the purpose of tenurial records tantamount to copy hold. A great deal of confusion resulted at St. John's when the old lists were burned in the fire of 1746. Disputes regarding land were usually decided orally by the officers of the navy, and written records of the evidence and decision of the commodore exist for only a few cases. After the departure of the ships in September the inhabitants had little to do until the following spring. Winter set in so early that fishing had to be abandoned and there were no other profitable enterprises in which the people could engage. Sealing, which was later to become an important industry during the off-season, was as yet undeveloped. The more enterprising Newfoundlanders occupied themselves in building boats, bringing in timber, and making repairs, but a great many did nothing but drink and enjoy themselves. There were no schools, the first grammar school being established at St. John's in 1799, and most of the inhabitants were illiterate. Stories of the exploits of hardy fishermen and other heroes were passed on from generation to generation in the form of ballads, and singing kept up the spirits of the people during the long cold months. Unlike the continental colonies, Newfoundland was late in becoming a center of activity for the churches. Although Erasmus Sturton, a Puritan minister, had been a member of Guy's original settlement and Calvert had brought a Roman Catholic priest to Ferryland, there had been no continuous ecclesiastical establishment in the island. Sometimes the naval chaplains held services ashore and conducted baptisms and marriages, but there was no resident priest of the Church of England until 1701. The Reverend John Jackson was commissioned chaplain of the garrison in that year with the understanding that he should also minister to the planters at St. John's. A number of Newfoundlanders and English shipmasters promised to build a church and provide him with a salary of over £300 a year, to be paid in fish. The Bishop of London and the Society for the Propagation of the Gospel were interested in the movement, but Mr. Jackson's character was such that he soon became involved in quarrels with officers of the garrison, Collin Campbell, agent for prizes, and some of the leading merchants and shipmasters. His successor, Jacob Rice, served until 1708 when St. John's was taken by the French and the garrison church destroyed. There was an interval until 1720, when a new church was built, but for many years it was provided only intermittently with a clergyman. No religious work was carried on in the outports until 1722, and the Church of England was not firmly established in Newfoundland until after the American Revolution. During these early years there were no dissenting clergy, although there were some Roman Catholics, Scottish Presbyterians, and New England Congregationalists. The Franciscans established a mission at Placentia in 1689 but this was discontinued after 1713. The influx of Irish in the early eighteenth century brought Roman Catholic priests to the island, but no formal ecclesiastical organization was erected until much later. The backwardness of education, the disorganized state of religious work, the debauchery, drunkenness and litigiousness of the people coupled with their poverty and oppression by English shipmasters and colonial traders account to a very considerable extent for the lack of initiative and enterprise which is so characteristic of the early Newfoundlanders.

The period from 1699 to 1713 saw the beginning of the decline of West Country influence. During these years three problems confronted the merchants and the government. These were: the enforcement of the act of 1699, the prevention of smuggling, and the elimination of the French. By the end of the seventeenth century the ineffectiveness of charter regulation had been clearly demonstrated. The western adventurers were still anxious to retain their control, but were beset by many difficulties. They finally had recourse to parliament, hoping that by means of statutory approval of their ancient fishing privileges they would be able to maintain a foothold in the fishery, which had become increasingly precarious with the passage of time. They were forced to take this step because of the constant pressure of their competitors, the exigencies of the war with France, 1689-1697, and the threatened reversal of royal policy under James II and William and Mary. Not only had the restrictions upon the participation of the inhabitants been removed, but on at least two occasions the crown had considered erecting a civil government in Newfoundland, and in 1696 had actually sent a military expedition to Newfoundland to chastise the French for their raids on the English settlements and to establish a fort and garrison at St. John's. Continuance of such policies and the extension of royal military protection to the Newfoundlanders spelled the ultimate extinction of West Country privileges. Recourse to parliament was made easy by the new importance which that body had assumed after the revolution of 1689. Temporarily, because of its desire to exercise its newly acquired authority and because of the critical situation with respect to France, parliament was concerning itself to a considerable extent with commercial and colonial matters. Moreover, as in the days of Charles I, the West Country was in a position to be heard, because of the large representation in the House of Commons from the western counties and boroughs. Consequently, the West Country was able to secure the passage of an "Act to Encourage the Trade of Newfoundland." This act is almost contemporary with the Navigation Act of 1696 and the abortive attempt of parliament to create a Board of Trade subject to its control. Undoubtedly, the investigation in parliament of the unsatisfactory joint military and naval expedition to Newfoundland of 1696-1697, which took place during the session of 1698-1699, helped to focus the attention of the legislature upon the affairs of the island and its fishery, and facilitated the passage of the Newfoundland Act.[9]

During the War of the League of Augsburg, 1689-1697, it became patent that the English fishery and settlements required better defense than that provided by the naval and mercantile marine in summer and by the ice in winter. As early as 1692 the crown considered the erection of fortifications but no action was taken and it remained for the French to demonstrate that an English military establishment was necessary if the island were to be held. In 1694 five French ships of forty and fifty guns attacked Ferryland but were repulsed. They returned with a greater force two years later and took St. John's, Ferryland, and several other important harbors. The joint military and naval expedition of 1696-1697 was organized with the threefold purpose of recovering the captured set-

[9] The joint military and naval expedition of 1696-1697 was investigated by the House of Lords from Feb. 23 to Apr. 17, 1699, Stock, *Proceedings and Debates*, II, 277-304; *House of Lords MSS.*, New Series, III, §1378. Cf. R. Usticke to Blathwayt, Jan. 12, 1697, Huntington Library, The Blathwayt Papers, Maryland, 1664-1701, no. 57; Ellesmere Papers (MSS. of the Earl of Bridgewater), nos. 9607, 9632, 9639, 9809, 9821, 9823, 9833.

tlements, driving the French from Placentia, and laying the foundations of a permanent military establishment. Although the first objective was achieved, the expeditionary force failed to oust the enemy from its position on the south coast owing to the inaction of its leaders, and most of the time was spent in building a fort at St. John's. When the forces were withdrawn a company of Colonel Sir John Gibson's Regiment of Foot was left behind to garrison Fort William together with a detachment of the Royal Artillery to man the guns. During the short interval between the conclusion of peace at Ryswick in 1697 and the outbreak of the War of the Spanish Succession in 1701 construction proceeded intermittently, but the stronghold was never fully completed. Colonel Roope of the Royal Engineers spent several summers in Newfoundland to direct the work, but he was handicapped by the difficulty of obtaining materials from England, the inefficiency and unskilled character of the local labor supply, and by his personal quarrels with the other officers of the garrison. The defenses were none too strong, and the garrison, surprised by the French in 1708, surrendered without offering much resistance. In 1698 the infantry detachment became an independent company, and three years later the commodore of the convoy was commissioned commander in chief of the land forces in order to straighten out the tangled affairs of the little garrison. Although this practice was continued throughout the war, the army officers resented being subordinate to an officer of the rival service, and the scheme never proved as satisfactory as was anticipated at Whitehall. The officers quarreled continually and violently among themselves. The enlisted men, long unrelieved, infrequently paid, poorly clad and undisciplined, were sullen and mutinous. They took sides in the squabbles of their superiors, which also involved Jackson, the chaplain, and many of the leading planters and merchants. The rivalry of Captain, later Major, Thomas Lloyd and Lieutenant, later Colonel, John Moody led to the temporary suspension of the former by Commodore Bridge in 1704, and both officers eventually were recalled to England to answer charges. Lloyd was reinstated, and Moody afterward procured a commission in Colonel Nicholson's regiment and reappeared in Newfoundland in 1713 as commander of the fort and garrison at Placentia. Lloyd remained in command from 1705 until his surrender to the French in 1708. Following French raids on Conception Bay in 1706, the commodore created a militia, distributing arms and ammunition to the planters. When, however, Lloyd proposed to retaliate against the French in the following year by marching on Placentia, the militia refused to coöperate and the plan had to be abandoned. During the entire war the garrison at Fort William comprised only about twenty-five or thirty men. The officers engaged in trade, and the enlisted men fished. Altogether the post was thoroughly disorganized and slipshod. In 1713 the fort at St. John's was abandoned and Placentia became the only military station until Fort William was repaired and garrisoned again during the War of the Austrian Succession. Occupied with more pressing military affairs during these years of war, the authorities in England paid little attention to the garrison at St. John's and a good deal of the ill-feeling and lack of discipline can be attributed to the failure of the government to recognize and provide for even the simplest wants of officers and men. The two wars with France from 1689 to 1713 proved fallacious the contention of the western adventurers that Newfoundland was naturally defensible. Thereafter British forces were always stationed there.

We know little about the actual consideration which

parliament gave to the Newfoundland Act. The bill was introduced seemingly without attracting much attention, and passed through the various stages without debate. It is significant that the committee of the House of Commons to which the bill was referred was composed of three members representing West Country constituencies, and it is also noteworthy that in general the bill conformed largely to the views held in the West of England. Because of the absence of any recorded debates in either house, it is difficult to ascertain, except by inference, the actual influence exerted by the western adventurers and their representatives to obtain its passage.[10] One suspects, however, that the rivalry between the crown and parliament for control of the commercial policies of the nation may have facilitated its passage. Fear that the crown might wish to extend its authority over Newfoundland, as it had done following the accession of William and Mary in several of the plantations, probably had some weight in determining the attitude of parliament.

The passage of the Act of 1699 was an apparent triumph for the western adventurers. For forty years they had urged the adoption of a permanent national policy which would maintain them in their ancient position and assure them of a large share in the administration of the fishery. An examination of the preamble indicates that the framers of the act accepted the current economic theory regarding the fishery. It states that the trade was commercially beneficial to the nation because considerable quantities of English provisions and manufactures were

[10] The crown considered establishing royal government in 1696 and 1697. See notes made by the Earl of Bridgewater at Board of Trade meetings, on representations from the West Country, Ellesmere Papers, nos. 9609, 9631, 9639, 9646. From these we learn that the opposition centered in the West of England as usual. The passage of the Newfoundland Act of 1699, took comparatively little time. Stock, *Proceedings and Debates*, II, 275.

consumed therein, thus bringing profit to tradesmen and work to artisans; and explains that the resultant foreign trade was also of indirect value to the nation because it helped offset imports of wine, oil, plate, iron, wool, and other useful articles, with the result that the royal revenue was increased and the trade and navigation of England encouraged.[11] An analysis of the sixteen articles, with the exception of the rider relating to the Greenland trade, shows that only slight changes were made in the old charter regulations. The act is the heir to the charters of 1634, 1661, and 1676, and in addition to repeating the customary rules, it perpetuated all the defects of the previous patents. In some respects it was more liberal, for the regulations applying to the inhabitants were much less strict than those in the charter of 1676. No severe restrictions were placed upon the Newfoundlanders, and the participation of both planters and byboatkeepers in the fishery was conceded. The act confirmed the position of the western adventurers as managers of the fishery, but it removed its administration from the hands of the West Country magistrates and mayors, placing it with the fishing admirals, who were to function under the supervision of the commodore as representative of the crown. The statute is merely a parliamentary confirmation of the practices which had been followed since the collapse of the charter of 1676, but it fixed the system of regulation so definitely that there was no opportunity to introduce flexibility into the administration. The rigidity of the statute was to cause endless trouble and embarrassment until the government finally discovered a way to nullify it.

[11] The preamble of "An Act to Encourage the Trade of Newfoundland," 10 & 11 William III, c. 25. The act was not repealed until 1867, after it had been inoperative for many years. Cited hereafter as the Newfoundland Act, 1699.

Scarcely had the law become effective when it was discovered that fishery regulations imposed by statute were quite as inoperative as those granted by royal prerogative. The defects were manifested immediately. Those concerned with its administration were soon found to be remiss in their duties either from indifference or ignorance, allowing themselves to be bribed or swayed by personal interest. Moreover, it failed to provide fines or penalties for violations, and contained no provisions for correcting many of the abuses still existing in the fishing industry.[12] The Board of Trade became gravely concerned over the general disregard of the act. In 1701, both Commodore Graydon, and George Larkin, who was sent to America as an investigator for the commissioners of customs, reported on the situation at Newfoundland, and agreed that all concerned in the fishery generally disobeyed the law.[13] In 1702, the Board of Trade, the Privy Council, and parliament all took up the question of providing more effective machinery. The board pointed out that neither the fishing admirals nor the inhabitants were observing the rules prescribed by the act any better than they had the charter regulations. It recommended that the commodore be instructed to exercise extreme diligence in preventing violations and proposed that on his return at the end of the current fishing season, he should recommend to parliament such additions to the act as

[12] Capt. Sir Stafford Fairborne to the Secretary of the Board of Trade [Popple], Sept. 11, 1700, *Cal. State Paps., Col.*, 1700, pp. 520-523; George Larkin to the Board of Trade, Aug. 20, 1701, *ibid.*, 1701, pp. 430-431; Board of Trade to Secretary Vernon, Apr. 23, 1701, *ibid.*, 1701, p. 171; Commodore Graydon's answers to inquiries, Mar. 13, 1702, *ibid.*, 1701, pp. 529, 530; Fairborne to Bridgewater, May 13, 1700, Ellesmere Papers, no. 9781.

[13] Capt. Graydon's answers to inquiries, *Cal. State Paps., Col.*, 1701, pp. 529-530; report of the Board of Trade to both Houses of Parliament, Feb. 16, 1702, *House of Lords MSS.*, New Series, IV, 447-450; order in council, Mar. 2, 1702, *Acts, Priv. Coun., Col.*, 1680-1720, pp. 397-398; Board of Trade representation, Mar. 17, 1702, *Cal. State Paps., Col.*, 1702, pp. 137-139.

might be necessary.[14] Further action was postponed awaiting the return of the commodore in the autumn of 1702.

Then, in November, 1702, the Board of Trade submitted a report to a committee of the House of Lords which embodied two definite recommendations: that the commodore be given authority by parliament to impose fines and penalties for violations of the act; and also the powers of a customs officer to search incoming vessels from New England and to take bonds from their masters not to carry away any more men than they had brought to the island.[15] A similar report was made in 1703, and again in 1706.[16] Receiving no support from Parliament, the board was obliged to fall back upon the expedient of reminding the shipmasters of their obligations and duties under the act.[17] Parliamentary interest in Newfoundland was momentary and occasional, so the real supervision of the fishery had to be carried on by the crown for many years thereafter upon the very inadequate basis of the act of 1699.

Meanwhile private individuals interested in the trade became active again in recommending various measures

[14] Order of the committee of the House of Lords, Nov. 24, 1702, *Cal. Stat. Paps., Col.*, 1702, p. 742; Board of Trade to the House of Lords, 1702, *ibid.* 1702, §1202 (i); Commodore Leake's answers to inquiries, 1702, *ibid.*, pp. 721 727; Stock, *Proceedings and Debates*, III, 4-6.

[15] Board of Trade report to the House of Commons, Jan. 16, 1706, Egerton MSS., 921, fols. 3-8; Stock, *Proceedings and Debates*, III, 6-7, 8.

[16] Some of the fishing admirals sent in reports to the Privy Council as required by the act of 1699, *Cal. State Paps., Col.*, 1701, §§102, 466; 1708 1709, §283; C.O. 194: 5, nos. 30, 30 (i). Information relative to the general violation of the statute is found in the following: *Cal. State Paps., Col.* 1702, §1154 (i); 1704-1705, §1373; Stock, *Proceedings and Debates*, III, 38 40.

[17] Board of Trade to Sir Charles Hedges, Secretary of State, July 13 1705, *Cal. State Paps., Col.*, 1704-1705, §1241; Colonel Roope, Royal Engineer at St. John's, to the Board of Trade, Feb. 12, 1706, *ibid.*, 1706-1708 §101.

LAW AND WAR, 1699-1713

for the better regulation of the fishery. In 1705, some of the western adventurers had so far altered their former attitude as to suggest the appointment of a civil governor, while others recommended the election of a chief magistrate for the island and constables for the various fishing harbors to maintain order during the winter. In view of the danger of French attack it was proposed that the fortifications be extended and a militia established.[18] Although it recognized the ineffectiveness of the act, the Board of Trade was reluctant to adopt these suggestions, taking the stand that the law provided all the civil government necessary for "so desolate a country." Thus, again the West Country point of view was reflected in a pronouncement of the official body charged with supervision of commercial and colonial matters.

Though the authorities insisted that the Newfoundland Act provided sufficiently for the government of the fishery, they were obliged to regulate many matters that were not mentioned in the statute, such as the prevention of piracy, the seizure and condemnation of prizes, the military and naval defense of the fishery and settlements during the war with France, prevention of illegal trade and the curtailment of the activities of the New Englanders who encouraged the emigration of able seamen and fishermen, none of which were covered by the act.

The prevention of piracy was entrusted to the commodore and the masters of merchant vessels of over 200 tons burden, who were commissioned to try pirates. George Larkin, who visited St. John's in 1701, was entrusted with the delivery of commissions and with the duty of establishing a prize court. He appointed George Newman, a merchant of St. John's, register, and delivered to the

[18] The western ports were heard from during the years 1703-1705. Petitions were received from Poole, Bristol, Bideford, *ibid.*, 1702-1703, §156; 1704-1705, §§1246, 1256.

commodore the commission constituting him a judge of prizes. It became the practice of each commodore to leave the commission at St. John's for the use of his successor in the following year. An agent for prizes was also appointed sometime previous to 1705. It is probable that the first agent was James Campbell, who was London representative for the Newfoundland garrison in the fort at St. John's, but Collin Campbell seems to have acted as his deputy. Archibald Cumming, later preventive officer, was appointed agent for prizes in 1706, although again it is possible that he was only James Campbell's deputy.[19] The appointment of these officers was due to the exigencies of the war with France but when taken together with the establishment of the permanent garrison at St. John's in 1696 and the appointment of the commodore as commander of both the land and sea forces, it is very evident that the crown was seeking to extend its authority over Newfoundland, not only to make the English zone secure against France, but also to create a nucleus of officials to form the basis of a permanent royal establishment.

Illegal trade and the enticement of seamen from Newfoundland to New England became increasingly important questions in the early years of the eighteenth century owing to the war. The authorities again took steps to in-

[19] Prize Officers to the Secretaries of the Lord High Admiral, Aug. 22, 1705, Admiralty 1: 3662, fols. 191, 192; the same to the Treasury, Aug. 14, 1705, *Calendar of Treasury Papers*, 1702-1707, §47. Cited hereafter as *Cal Treas. Paps.;* Savage to Popple, Nov. 21, 1707, Aug. 26, 1708, *Cal. State Paps., Col.*, 1706-1707, §1206; 1708-1709, §119; representation of Archibald Cumming to the House of Commons, *ibid.*, 1706-1707, §138; presentment of Cumming by the Commissioners for Prizes, Mar. 19, 1706, *Cal. Treas. Paps.* 1702-1707, §106; report of Prize Officers to the Lord High Treasurer, *ibid.* 1702-1707, §29; Board of Trade representation to the king, June 5, 1700 *Cal. State Paps., Col.*, 1700, §498; the same to the Admiralty, Feb. 5, 1701 *ibid.*, 1701, p. 131; George Larkin to the Board of Trade, *ibid.*, 1701, §756 Popple to William Lowndes, Jan. 23, 1705, *ibid.*, 1704-1705, §§818, 818 (i)

LAW AND WAR, 1699-1713 219

vestigate smuggling, for, as Edward Randolph remarked in 1701, Newfoundland had always been and still continued to be "a great staple" for European and plantation commodities. He reported that Scottish merchants had recently established a factory there and were sending enumerated commodities to Scotland, Holland, and other prohibited places; while English vessels brought in European goods which were dispersed in small quantities throughout the colonies.[20] The authorities tried to put a stop to the traffic. George Larkin went out in 1701 as a representative of the commissioners of the customs to look into smuggling in Newfoundland and the plantations. His investigation accomplished little with respect to Newfoundland because of his own lack of originality and because of the stereotyped form of his instructions, which were nothing more than the usual heads of inquiry given the commodores. At about the same time the Board of Trade cautioned the commodores to be particularly careful to prevent illegal trade through Newfoundland.[21] These efforts were very insignificant and ineffective in view of the rapid expansion of the traffic, and produced no results whatsoever.

More specific information is available regarding the foreign commodities brought to Newfoundland after 1700. Illegal imports came regularly from Portugal, the

[20] Edward Randolph to the Board of Trade, Feb. 19, Mar. 17, 1701, *Cal. tate Paps., Col.*, 1701, §§180, 259, especially pp. 92, 133; Popple to Lowndes, Jan. 23, 1705, *ibid.*, 1704-1705, §§818, 818 (i).

[21] Board of Trade journal, May 30, 1699, *Cal. State Paps., Col.*, 1699, p. 59; Capt. Fairborne's answers to inquiries, 1700, *ibid.*, 1700, §774, especially clauses 20, 21; Board of Trade to the Admiralty, Apr. 11, 1701, *ibid.*, 701, §326; the same to the king, Apr. 10, 1701, *ibid.*, §324; order in council, pr. 10, 1701, *ibid.*, §325; Board of Trade to Larkin, Apr. 19, 1701, *ibid.*, 354; heads of inquiries for Larkin, *ibid.*, §354 (i); Board of Trade journal, pr. 19, 1701, *ibid.*, p. 170; Larkin to the Board of Trade, Aug. 20, 1701, *id.*, §756.

Azores, and Madeira in English salt ships. The commodities sent from Portugal consisted chiefly of wine, brandy, oil, French linen, Levantine silk, and bacon. Owing to the French war, trade from Spain and France was irregular. That from Spain was carried on either by the English salt ships or by Spanish vessels which went to Newfoundland with British passes to load fish. It consisted of wine, brandy, and large quantities of other illegal imports. A good many French articles were obtained from prize cargoes condemned at Newfoundland. Besides yielding salt which was a legal import, the prize ships also provided ladings of wine, brandy, dowlas, linens, silks, ala modes, lustrings, sarcenets, canvas, hats, and paper. English ships not only brought in goods from southern Europe but also carried goods direct from Holland, such as canvas, linen, cordage, iron poles, pots, and other articles. The clandestine traffic was largely conducted by English shipping, although some foreign ships also engaged in the trade.[22] Most of the foreign goods were bartered or sold to the New England traders.

The New Englanders still continued to dispose of considerable amounts of plantation goods in Newfoundland either for local consumption or for reshipment to Europe. The chief items shipped from the colonies to foreign countries *via* Newfoundland at this time were sugar, tobacco, pitch, and tar. The island could consume only limited quantity and variety of goods, but articles which could not possibly have been used locally passed through its ports. For example in 1712, Archibald Cumming, the

[22] Information relative to illegal trade is derived from the commodore answers to inquiries, instructions to these officers, communications from Archibald Cumming, Board of Trade representations, and orders in council for the period from 1701-1713. See especially: *Cal. State Paps., Col.*, 1700 §879; 1706-1708, §§139, 588 (i); 1710-1711, §§511 (i), 558, 815; 1711-1712 §149 (i); *Acts, Priv. Coun., Col.*, 1680-1720, §§865, 1063; C.O. 194: 5, no. 17, 26, 34, 59 (i).

preventive officer, reported a sloop fully laden with logwood bound for the Straits.[23] It is difficult to tell whether there was any actual expansion of the trade after 1701. Statistical information is lacking, but probably the War of the Spanish Succession raised prices and created shortages, which were sufficiently enticing to encourage illegal trading. It is impossible to know whether the more frequent reports from the resident officers in Newfoundland, or an actual improvement is responsible for the greater number of records concerning smuggling which are extant for these years.

One thing which is apparent at the beginning of the eighteenth century is that the New England merchants had developed an extensive organization in Newfoundland to further their trade with the inhabitants and foreign countries. Business was sufficiently extensive by 1700 to warrant the appointment of agents in the various harbors, and the petty New England traders of former days had given way to mercantile establishments of considerable size, which through their agents kept themselves informed as to what foreign commodities were available. As formerly, they obtained these goods by barter or trade with the inhabitants or English shipmasters. The New England vessels made two or three voyages a year bringing in bread, flour, pork, beef, onions, apples, sheep, pine boards, shingles, tobacco, molasses, sugar, lime juice, and rum, which were disposed of either for bills of exchange or for foreign goods. There was scarcely a New England vessel arriving at Newfoundland whose bill of lading mentioned half her cargo, or whose lading on return to New England would not have made her liable to seizure. Although they carried home considerable quantities of

[23] Cumming to the Board of Trade, Oct. 21, 1712, C.O. 194: 5, no. 17. Cf. Graydon's answers to inquiries, 1701, *Cal. State Paps., Col.*, 1701, §879 (xii), clauses 18, 20, 27.

European goods and disposed of enumerated plantation goods for shipments to Europe, the New Englanders did not import foreign goods into Newfoundland, but confined themselves to commodities from the continental colonies and the British West Indies. The inhabitants preferred to buy from the colonial traders, since their provisions were fresher and cheaper, and were supplied more frequently than the expensive salt provisions and other English goods brought from a longer distance.[24] This preference was gall and wormwood to the English merchants who complained bitterly of the New England trade, which they held responsible for the decline of their own business. The colonial traders were reported as making about thirty-five per cent profit on the resale of bills of exchange obtained there. The trade was of vital importance to New England because it partly offset the unfavorable balance of trade of that group of colonies, which amounted to about £100,000 annually. The colonial merchants stated that they made more from their dealings with Newfoundland than the entire annual profit from their own region, and pointed out that if they were barred from trading there they would be unable to meet their obligations in England.[25] Recognition of this factor, possibly accounts for the lack of unanimity of opinion among merchants regarding the necessity of enforcing the acts of trade at Newfoundland.

[24] A great deal of information relative to the plantation trade is contained in the answers to inquiries and covering letters of the commodores, communications from Larkin, Cumming, Roope, Jackson, etc. See *Cal. State Paps., Col.*, 1700, §§774, 774 (i); 1701, §879 (xii), pp. 431-432; 1702, §1154 (i); 1706-1708, §§56, 74, 588 (ii); 1708-1709, §233 (i); 1710-1711, §511 (i); 1711-1712, §§149, 149 (ii); C.O. 194: 5, nos. 16 (i), 17, 26, 34, 59 (i).

[25] Several Board of Trade representations submitted either to parliament or the crown contain comment on the plantation trade and smuggling at Newfoundland. See *Cal. State Paps., Col.*, 1702, §207; 1706-1708, §32 (i); 1710-1711, §588 (i); Huntington Library, HM 821; *Acts, Priv. Coun., Col.*, 1680-1720, §§865, 1063, 1134.

Had the New Englanders confined their evasions of English law to violations of the acts of trade, or had they contented themselves with the provision trade, they would have aroused less resentment in England than was actually the case. But their persistence led them to interfere constantly with the nursery for seamen. The training of sailors, always important to a maritime nation, became increasingly vital to England during the wars with Louis XIV. Though the practice of enticing sailors and fishermen to New England had existed for some time, the promises made to the men of higher wages and better living conditions in the colonies had a very injurious effect when made in wartime, not only because labor had become scarce in the West of England but also because trained men were required for service in the navy more than ever before. Both the western adventurers and the British government felt that the New Englanders and those English fishermen and Newfoundland planters who failed to employ the legal quota of green men as prescribed by the Act of 1699, were injuring the naval service.[26] Although the colonial shipmasters and merchants had not been responsible for the situation, they had always been willing to take advantage of it in order to supplement their earnings by the development of passenger traffic between Newfoundland and the continental colonies, and also in order to obtain labor suitable for their own fishery and shipping trade.

[26] A quota of green seamen and fishermen was prescribed, and the discharge of men at Newfoundland prohibited by clauses 9 and 10 of the Act of 1699. From time to time attempts were made to enforce these provisions, the most notable being in 1708, when a royal proclamation was issued calling for general observance of the statute. Order in council, May 20, 1708, *ibid.*, 1680-1720, pp. 558-559; the proclamation, June 26, 1706, Clarence S. Brigham, *British Royal Proclamations Relating to America*, pp. 163-167. Cited hereafter as Brigham, *Royal Proclamations*. The difficulties of enforcing these provisions are emphasized in the following: *Cal. State Paps., Col.*, 1710-1711, §§91, 91 (i), 227, 244; 1711-1712, §149; C.O. 194:5, no. 16 (i).

224 BRITISH FISHERY AT NEWFOUNDLAND

Emigration to New England was encouraged by the general disregard of the provisions of the Act of 1699 designed to further the training of seamen.[27] According to law the shipmasters, byboatkeepers, and planters were all supposed to employ a quota of green men annually, comprising one-fifth of the persons engaged. Some of the seamen were paid off in Newfoundland and were free to go where they pleased, while others deserted and hid in the woods or worked temporarily for the planters until they had an opportunity to go to New England. The home government was much concerned about the situation, as several hundred left the island annually, but no one had sufficient authority to stop it. The law provided no fines and penalties, and the expedients adopted to check the migration failed to work successfully. Most of it took place before the arrival or after the departure of the convoy, or in the outports where the supervision of the commodore did not usually extend. Although the Board of Trade tried several times to procure for the commodore and the fishing admirals the authority of customs officers to search New England vessels, and proposed that the masters be required to comply with the terms of their bonds under penalty of the law, parliament never granted these requests.[28] Many years passed before the general exodus was brought under control.

The steady migration to New England and the smuggling at Newfoundland made the situation serious enough to warrant at least the appointment of a few officers to

[27] The migration to New England is mentioned frequently by the commodores, *Cal. State Paps., Col.,* 1700, §774 (i); 1701, pp. 529-530; 1702, §1154 (i); 1708-1709, p. 526; 1710-1711, §149 (ii); C.O. 194:5, nos. 16 (i), 59 (ii).

[28] Board of Trade representation to the House of Lords, Nov. 28, 1702, *Cal. State Paps., Col.,* 1702, §1202; Sir Charles Hedges to the Board of Trade, July 5, 1705, *ibid.,* 1704-1705, §§1228, 1228 (i); Board of Trade to Hedges, July 13, 1705, *ibid.,* 1704-1705, §1241.

report upon, if not prevent, these injurious practices. Archibald Cumming was appointed preventive officer in August, 1708, to act under the direction of the customs commissioners.[29] Though he occupied the office for many years, it is safe to say that as he enjoyed little real authority he never prevented any illegal trade, and acted chiefly as an informer for the customs and the Board of Trade on trade conditions. At the time when Cumming was first appointed, the government also decided to erect a court of vice-admiralty at Newfoundland composed of a judge, register, and marshal, thus creating a vice-admiralty district there in accordance with the practice followed elsewhere in America. This course did not require the appointment of a vice-admiral for the district. James Smith was appointed judge, Bryan Rushworth, register, and Thomas Hayne (or Heaynd), marshal. These officers received their warrants from Prince George of Denmark, lord high admiral, on September 1, 1708, but none of them, except Smith, appears to have made use of them. Smith found it impossible to enforce the acts of trade without a governor and naval officer to assist in their execution, and declared it was useless to maintain a preventive officer there in the absence of other officials. He did not remain in Newfoundland very long, and received no salary for his work there.[30]

[29] Memorial of James Campbell to the Lord High Treasurer, Aug. 23, 1714, *Cal. Treas. Paps.*, 1714-1719, §7. According to Henry McCulloch, Jr., Cumming held the post of preventive officer until 1716. McCulloch to the Lords of the Treasury, no date, State Paps. Dom., George II: 156, no. 272. The appointment of such an officer was approved by order in council, May 20, 1708, *Acts, Priv. Coun., Col.*, 1680-1720, p. 559.

[30] Hedges to Burchett, Aug. 7, 1708, Admiralty 1: 3667, fol. 129; warrant of Prince George of Denmark, Lord High Admiral, appointing officers for the vice-admiralty court of Newfoundland, Sept. 1, 1708, High Court of Admiralty, Misc.: 824, no. 467; William Brown, Deputy Register of the High Court of Admiralty, to Thomas Corbett, Secretary to the Admiralty, May 1, 1736, Admiralty 1: 3673; memorial of James Smith to Viscount Townshend, Secretary of State [1714], C.O. 194: 5, no. 57 (i).

226 BRITISH FISHERY AT NEWFOUNDLAND

As has already been said the powers of the commodore over the fishery and settlement were weak and incomplete. Although no assistance was forthcoming from parliament, that body continued to manifest a small amount of interest in Newfoundland by calling for annual reports from the Board of Trade relative to the fishery, and concerning itself occasionally with the military program, the condition of the industry, and the violations of both the Newfoundland Act and the acts of trade. It is evident, however, that parliament was not convinced of the impossibility of enforcing the Act of 1699, for in 1708 it asked the queen to take steps to have the royal officers do so, but refused to pay any heed to the repeated requests of the Board of Trade for necessary amendments to the act. The board, despairing of legislative sanction, after consulting the solicitor general, finally recommended court action against violators of the statute, justifying this step by alleging that offenses were *contra formam statuti*.[31]

Assuming that such action was legal, the Board of Trade proceeded to recommend that the commodore be commissioned "to command at land" during his stay at Newfoundland, and that he be directed to redress and punish all abuses and offenses contrary to the act "according to the known usage and customs there." Furthermore, in order to strengthen the administration of the fishery regulations in England, it recommended that the customs officers in the outports be instructed to keep lists of the shipmasters and byboatkeepers who were delinquent in employing the legal quota of green men. In order

[31] Order of the House of Commons to the Board of Trade, Jan. 16, 1707, *Cal. State Paps., Col.*, 1706-1707, §720; Board of Trade report to the Commons, Jan. 23, 1707, *ibid.*, 1706-1707, §726; order of the House of Lords to the Board of Trade, Nov. 12, 1707, *ibid.*, 1706-1707, §1191; Board of Trade report to the Lords, Nov. 19, 1707, *ibid.*, 1706-1707, p. 606. See also the same to the same, no date, Huntington Library, HM 821.

LAW AND WAR, 1699-1713

to warn all persons of the government's intention to enforce the Newfoundland Act, it proposed that a royal proclamation be issued. As these recommendations were contemporary with the abortive attempt to erect a court of vice-admiralty and with the appointment of Archibald Cumming as preventive officer, it can readily be seen that the Board of Trade, the Admiralty, and the Commissioners of the Customs were all anxious to strengthen the royal administration in Newfoundland. However, the Privy Council endorsed only a very small part of the recommendations of the Board of Trade, for it disapproved of extending the commodore's authority over the settlement. It also rejected the plan for bringing violators of the statute to justice in the royal courts. Though it agreed to the report in general, it was content to recommend that a proclamation be issued calling upon all persons to observe the act. The proclamation was subsequently issued, but as might have been expected, it had no effect whatsoever, as everyone continued to disregard the law.[32] Thus, the attempt made by the Board of Trade to strengthen the Newfoundland Act, to provide some means for checking smuggling, and to prevent emigration to New England during the critical times of the War of the Spanish Succession, failed because the higher authorities in England would not accept any radical changes in the law as it stood.

The long war with France from 1701 to 1713 left its scars upon the English fishery at Newfoundland. The period marks a turning point in the development of the industry because all concerned were brought to the brink of ruin. The western adventurers, planters, and byboat-keepers all suffered greatly, but the West Country fisher-

[32] Order in council, May 20, 1708, and Board of Trade report, May 19, 1708, *Acts, Priv. Coun., Col.*, 1680-1720, pp. 553-559; royal proclamation, June 26, 1708, Brigham, *Royal Proclamations*, pp. 163-167.

men suffered most of all. Owing to the risk of losing their ships to the enemy or having their fishery despoiled by French attacks, only the most daring merchants sent their ships to sea. Even those willing to trust to chance were discouraged either by the tardiness of the British government in furnishing the usual naval escort, which made them lose the better part of the fishing season, or by the impressment of their men, which made it difficult to procure crews. Consequently, the number of fishing ships from England was greatly reduced, and although a few English byboatkeepers continued to go to Newfoundland even in the worst years, this branch of the industry, too, became very much impoverished. The lessening number of fishing ships and byboatmen left a proportionately larger share of the fishery to the inhabitants, but because of the unfavorable circumstances produced by the war the latter were unable to profit by the misfortunes of their competitors. Many of the planters lost heavily in the French attacks on St. John's, Ferryland, Trinity, Conception, and Bonavista bays, while others became heavily indebted to English and colonial traders for provisions, and were unable to pay their creditors because the war had closed the Spanish and Mediterranean markets.[33] The only assured market left was Portugal, although occasionally sales were made in those parts of Spain favorable to the Grand Alliance. Furthermore, the war dis-

[33] Conditions in the fishery from 1699 to 1713 are described in the reports and letters of the commodores, the memorials of the western adventurers, planters, and other interested persons. See the answers to inquiries and schemes of the fishery, 1702-1713: *Cal. State Paps., Col.*, 1702, §1154 (i); 1706-1708, §§19, 588 (i); 1708-1709, §§223 (i-iii), 859 (iii); 1711-1712, §§149 (i, iii); C.O. 194:5, nos. 16 (i), 59 (i, ii). Cumming sent in some information to the Board of Trade. C.O. 194:5, nos. 26, 34. The western adventurers and others submitted a number of reports and complaints. *Cal. State Paps., Col.*, 1710-1711, §§74-80, 91, 122. The most important reviews of conditions were those submitted to parliament by the Board of Trade in 1706 and 1707, Egerton MSS., 921, fols. 3-8; Huntington Library, HM 821.

rupted the normal shipping services to such an extent that frequently there were not enough sack ships calling at Newfoundland to carry away the greatly reduced catch, and the inhabitants suffered additional losses through accumulated stocks which they were obliged to carry over from season to season. Faced with these obstacles the recovery of the Newfoundlanders was bound to be slow, but it was still further held up by the trading activities of the military officers and the prize officer, who carried on an extensive local business and used their official positions to engross commodities and forestall the market. In spite of protests from the Newfoundland merchants and in spite of recommendations by the Board of Trade that the officers be prohibited from engaging in business, they continued to operate throughout the war period.[34] In spite of all these handicaps, the inhabitants of Newfoundland during the War of the Spanish Succession were able to entrench themselves more securely in the management of the fishery because of the decline of West Country activity. Henceforward they were to play a much more important part than before in the economic development of the island.

The commercial competition and active enmity of France were considered the fundamental reasons for the discouragement suffered by the English fishery from 1699 to 1713. During these years the English became more intensely aware of the French menace and began to demand the total elimination of their rival from the fishery. That which heretofore had only been of interest to those actively engaged in the industry now became of national importance. France was accused of having adopted

[34] [James] Campbell's memorial to the Board of Trade, continued [1715], B.T. Newfoundland, 7, L67 (Transcript in the Public Archives of Canada, Ottawa, Ont.); Cumming to the Board of Trade, Oct. 10, 1713; Apr. 7, 1715; Oct. 10, 1715, C.O. 194: 5, no. 34, fol. 327, no. 100.

a deliberate policy of expansion in the North American fisheries in order to gain commercial supremacy and to increase her naval power. English sentiment is well expressed by the pamphleteer Peck, writing in 1702: "Their Newfoundland Fishery, by Encroachment on us, has considerably increas'd their Seamen and Shipping; and it is a certain Maxim, that all States are powerful, as they flourish in the Fishing Trade."[35] Fear of the French menace to the English fishery and commerce was constantly reiterated in the writings of the time, and as the war progressed the desire of the British to eliminate their rivals was frequently expressed. Sir Francis Brewster said that England would be justified in taking the Newfoundland and Greenland fisheries from the enemy as these were the chief nurseries for French seamen, and that England would not be safe until this was accomplished. France's gain in shipping had taken place at England's expense, for now she controlled not only her own domestic market for fish but was also supplying Spain with fish from Ireland and Newfoundland. On the other hand some Englishmen felt that their rival's fishing industry was not as prosperous as was generally supposed. Both Daniel Defoe and an anonymous commentator of 1702 pointed out that the Norman fishery had declined owing to excessive taxes, and that Fécamp, formerly an important port for the Newfoundland trade, had greatly curtailed its activity, many French fishermen having emigrated to Holland.[36] In spite of these dissenting opin-

---

[35] The South Sea Company seems to have tried to get control of the fishery. Peck, *Some Observations for the Improvement of Trade, by Establishing the Fishery of Great Britain; as a proper Means to Obtain the Ballance of Trade, employ the Poor, and promote the Interests of the South Sea Company* (London, 1702), p. 8.

[36] Sir Francis Brewster, *New Essays on Trade* (London, 1702), pp. 5-6, 15, 61-62; William Cary, on the trade and fishery of Newfoundland [Jan. 10, 1705], *Cal. State Paps., Col.*, 1704-1705, §798; Joshua Gee, *The Trade*

ions, there was an increasing tendency to emphasize the expansion of the French fishery.

The extent of the French fishery on the coasts and banks of North America during the period from 1699 to 1713 is most impressive. It was estimated by contemporaries that there were from 400 to 800 French ships engaged in fishing at Newfoundland, Nova Scotia, on the Banks, and in the Gulf of St. Lawrence. This far-flung industry employed from 16,000 to 30,000 seamen annually. At Newfoundland there were only 5,000 Englishmen, and the French controlled three-quarters of the coastline. Of course the inclusion of the New England fishery in the British estimates would have lessened the discrepancies between the two nations but as this was controlled exclusively by the colonists and competed actively with the Newfoundland fishery, it was usually omitted from the calculations of British writers.

Occupying the southern, western, and northern shores of the island, the French had developed an extensive fishery at Newfoundland. At Placentia alone they had from fifty to fifty-five fishing ships; a considerable number of large vessels from St. Malo fished as usual north of Cape Bonavista, and there were about 100 more Frenchmen on the Grand Bank. The French carried on their fishery largely by ships, which also carried the catch to market, although there were a few resident fishermen and some byboatkeepers at Placentia, St. Mary's, and St. Pierre. The French had certain advantages over the English. In

and Navigation of Great Britain Considered (London, 3d ed., 1731), pp. 10, 11, 13-14; [Thomas Thompson], *Considerations on the Trade of Newfoundland* [printed in 1710 or 1711?], Additional MSS., 13,872. Cumming was either the actual author or else he later copied it and submitted it as a report. C.O. 194:5, no. 62 (i). Daniel Defoe, *A Plan of the English Commerce* (London, 2d ed., 1728), p. 170; Anonymous, *A Political Account of the Diminutions of the Revenues and Trade of France . . . from . . . 1660 to 1699 . . .* (London, 1702), p. 9.

the first place, their fishery was so extensive that fewer ships fished in each harbor, and congestion was avoided without the consequent exhaustion of the supply as was the case in the English zone. In the second place, owing to the absence of ice on the southern coast the fish arrived there earlier, which made it possible for the French to begin operations considerably in advance of the English and to complete curing before the sultry summer weather began. In the third place, the excellent beaches on the southern and western shores obviated the use of staging. Besides the profitable fishery south and west of the British, the French also obtained enormous catches north of Cape Bonavista. Though the rocky coast made staging necessary there were no inhabitants there to destroy it nor to distract the fishermen by trade. The bank fishery was also extremely profitable to the French, and was unique because it could be carried on during the entire year. In general, the British considered their rivals to have a much better strategic position than they had themselves. Because of geographic and climatic advantages and because they obtained a better grade of salt than the English, the French were able to cure their fish more rapidly and better. Moreover, the French fishery was relatively free from the disputes which continually disrupted the English industry, for even in the harbors where inhabitants and fishermen were found together the relations between landsmen and seamen were excellent. At Placentia a good deal of the friction was eliminated by the presence of a resident governor. There the distribution of the beaches was made by the fishing admirals under his direction, and space allotted to each in proportion to his individual needs. The French fishery was so supervised by the royal governor that there was less opportunity for quarreling and violence. Besides making fish of a superior quality under more harmonious working conditions, the French had the added advantage of a lower cost of production owing to a lower standard of living and lower wages. Their natural advantages, their seemingly greater efficiency, and their lower production costs enabled the French to compete on excellent terms with the English in European markets. Their home market was protected by a high tariff and they were able to make deliveries to foreign markets six weeks or two months earlier than the English. Consequently, they had gained a large share of the Spanish and Italian trade.[37]

As the War of the Spanish Succession drew to a close British demands for the elimination of France from the Newfoundland fishery increased. Many felt that the English industry could not be improved as long as France threatened to seize the entire island, or as long as the establishment of a Bourbon dynasty in Spain implied exclusion from that important market. The continued congestion in the zone between Cape Race and Cape Bonavista made the English cast covetous eyes in the direction of Placentia and the neighboring fishing harbors of the south coast. Many hoped that the conclusion of peace would see the entire island of Newfoundland in British hands. The eradication of the French meant not only the removal of the danger of military aggression, but also the eventual dominance of all the North American fisheries by Great Britain, a British monopoly of the trade to Spain, Portugal, and Italy, and French dependence upon

[37] Information concerning French activities and the effect of their competition is found in the reports of the commodores and of Cumming. For the commodores' reports see: *Cal. State Paps., Col.*, 1700, §774 (i); 1701, §756 and pp. 531, 532; 1702, pp. 724-727; 1706-1708, §588 (i); 1708-1709, §§589 (i, iv); C.O. 194:5, no. 16 (i). Cumming to the Board of Trade, Dec. 11, 1713, C.O. 194:5, no. 26. There is a great deal of information regarding the war and its effect on the fishery and settlement contained in the *Calendars* from 1702 to 1712-1714, inclusive. French designs and attacks are described in considerable detail, and British plans and counter-attacks are also given much space.

234 BRITISH FISHERY AT NEWFOUNDLAND

Great Britain for fish. Moreover, Great Britain would have exclusive control over the trade in provisions and supplies for the fishery and settlement, and the British training school for seamen would be increased at the expense of France. One enthusiast considered complete control of Newfoundland as a greater benefit to the nation than possession of either the East or the West Indies.[38] The realization that France was the great commercial rival was expressed more emphatically during the War of the Spanish Succession than at any previous time. Demands for her elimination, though only partly realized, henceforward formed the keynote of British fishery policy in the mid-eighteenth century.

In spite of the hopes so frequently expressed in Great Britain during the war that all Newfoundland and Nova Scotia might fall to her at the conclusion of a peace with France, such was not to be. The views of the extremists who demanded the total elimination of France from northeastern America received scant attention from the diplomats concerned in the peace negotiations which lasted from 1711 to 1713, because compromises with France were necessary in order to terminate the conflict. During the general negotiations at Gertruydenberg the British diplomats demanded the complete cession of all Newfoundland, Hudson Bay, and the island of St. Christopher in the West Indies.[39] Upon the collapse of this con-

---

[38] Reflections on the present settlement of Newfoundland, 1704, *Cal. State Paps., Col.*, 1704-1705, §69 (i); masters of ships at St. John's to the Board of Trade, Oct. 15, 1705, *ibid.*, 1704-1705, §1373; Board of Trade report to the House of Commons, 1706, *ibid.*, 1706-1708, §32 (i); Col. Roope's memorial to the committee of the House of Commons, 1706, *ibid.*, 1706-1708, §56; answers to inquiries, 1706, *ibid.*, 1706-1708, §588 (i).

[39] Instructions to the Duke of Marlborough and Viscount Townshend, May 2, 1709, *British Diplomatic Instructions, 1689-1789*, II, *France, 1689-1721* (Royal Historical Society, 1925), p. 11. Cited hereafter as *Brit. Diplo. Instrucs.*, II, *France*. Henry Boyle to Marlborough and Townshend, May 18, 24, 1709, *ibid.*, pp. 12, 14; the same to Townshend, May 30, 1710, *ibid.*, pp. 21-22.

ference, Great Britain and France entered into separate preliminary negotiations in 1711, wherein the British government insisted upon the cession of both Newfoundland and Acadia, including Cape Breton Island, and the relinquishment by France of all her fishing rights at Newfoundland. These demands were too extreme to be acceptable to France, but served as the basis for further negotiations wherein compromises relating to Newfoundland and Acadia were eventually made. According to these terms the British would have obtained considerable territory but France too would have been left in a decidedly favorable position in northeastern America. A treaty was drafted which permitted the French to continue fishing and curing fish at Newfoundland and allowed them to settle and fortify Cape Breton Island. On the other hand the draft treaty acknowledged British sovereignty over continental Nova Scotia and Newfoundland including Placentia.[40] Such liberal concessions to France were not received favorably in England. The Board of Trade opposed the suggested arrangement on the ground that the cession of Newfoundland to Great Britain without the complete exclusion of the French from the fishery would nullify British sovereignty, because the French would retain tangible advantages, while Great Britain would gain nothing but nominal recognition of her title to the island. The draft treaty proposed that all fortifications at Newfoundland should be demolished, but that France was to be permitted to retain and fortify Cape Breton Island. The board objected to these provisions on the ground that Cape Breton should be con-

[40] Report from the Committee of Secrecy, June 9, 1715, *Reports from Committees of the House of Commons*, I, 3-98. See especially pp. 26-29. Project of a treaty with France, received from the plenipotentiaries in their letter of April 8, 1712, S.P., Foreign, France: 154, clauses 11, 12; Matthew Prior to Bolingbroke, Versailles, Dec. 12/23, 1712, S.P., Foreign, France: 154.

sidered part of Acadia and hence included in the cession of that region, and that the fortification of that island would require Great Britain to undertake the military defense of Newfoundland, which the proposed treaty forbade.[41]

The British plenipotentiaries were little influenced by the opinions of the Board of Trade or by mercantilist opinion at home. When the negotiators became deadlocked at Utrecht over the question of British sovereignty at Newfoundland and French control at Cape Breton, the two governments undertook direct negotiations. Bolingbroke sent Matthew Prior to Paris to deal directly with Desmarais with the hope that compromises of a political nature might be made in exchange for a commercial treaty favorable to Great Britain. Prior proposed that France retain Cape Breton with the right to fortify it, and that continental Acadia should be ceded to Great Britain; while Newfoundland and adjacent islands were to be granted to the British with reservations favorable to France which would allow her subjects to fish on the coasts of the island from Cape Bonavista northward to the Strait of Belle Isle and thence southward down the west coast as far as Point Riche. Prior also conceded to France possession of all the islands in the Gulf and River St. Lawrence.[42] These concessions were decidedly favorable to France and considerably more than the Board of Trade would have granted had it had a voice in the nego-

[41] Secretary St. John to the Board of Trade, Apr. 2, 1712, *Cal. State Paps., Col.*, 1712-1714, p. 254; Board of Trade to St. John, Apr. 5, 1712, *ibid.*, p. 256; Solomon Merrett's memorial to the Board of Trade, Apr. 5, 1712, *ibid.*, §373; Filson to the same, Dec. 9, 1712, C.O. 194: 5, no. 15; Popple to Filson, Dec. 10, 1712, *Cal. State Paps., Col.*, 1712-1714, §166.

[42] Instructions to the Duke of Shrewsbury, Dec. 11, 1712, *Brit. Diplo. Instrucs.*, II, *France*, pp. 39-40; Matthew Prior to Bolingbroke, Versailles, Dec. 12/23, 1712, S.P., Foreign, France: 154; proposition relative to Cape Breton, Acadia, and the fishery of Newfoundland, same date, S.P., Foreign, France: 154.

tiations. Prior ran counter to the contemporary demand for the total elimination of France from the Newfoundland fishery and thereby earned the enmity of the influential mercantilists at home.

The Prior-Desmarais draft agreement also included adjustments relative to Hudson Bay and the West Indies. It was finally accepted by both governments, but only after the British had insisted upon a definite limitation of the French fisheries at Nova Scotia and Newfoundland, demanding that their boundaries should be set forth explicitly in the treaty of peace. Though the French were willing to accept the principle of limitation, they desired that the bounds of their Nova Scotia fishery should be expressed only in the broadest terms. Great Britain, seeking to have her rights at Newfoundland clearly defined, desired to place restrictions upon the French, which had they been incorporated in the final treaty would have been decidedly onerous to France. Besides limiting the French fishery at Newfoundland to the indefinite region between Cape Bonavista and Point Riche, the British proposed that the French be prohibited from going to Newfoundland before April and that they be required to leave before the autumnal equinox. This provision was intended to limit the French fishery to the same period as that of the English and thus remove some of France's advantages in the matter of deliveries. The French, as might be expected, refused to consider these proposals, explaining that their shipping would be exposed to danger and inconvenience, and pointing out that such terms would lead to misunderstandings between the two powers. France stated that were these conditions imposed she must insist on retaining her fishing rights from Fortune Bay westward and northward to the Strait of Belle Isle and thence down the eastern coast of the Petit Nord to Cape Bonavista. The British then fell back on their earlier proposal, that the French retain fishing rights from Point Riche around the northern part of the island to Cape Bonavista, but they dropped the proposed time limit for the fishing season. The reason for the British retreat from these excessive demands is probably to be found in the anxiety of the Bolingbroke ministry to obtain commercial concessions from France. Finally, the right of the French to fish without restriction from Point Riche to Cape Bonavista was granted in exchange for the right of Great Britain to export British-caught fish to France on the basis of the moderate French tariff of 1664.[43] There is no evidence that Bolingbroke and his advisors were influenced to any extent by the demands of the mercantilists. It is more probable that the commercial concessions obtained from France were better liked in the West Country than among those who sought to nationalize the fishing industry at Newfoundland.

According to the ninth article of the commercial treaty which was concluded at Utrecht on March 31-April 11, 1713, British-caught fish packed in barrels might be imported into France on paying the duty prescribed in the French tariff of 1664. As most of the fish caught by English fishermen was shipped in bulk, the merchants quickly discovered that they could not comply with the requirements and engage profitably in furnishing fish to France. This particular article was severely criticised both in parliament and out because the conditions imposed would

[43] Points in dispute upon the project of a treaty of peace, Jan. 1, 1713, S.P., Foreign, France: 154; Prior to Bolingbroke, Jan. 6/17, 1713, S.P., Foreign, France: 154; Shrewsbury to Dartmouth, Jan. 14, 1713, S.P., Foreign, France: 157; memorandum from de Torcy, Jan. 14 and 17, 1713, S.P., Foreign, France: 157; Shrewsbury to Dartmouth, Feb. 7, 1713 (new style), S.P., Foreign, France: 157; Prior to the same, Feb. 13, 1713, S.P., Foreign, France: 157; report from the Committee of Secrecy of the House of Commons, *Reports from Committees of the House of Commons*, I, 3-98. See especially pp. 26-29.

make the trade unprofitable.⁴⁴ The unpopularity of this section of the treaty had a good deal to do with its rejection. The refusal of parliament to accept the commercial treaty left Great Britain in a very disadvantageous position with respect to the provisions of the treaty of peace. Political concessions to France had been made in it in order to obtain commercial advantages which parliament proceeded to reject. As the peace agreement did not have to be approved by parliament it became effective at once. As soon as it was published, the disastrous effect of Prior's negotiations was apparent to all, and the general dissatisfaction which it produced contributed to the increasing unpopularity of the Bolingbroke ministry and ultimately to its fall.

The definitive treaty of peace with France concluded at Utrecht in 1713 contains three clauses relative to the adjustment of Anglo-French affairs in northeastern America. According to the twelfth article, France ceded all of Nova Scotia with its ancient boundaries, except Cape Breton Island and Ile St. Jean, and agreed to a limitation of her fisheries on the coast of that province to the east and south east at a distance of thirty leagues from the shore beginning at Sable Island. According to the thirteenth article, Newfoundland and adjacent islands were ceded to Great Britain, arrangements being made for the transfer of Placentia within seven months after the conclusion of the peace. In the same article, France abandoned all claims to sovereignty over Newfoundland and its dependencies, agreeing not to erect fortifications anywhere or to erect permanent buildings, temporary huts and stages used in the fishery being excepted. The

---

⁴⁴ The Treaty of Navigation and Commerce between Great Britain and France, Utrecht, Mar. 31/Apr. 11, 1713, George Chalmers, *A Collection of Treaties between Great Britain and other Powers*, II, 390, 397, 422-423. Cited hereafter as Chalmers, *Treaties*.

French were conceded the right to fish and dry their catch ashore along the coast from Cape Bonavista to the northern point of the island, and thence down the western shore to Point Riche. The fact that no one was certain of the exact location of Point Riche caused no concern at the time, although it was to cause endless controversy in the future. The fourteenth article provided that French subjects dwelling in the territories ceded to Great Britain should have liberty to withdraw and take such personal effects as they should choose. Those willing to remain and become British subjects were to be permitted to enjoy the free exercises of their religion according to the usage of the Church of Rome, "as far as the laws of Great Britain do allow the same."[45] Altogether the treaty was a compromise with the advantage in favor of France. Great Britain obtained sovereignty over the island and the removal of the French from their fort, settlements, and fishery on the southern coast, but gave in exchange the right to fish from Point Riche around the island to Cape Bonavista, and most important of all the right to colonize and fortify Cape Breton Island, a place of far greater strategic importance than Placentia. Neither side was satisfied with the arrangement, and discord and friction between the two nations continued for many years. The problem of French fishing rights at Newfoundland was not settled satisfactorily until 1904.

The reaction of the mercantilists to the provisions of the peace treaty and the commercial agreement was immediate and unfavorable. They were especially critical of the arrangements made relating to Newfoundland, Cape Breton, and Nova Scotia. One contemporary propagandist, writing before the final peace was signed but when its provisions were generally known, felt that Great Britain would have made a better bargain had she offered to cede

[45] Treaty of peace, Articles XII-XIV, Chalmers, *Treaties*, II, 380-382.

the old English fishing zone in exchange for the region about Placentia. Daniel Defoe was fearful of the effect of permitting France to continue in the Newfoundland fishery and disapproved of granting Cape Breton to the rival power. He was convinced that the English fishery had "most miserably decayed" as a result of allowing France to fish in those regions, and that conditions would become worse if the French were given Cape Breton and allowed to fish at Newfoundland. Such an arrangement, Defoe thought, could only result in the British fishery being entirely surrounded and stifled by France. Criticism of the commercial treaty was also severe and much more effective. Many considered the concessions given by France to be worthless in comparison with the advantages which Great Britain would have gained by insisting on exclusive fishing rights at Newfoundland. Some doubted whether Great Britain would gain much from the commercial treaty as long as France had obtained rights at Newfoundland and had been put in a position to develop other fisheries also. These critics all felt that had the French been entirely excluded they would eventually have become dependent upon Great Britain for fish.[46] In fine, there was general disappointment and disgust in mercantilist circles at the provisions of the treaty of peace with France, and a feeling that the Bolingbroke ministry had played into the hands of the ancient rival.

The feeling that Great Britain had not made the best of

[46] Anonymous, *The Offers of France Explained* (London, 1712), p. 9; Daniel Defoe, *Some Further Observations on the Treaty of Navigation and Commerce* (London, 1713), p. 8; John Egerton, *A Vindication of the late House of Commons in Rejecting the Bill for Confirming the Eighth and Ninth Articles of the Treaty of . . . Commerce* (London, 1714), pp. 29-30; Anonymous, *A Letter to a West Country Clothier and Freeholder* (London, 1713), pp. 13-14; William Wood, *A Survey of Trade* (London, 1718), p. 211; Anonymous, *A Letter to the Hon. A[rthur] M[oo]re Com[missio]ner of Trade and Plantations* (London, 1714), pp. 12, 28; Anonymous, *Rejection of the French Treaty of Commerce* (no date), pp. 13-14.

her opportunities as a victor in the war with France became more acute as soon as the Treaty of Utrecht became operative. Joshua Gee felt that the French had proved themselves more clever than the English in commercial matters, and had gained from Great Britain "a Treasure equal to a Mine of Gold." William King emphasized the danger of allowing the French to continue in rivalry with the British at Newfoundland and Cape Breton. He drew attention to the activity of France at Cape Breton which had taken place as soon as her occupancy of that island had been confirmed, and prophesied that she would become a more formidable rival in the fishery than before. Although the French had yielded Placentia, nevertheless, they had acquired a much better fishery, and had thus become Great Britain's rival with the consent of the British people, who had obtained no advantages from the so-called concessions in the commercial treaty. Persons more intimately concerned with the Newfoundland trade felt that France had obtained a tremendous advantage by securing control of Cape Breton and by retaining fishing rights at Newfoundland. At Cape Breton the ships could begin fishing as early as February and carry on their activities as late as Christmas. The French also enjoyed similar advantages on the coasts of New France and in the Laurentian region generally. It was also felt that a great mistake had been made in allowing France to fortify Cape Breton.[47] British merchants and fishermen be-

[47] Joshua Gee, *The Trade and Navigation of Great Britain Considered* (London, 4th ed., 1738), p. 8; William King, *The British Merchant, or Commerce Preserv'd* (London, 2d ed., 1721), II, 286-297; [William Taverner], "A True Account of the Island of Gaspey [Cape Breton] and how Advantageous it will be to the French" [1715], C.O. 194: 23, no. 17. See also: Cumming to the Board of Trade, Dec. 11, 1713, C.O. 194: 5, no. 26; his memorial relating to Newfoundland and the French at Cape Breton, received Feb. 4, 1715, C.O. 194: 5, no. 62; memorial of James Campbell to the Board of Trade, Feb. 1, 1715, C.O. 194: 5, no. 65; his memorial continued, B.T., Newfoundland, 7, L67 (Ottawa transcript).

came increasingly alarmed as it became apparent that France was developing an important military and naval base as well as a trading center and fishing resort at Louisbourg. Many of the ideas expressed so emphatically by critics at the conclusion of peace with France were reiterated in the years immediately following, particularly at those times when relations between the two nations were strained or when open hostilities existed between them. Above all, British merchants realized that Bolingbroke, Oxford, and Prior, who had had the most to do with the peace negotiations had not been influenced in the least by mercantilist theories. The balance of trade had been sacrificed to the balance of power.

During the fourteen years between the passage of the Newfoundland Act in 1699 and the end of the War of the Spanish Succession in 1713 the fishing industry had suffered many vicissitudes. During the war the western adventurers had lost a good many ships or were obliged to lay up others to await the return of security and prosperity. The Spanish market was practically closed for many years and the risks involved in delivering fish to Portugal and Italy were great enough to offset the prevailing high prices. The ineffectiveness of the Act of 1699 had been amply demonstrated by 1713, and as before, disorder and confusion were general. The inhabitants, who obtained some advantage from the temporary embarrassment of the western adventurers, were unable to make the most of their opportunities, for they too suffered from the attacks of the enemy, and their fishery was demoralized. Only the Irish, New England, and foreign traders who continued to carry on their legal and illegal affairs seemed to profit from the war. Probably the fishery was in worse plight during the War of the Spanish Succession than at any other time in its history.

The government offered comparatively little help to

the struggling community and its staple industry. The military and naval defense provided proved utterly inadequate to meet the aggressive tactics of the French, and although the enemy was technically beaten, the peace left England's rival in a strong position in the North Atlantic fisheries. The indifference of the Bolingbroke ministry to the pleas of the British mercantilists was matched only by the lack of interest displayed in the West Country over the arrangements made in 1713 with France. Indeed, it is safe to conclude that up to that time the British government had not as yet adopted a truly mercantilist policy with respect to Newfoundland.

CHAPTER VIII

PEACE AND DISORDER, 1713-1729

AS might be expected, Newfoundland was in a state of confusion following the peace of 1713, and the period which came after was one of economic change and social disorder, owing partly to the effects of the war and partly to the complete collapse of the fishery regulations. The western adventurers, who had suffered severely during the war, were a long time in recovering from the blow dealt them, and never regained their dominant position. The scarcity of labor in the West of England, first apparent during the war, continued for some years thereafter, while the steady drain of trained seamen from Newfoundland to New England also persisted. An added discouragement was the continued failure of the fish for several years in succession. After 1713 the adventurers turned their attention increasingly to participation in general shipping, performing the functions of carriers and traders, and becoming less directly interested in the fishery, although English fishing ships always continued to go to Newfoundland. The shipmasters, however, devoted more time to barter, the collection of debts, the monopolization of local markets, and smuggling.[1]

The war with France left the inhabitants in almost sole control of the fishery for a number of years, and by 1713

---

[1] Answers to inquiries, 1708, 1713, 1715, 1717, 1722, *Cal. State Paps., Vol.*, 1708-1709, §233 (i); C.O. 194:5, nos. 59 (i), 99 (ii); C.O. 194:6, nos. 40 (i), 50; C.O. 194:7, fols. 99-120; Cumming to the Board of Trade, Dec. 1, 1713, C.O. 194:5, no. 26; his memorial, received Feb. 4, 1714, C.O. 194:5, no. 62.

246 BRITISH FISHERY AT NEWFOUNDLAND

they were making from two-thirds to three-quarters of the annual catch. They too, suffered from losses incurred during the war and from the succession of poor seasons. In order to better their wretched condition, proposals were offered to the Board of Trade suggesting either the removal of the inhabitants to the newly acquired southern coast or the offer to them of a bounty on fish. The settlement of the southern harbors would have relieved congestion in the old British zone, but many were too poor to move, and the western adventurers were unwilling to establish themselves on a strange coast. When it was seen that neither the Newfoundlanders nor the English fishermen were willing to occupy the south coast it was suggested that orphans or disbanded soldiers be located there. Besides the inherent conservatism of the inhabitants and West Countrymen, settlement was handicapped by the presence of French who had taken the oath of allegiance, and by others who had accepted British citizenship in order to retain title to their property, but who had removed to Cape Breton. None of the plans to settle the region or to free it from the influence of the French at Louisbourg were carried out, and the southern coast filled up gradually by infiltration from the old British settlements and from Ireland, although it never became as populous as the eastern coast.[2]

The influence of the western adventurers was strong enough to prevent the government from adopting a policy

[2] Solomon Merrett to the Board of Trade, Feb. 4, 1714, C.O. 194:5, no. 63; Col. John Moody to the same, Sept. 9, 1714, C.O. 194:5, no. 77; William Bromley to Moody, Sept. 8, 1714, C.O. 194:5, no. 64 (iv); Campbell's memorial, Feb. 23, 1715, C.O. 194:5, nos. 66, 66 (i); Board of Trade to Secretary Stanhope, Mar. 10, 1715, C.O. 195:6, pp. 50-52; merchants of Bideford to the Board of Trade [Mar. 31, 1715], C.O. 194:5, no. 83; answers to inquiries, 1716, C.O. 194:6, no. 29 (i); Capt. William Taverner's reports, Feb. 15, 1716 [May 20, 1718], C.O. 194:6, nos. 12 (i), 48 (iv); Col. Samuel Gledhill to the Duke of Newcastle, Sept. 30, 1731, C.O. 194:24 P 60.

of colonization for Newfoundland. They asserted that the settlement of the southern coast would bring great losses to English commercial interests, because much of the profit made by the inhabitants was spent either locally or in New England. Moreover, were colonization encouraged it would aid the planters in accumulating enough capital to control the entire industry. Consequently, the Board of Trade not only frowned upon colonization, but even went so far as to suggest the removal of all the Newfoundlanders to Nova Scotia, ostensibly to strengthen the British there, but actually in order to return the fishery to the western adventurers. Organized settlement was a policy never favored officially at Whitehall throughout the entire period.[3]

Important and far-reaching changes took place in the management of the fishery between 1713 and 1729. The scarcity of skilled labor in England and Newfoundland continued, and by 1717 green men, too, were hard to procure in the West Country, owing to the higher wages offered in New England. Costs of operation rose during these years also. Byboatkeepers were obliged to offer from £16 to £18 per season to skilled fishermen, while the cost of fitting out a fishing ship of 100 tons, manned by fifty men, and equipped with ten fishing boats amounted to about £1,000 annually. High wages, scarcity of skilled workers, coupled with the losses in the recent war, caused many operators to become insolvent. Many felt, however,

[3] Communications from the mayors and merchants of the West Country, received during 1715 and 1716, C.O. 194: 5, nos. 78 (i), 83, 107 (i); C.O. 194: 6, nos. 4, 7, 8. Relative to the removal of the inhabitants see the Board of Trade representation to the king, May 30, 1718, C.O. 218: 1, pp. 262-375, especially pp. 362-364. See also the representation, Dec. 19, 1718, C.O. 195: 6, pp. 416-464, especially p. 464; *Acts, Priv. Coun., Col., The Unbound Papers*, §277; the Board of Trade to Lord Carteret, Apr. 1, 1721, C.O. 195: 7, pp. 76-78; the same to the king in council, Dec. 20, 1728, C.O. 195: 7, pp. 158-174; *Acts, Priv. Coun., Col.*, 1720-1745, pp. 215-218; the same to Newcastle, Apr. 24, 1734, C.O. 195: 7, pp. 340-342.

that there were other reasons for the decline of the West Country fishery, namely, the discontinuance of two old customs: that of hiring men on shares; and that of apprenticing inexperienced men for a period of seven years. These practices were accounted the chief reasons for the success of the western adventurers in former times, but since they had been abandoned, probably between 1675 and 1699, the West Country operators had been forced to compete for labor with the Newfoundland planters who offered higher wages. Those interested in the recovery of the industry in the West of England considered the wage system harmful, and advocated a return to the former method of paying men in shares, some enthusiasts even going so far as to propose that fishing on any other basis should be made illegal. New England's success was attributed to the continuance of the share arrangement. If the Newfoundland industry continued on a wage basis, some thought that within a short time the New Englanders would not only be able to undersell Newfoundland traders abroad, but would also be able to invade the northern fishery itself. The chief objection to the wage system was that the fisherman lost interest in his work as soon as he had caught fish enough to equal his wages, while at the same time he became resentful if any fault were found with his work. By 1720 there were only a few ships which followed the old custom of fishing on shares, and within a few years it had been generally abandoned, except by vessels from Bideford and Barnstaple. Indeed, these ports were the only ones which sent out vessels for the sole purpose of fishing. Most of the others engaged in general trade, or carried on their fishing operations by employing cheap labor from southern Ireland.[4] The old

[4] Answers to inquiries, 1715, 1717, 1722, 1723, 1724, C.O. 194:5, no. 99 (ii); C.O. 194:6, no. 50 (i); C.O. 194:7, fols. 99-172, 205-214, 231-241; the same for 1725, inclosed in Bowler to the Board of Trade, Oct. 13, 1726

coöperative system had been discarded largely because the West Country operators had become interested in general trading as well as shipping. Now become capitalists, they sought to industrialize the fishery completely. The development of the offshore fishery on the Newfoundland Banks was the most significant change which occurred after 1713. It arose from the frequent failure of the fish in the harbors and on the inshore ledges, which was attributed to congestion in the zone between Cape Race and Cape Bonavista. Conditions were made worse by many thoughtless practices. The spawning places of the cod were destroyed by throwing refuse into the harbors, and by the careless use of seines to take bait, which killed many of the cod fry. Probably temporary changes in the nature of the sea also brought about poor seasons, although contemporaries did not put these forward as reasons for the decline. At any rate the failure of the inshore fishery caused fishermen to look to the banks. The new fishery was developed by the inhabitants who had apparently learned of it from their French neighbors. The methods employed were quite different from those followed in the harbors. Instead of using the dory, the fishermen employed decked shallops, sloops, or other small vessels such as shalloways. These boats remained at sea for five or six days until fully laden, when they returned to the harbors to deliver the fish for curing to persons who remained ashore for that purpose. As soon as the shallop had discharged its cargo, it returned to the banks for another catch. This procedure was kept up during the entire fishing season with the result that there was an appreciable increase in the annual catch. Measured in terms of the fishing dory, the bank fishery produced about 200 quintals per boat as against 150 quintals,

C.O. 194: 8; the same for 1729, inclosed in Beauclerk to the same, Oct. 14, 1729, C.O. 194: 8, O 63.

250 BRITISH FISHERY AT NEWFOUNDLAND

which was considered a good catch inshore. The latter fishery was still carried on extensively, but that on the banks proved so advantageous that many people were attracted to it. A bark, employing ten hands, cost about £70 per season, including labor and maintenance, and caught about 600 quintals. This was much better than could be done inshore, by seven boats, employing thirty-five hands, which cost for wages alone, nearly £400 for the season. The new method was considered more efficient and economical than the old, although the cod caught inshore yielded a larger quantity of oil than the bank fish.[5] The more conservative West Country operators did not venture into bank fishery immediately, because they found English labor unwilling to work offshore. However, by 1720 they were sending their ships to Newfoundland manned only by sailing crews. These vessels, already accustomed to call at ports in southern Ireland for labor, provisions, and saleable goods, hired Irishmen or picked up some of the poorer Newfoundlanders on arrival as extra hands. When the inhabitants were employed each was allowed a third of the fish he caught, apparently not as a share, but as a wage payment in cod, the principal circulating medium. Gradually, the bank fishery gained in favor among the western adventurers, so that by 1736 the fishing admirals were so busy on the banks that they had no time to attend to the regulation of the fishery in the harbors.[6]

In spite of the development of the bank fishery, trade continued to be poor or uncertain for many years, and prosperity did not return until about 1722. The fishing

[5] Answers to inquiries, 1715 (two separate reports), C.O. 194: 5, no. 99 (ii); C.O. 194: 6, no. 10 (i).

[6] Memorial from Capt. Caleb Wade, 1715, C.O. 194: 5, no. 82 (1); answers to inquiries, 1720, C.O. 194: 7, fols. 3-10; Gov. Lee to the Board of Trade, Sept. 25, 1736, C.O. 194: 10, P 7.

ships regained some of their former advantages in the industry and by 1724 there were about 200 ships from Great Britain, including fishing vessels and carriers, with a total tonnage of about 15,000, and employing over 3,000 hands. However, by this time most of the fish were caught by the inhabitants and byboatkeepers. The season of 1728 was the best in fifteen years. The catch was so good that the only complaint heard from those engaged in the industry was that there were not enough sack ships available to carry it to market.[7]

Compulsory contraction of debts and their forcible collection were great burdens upon the industry. The Act of 1699 provided for the settlement of disputes relating specifically to the fishery, but it prescribed no method for the adjustment of ordinary commercial disagreements. By 1713 the need of this sort of regulation was very apparent. Every year disputes arose between the inhabitants and the visiting New England traders, the factors representing English, Scottish, Irish, and colonial merchants, and the masters of sack ships and fishing ships interested in trade. The Newfoundlanders exchanged with these traders their fish and oil for the necessaries of life, as

[7] Conditions in the fishery are described in the answers to inquiries, 1715-1720, 1722-1724, C.O. 194: 5, no. 99 (ii); C.O. 194: 6, nos. 10 (i), 29 (i), 50 (i), 71; C.O. 194: 7, fols. 3-10, 99-172, 205-214, 231-241; the same for 1725, inclosed in Bowler to the Board of Trade, Oct. 13, 1726, C.O. 194: 8; the same for 1728, inclosed in Beauclerk to the same, Oct. 14, 1728, C.O. 194: 8, O 63. See also papers received from Cumming in 1715, C.O. 194: 5, nos. 62, 86, 86 (i), 100, 100 (i); Campbell's memorials, C.O. 194: 5, nos. 65, 66, 66 (i); Jeremy Dummer, Agent for Mass. Bay, to the Board of Trade, Oct. 13, 1713, C.O. 5: 866, no. 7; letters from West Country mayors and merchants, 1715-1719, C.O. 194: 5, no. 61; C.O. 194: 6, nos. 55, 56; Board of Trade representation to the king, Dec. 19, 1718, C.O. 195: 6, pp. 416-464; *Acts, Priv. Coun., Col., The Unbound Papers*, §277. See also the schemes of the fishery, 1723, 1724, C.O. 194: 7, fols. 195, 216, 242, the same for 1725, inclosed in Bowler to the Board of Trade, Oct. 10, 1725, C.O. 194: 8; the same for 1726, inclosed in Bowler to the Board of Trade, Oct. 13, 1726, C.O. 194: 8.

well as for liquor and a few luxuries. When the season proved a bad one, the inhabitants had to be trusted by the traders until the following year, when they were expected to pay. Several bad seasons in succession soon got the Newfoundlanders hopelessly in debt and unable to extricate themselves. Closely connected with this evil, was the pernicious practice, followed especially by the masters of the salt ships, of forcing people to buy goods they did not want. This unfair dealing was usually employed against the poor, although sometimes the well-to-do also were victims. The masters, who brought in considerable quantities of wine and brandy, forced the planters to accept a butt or a quarter-cask with every ten hogsheads of salt, especially during seasons when that essential commodity was scarce. As this amount of salt would only make a hundred quintals of fish, and as the salt and wine or brandy cost the planter between seventy-five and eighty quintals, he was usually left with little to meet his other obligations, and often without enough to finance his next fishing season. Those acquainted with conditions felt that if it were not stopped it would ruin some of the most enterprising Newfoundlanders. Owing to the absence of courts of law debts were collected by force, with the result that the strongest usually succeeded in seizing all the property of his debtor, sometimes before the season was half over, thus ruining the debtor's chances of recouping his losses and preventing him from paying his other creditors. Often the planter became so reduced in circumstances that he had to become a common laborer in the fishery in order to support his family. Captain William Taverner felt that the forcible collection of debts was more responsible for the decline of the fishery than the French war; and Commodore Kempthorn held it responsible for a great deal of disorder. The fishing admirals could not or would not interfere, as they themselves were often creditors. A great deal of time was lost during the season, because the creditors busied themselves during August in tricking and watching each other, and in making seizures. There were frequent quarrels, and recourse was made to "Clubb-law" in order to determine which creditor was to get the property and fish. Kempthorn felt that if it had not been for the time wasted in disorderly quarreling, the fishery could have been carried on into September. Forcible collection of debts encouraged the emigration to New England, because the unpaid employees of the unfortunate planters sought the first opportunity to leave the island rather than run the risk of starving to death. Exploitation did not end with the close of the fishing season, but was carried on throughout the winter by the resident Irish and New England factors, who had a great deal of power, "so that the planter was little better than a slave to them." Trickery and unfair dealing on the part of these agents was responsible for a good deal of the disorder which occurred after the departure of the convoy, and led to the emigration to New England of some of the more enterprising planters who were unwilling to enter "slavery for life."[8] The ill effects of forced buying and forcible collection of debts were to endure for many years, and to contribute to the retardation of the economic and social development of Newfoundland. Harmful as these practices may be in a society based upon a diversified industrial economy, they were disastrous to the well-being of a people whose only means of livelihood was the production of a single commodity.

The unregulated sale and universal consumption of large quantities of liquor also interfered with the effi-

[8] Answers to inquiries, 1712, 1713, 1719, 1720, C.O. 194:5, nos. 16 (i), 59 (i), 99 (ii); C.O. 194:6, no. 71; C.O. 194:7, fols. 3-10; Capt. Taverner's remarks [Mar. 19, 1715], C.O. 194:5, no. 35; Cumming to the Board of Trade (1715), C.O. 194:5, nos. 34, 62, 100.

ciency of the industry. Intemperance contributed considerably to the prevalence of disorder and violence. The problem was not new, but after 1713, it received more attention than formerly. All the shipmasters engaged in the liquor traffic. Spirits were retailed from stores and homes and from the decks of New England sloops. There was always a considerable amount of drunkenness, particularly on Sunday when the men were idle, but on week days too when it interfered with the regular work. According to Commodore Bowler, sixty-five of the 420 families resident in Newfoundland in 1726 kept public houses, of which forty-six were in St. John's and vicinity, ten at Ferryland, four at Bay of Bulls, four at Trepassey, and one at Bonavista.[9] In the absence of any regulations controlling the sale, the traffic in liquor contributed greatly to the piling up of indebtedness, forcing many into virtual peonage, and obliging others to emigrate in search of relief.

After 1713 the planters and local merchants occupied much more influential places in the fishery and in general commerce than before. In spite of discouragements, some Newfoundlanders had managed to accumulate a considerable amount of capital. By 1715 St. John's, where many of them lived, was regarded as the commercial center or "metropolis" of the island. There were few ships which either did not fish in its vicinity or make that place a port of call. The prices of provisions, liquor, fish, and oil were fixed by the St. John's merchants, the outports basing theirs upon the St. John's figures.[10] The attempt of the

[9] Cumming's memorial to the Board of Trade, received Feb. 4, 1715, C.O 194: 5, no. 62; answers to inquiries, 1715, C.O. 194: 5, no. 99 (ii); the same 1722, C.O. 194: 7, fols. 99-120; Capt. Taverner's remarks [Mar. 19, 1714] C.O. 194: 5, no. 35; mayor of Dartmouth to the Board of Trade, Jan. 9 1716, C.O. 194: 6, no. 4; scheme of the fishery, 1726, inclosed in Bowler to the same, Oct. 13, 1726, C.O. 194: 8.

[10] Answers to inquiries, 1715, C.O. 194: 5, no. 99 (ii).

PEACE AND DISORDER, 1713-1729 255

property owners there to establish a local government in 1723 to cope with disorder and crime, shows that by that time there were enough persons of property living within the town to require protection. The activities of William Keen, in advocating the erection of civil government imply that the local capitalists desired protection against the incessant lawlessness.[11] In spite of these indications of a more mature capitalist society, life in Newfoundland was still much more primitive than in corresponding seaboard regions in the neighboring continental colonies.

Illegal trade continued to flourish after 1713 as before. The acquisition of the southern coast with a resident population of Frenchmen complicated the situation. The new subjects continued to carry on open and direct trade with France and Cape Breton, or an indirect trade by way of the Channel Islands. French merchants invented various subterfuges in order to maintain commercial contact with their former compatriots. Shipmasters of French parentage but born in England were sometimes employed, or the shipping of the Channel Islands was engaged secretly. The officers of the British garrison at Placentia, and Captain William Taverner, the royal surveyor of the southern coast, were accused of trading illegally with the French. Colonel John Moody, lieutenant governor and commander of the garrison at Placentia, was obliged to prohibit French ships from disposing of goods in any way within his jurisdiction.[12] Various

---

[11] Account of the proceedings at the harbor of St. John's in erecting a civil government there in 1723, C.O. 194: 7, fols. 246-252. The preamble of the articles of association is a quotation from John Locke, *Essay Concerning the True Original, Extent and End of Civil Government* (London, 46th ed., 1713), p. 256. The agreement was signed by John Jago, the Church of England chaplain and fifty others. The association held courts almost weekly during the winter of 1723-1724.

[12] Taverner's remarks, received Mar. 19, 1714, C.O. 194: 5, no. 35; Lowndes to [the Board of Trade], inclosing a paper from [Moses] Jacqueau,

schemes were brought forward as before to put a stop to the traffic, but none received official support. James Campbell, a former merchant of St. John's, proposed the enforcement of the acts of trade at Newfoundland, and Captain Caleb Wade submitted an elaborate plan of seizure and confiscation of foreign ships and goods. However, the reports received from the commodores were usually indefinite and sometimes contradictory with respect to smuggling.[13] The uncertainty of the information received by the home authorities probably went a long way toward preventing the adoption of any drastic enforcement of the laws against clandestine trade.

In spite of the indefiniteness of much of the information received by the Board of Trade, there is every indication that smuggling was more highly organized and carried on more systematically than before the War of the Spanish Succession. British ships, particularly those whose masters became fishing admirals, sailed to France, Spain, Portugal, and the Cape Verde Islands, ostensibly for salt but also to obtain wine, brandy, and other foreign products and manufactures. Moreover, by 1717 there were merchants in Newfoundland with sufficient capital to buy entire cargoes, which they sold at retail to the planters and the New Englanders. Some of the British shipmasters also engaged directly in retailing, renting

May 8, 1714, C.O. 194:5, nos. 44, 44 (i); order in council, Dec. 21, 1716, C.O. 194:6, no. 19; inclosures with the above order referring to the representation of Ambrose Weston and William Cleeves, C.O. 194:6, no. 19 (ii-vii); George Lewen to the Board of Trade, Mar. 29, 1717, C.O. 194:6, no. 26; petition of Arbuthnot, Young, and Cleeves to the same, received Feb. 17, 1716, C.O. 194:6, nos. 11, 11 (i, ii); Taverner's reply to the above petition [Mar. 6, 1716], C.O. 194:6, no. 35; Col. Moody's order to French ships not to break bulk at Placentia, June 30, 1714, C.O. 194:5, no. 57 (iv).

[13] Memorial of James Campbell to the Board of Trade, Feb. 1, 1715, C.O. 194:5, no. 65; Capt. Caleb Wade to the same, Mar. 16, 1715, C.O. 194:5, nos. 82 (i, ii); Capt. Fotherby to the same, May 30, 1715, C.O. 194:5, no. 98.

storehouses, and offering their goods on credit to the inhabitants. If unable to dispose of all their stock during the fishing season, they left trustworthy men in charge who were authorized to dispose of the goods during the winter. The British masters often bartered goods to the New Englanders for plantation commodities, but the illegal traffic was so well organized that it was difficult for the commodores to ascertain its true extent. The smugglers were naturally reticent and took care to hide the forbidden goods from the officers of the navy. No one could really make any serious attempt to stop smuggling because as before, no one was certain whether the Navigation Act of 1663 applied to the Newfoundland trade. Attempts on the part of officers of the navy to check it were not supported by the legal authorities in England, and only the increased vigilance of the customs officers in the continental colonies had any effect at all.[14]

The colonial traders became increasingly active after 1713, competing with the English and Irish in supplying Newfoundland with a considerable part of its provisions, as well as livestock, clothing, and fishing gear. Most of the sloops engaged in the trade were owned in New England, but there were some vessels from New York and Pennsylvania, while ships from Virginia, Bermuda, Jamaica, Barbados, and Antigua appeared occasionally. The largest number of vessels hailed from Boston, and these carried the greatest variety of goods, consisting of livestock, general supplies, woodenware, lumber, bricks, naval

[14] Answers to inquiries for 1717-1720, 1722, 1723, 1724, C.O. 194: 6, nos. 50, 59, 71; C.O. 194: 7, fols. 3-10, 99-172, 205-214, 231-241; Francis Fane's opinion on the carriage of Canary wine to the Plantations, George Chalmers, *Opinions of Eminent Lawyers*, II, 275-276; Capt. St. Lo to the Board of Trade, inclosing answers to inquiries for Placentia, Sept. 30, 1727, C.O. 194: 8; Gov. Henry Osborn's instructions to Lieut. Gov. Gledhill, Sept. 6, 1729, C.O. 194: 8; Lord Vere Beauclerk's answers to inquiries for 1729, C.O. 194: 8.

stores, rum, sugar, molasses, tobacco, logwood, braziletto, chocolate, and rice. The New Hampshire sloops confined their cargoes to lumber, cattle, and apples; and those from Rhode Island brought provisions, lumber, rum, molasses, and pitch. The New Yorkers carried similar commodities, besides wheat, Indian corn, straw-work, cordage, leather, lime juice, and European goods. Those from Philadelphia had the same general cargoes as the others, but also carried shipments of flour, ham, venison, and lime juice. Virginia sent some tobacco direct, but most of it passed through other colonies before arriving at Newfoundland. The West Indian vessels carried only rum, sugar, molasses, and lime juice.[15] General supplies for the Newfoundlanders and enumerated commodities for reexport to Europe comprised the bulk of the colonial trade.

The trading methods of the colonials, though probably no worse than those of the British and Irish, were the cause of general complaint. The practice of selling fish to the sack ships for bills of exchange or cash provoked bitterness, because the colonials discounted the fish they had received in barter from the Newfoundlanders by one or two *reales* per quintal, and not only undersold them, but also cornered the money market. The local traders who could not afford to sell below the current price were sometimes forced to carry their stock over from one season to the next, while those who paid their employees in bills of exchange, or used them to finance the next year's fishing, were seriously handicapped by the financial monopoly of

[15] Answers to inquiries for 1713 and 1715, C.O. 194: 5, nos. 59 (i), 99 (ii); C.O. 194: 6, no. 10 (i); Taverner's remarks [Mar. 19, 1714], C.O. 194: 5, no. 35; Cumming's memorial [received Feb. 4, 1715], C.O. 194: 5, no. 62; his letter to the Board of Trade, Oct. 10, 1715, C.O. 194: 5, no. 100; Charles Carkesse, Secretary to the Commissioners of Customs, June 4, 1717, inclosing papers relating to the plantation trade with Newfoundland, C.O. 194: 6, nos. 33, 33 (i-iv).

the New Englanders. By 1725 the colonial traders had developed an organized system of distribution in Newfoundland. Goods were handled by resident factors who had an advantage over their British competitors. Permanent residence enabled them to demand high prices in the winter to offset the price cutting necessary during the summer when the British ships were there. Sometimes the factors made profits of from 200 to 400 per cent on retail transactions. The New England factors also acted as transportation agents, supplying their own colonies with capable but insolvent laborers. They regularly enticed workers to contract large debts and then sold them as indentured servants to the masters of sloops bound for New England.[16]

The New Englanders owed a good deal of their unpopularity to the development of the extensive passenger traffic in indentured labor. Both the British government and the English fishing interests were as anxious as ever to stop the traffic, but their feeble efforts were marked with no success. Sometimes as many as sixty men were carried away in a single sloop, and contemporary estimates place the annual exodus at between 1,200 and 1,400 men, except when the commodore happened to be unusually vigilant. The migration was due not only to the oppression of the trading shipmasters and factors, but also to the practice of English masters of fishing ships of leaving their men at Newfoundland at the end of the season, thus increasing the number of unemployed and indigent. Men who could pay their passage to New England were charged £3 for the voyage, while the penniless were sold

[16] Taverner's remarks [Mar. 19, 1714], C.O. 194:5, no. 35; Cumming's memorial [received Feb. 4, 1715], C.O. 194:5, no. 62; his letter to the Board of Trade, Oct. 10, 1715, C.O. 194:5, no. 100; answers to inquiries for 1713 and 1715, C.O. 194:5, no. 59 (i); C.O. 194:6, no. 10 (i); Kempthorn to Burchett [Oct. 12, 1715], C.O. 194:5, no. 99 (i).

as indentured servants for a period of four years to work off their passage on arrival in New England. Many of these men left their families in England where they became public charges. Legislation to curb the traffic was never seriously considered by the authorities, and the old and feeble expedient of bonding the colonial masters not to carry away more men than they had brought to Newfoundland, was continued with the same want of success as before. Similarly the requirement that the colonial sloops sail with the convoy was equally ineffectual. Some returned as soon as the king's ships were well out at sea, while others, more insolent, openly defied the commodores. Sometimes men were headed up in casks in order that they might escape the watchful eyes of the officers of the navy. The Board of Trade besought the commodores to make strenuous efforts to suppress the traffic, and tried to find legal support for its total prohibition. The advice of learned counsel was asked concerning the effectiveness of the bonds taken from the colonial traders, but the law officers made no reply to the inquiry. No legal action was ever taken, the bonds were worthless, and the efforts of the commodores to suppress the traffic were useless.[17]

[17] Capt. Fotherby to the Board of Trade, May 30, 1715, C.O. 194: 5, no. 98; Kempthorn to Burchett [Oct. 12, 1715], C.O. 194: 5, no. 99; answers to inquiries for 1715, C.O. 194: 5, no. 10 (i); Capt. Passenger to the Board of Trade, Oct. 1, 20, 1717, C.O. 194: 6, nos. 37, 39; his answers to inquiries for 1717, C.O. 194: 6, no. 50; his order to the New England ships to sail before the convoy [Sept. 23, 1717], C.O. 194: 6, no. 39 (i); Cumming to the Board of Trade, Oct. 10, 1715, C.O. 194: 5, no. 100; merchants of Bideford to the same [received Dec. 10, 1715], C.O. 194: 5, no. 107 (i); mayor of Dartmouth to the same, Jan. 9, 1716, C.O. 194: 6, no. 14; Popple to Burchett, Mar. 3, 1718, C.O. 195: 6, pp. 375-376; Burchett to Popple, Mar. 12, 1718, C.O. 194: 6, no. 43; answers to inquiries for 1718-1720, 1722, and 1723, C.O. 194: 6, nos. 59, 71; C.O. 194: 7, fols. 5-10, 92-172, 205-214; Bowler to the Board of Trade, Oct. 10, 1725, Oct. 13, 1726, C.O. 194: 8; Capt. St. Lo to the same, Sept. 30, 1727, C.O. 194: 8. Scott's order to the New England traders, Sept. 20, 1718, C.O. 194: 6, no. 59 (ii); bonds taken by Percy from the New England traders, 1720, C.O. 194: 7, fols., 13-17; affidavit of persons

PEACE AND DISORDER, 1713-1729 261

Though by 1713 the Act of 1699 had proved ineffective, yet for a number of years thereafter the British government was occupied with the hopeless task of trying to enforce it. Already defective, the fishery regulations were to demonstrate their utter worthlessness in the face of the changing conditions. By 1715 the fishing admirals had proved themselves wholly incompetent to deal even with disputes relating to the fishery. They often ignored their duties entirely, unless it served their own interest or that of friends to exercise their legal functions. If self-interest ruled, persons occupying coveted fishing places were ejected on the ground of faulty title, even when such action ran counter to previous judgments by the commodores. Sometimes the fishing admirals were shipmasters making their first voyage to Newfoundland and, therefore, unacquainted with their obligations or the customs of the place. These newcomers were controlled by the old, experienced West Country shipmasters who used them as catspaws in dealing with offending rivals. Some of the old masters had become so powerful locally that they were called "kings" by the Newfoundlanders. These bosses were entirely selfish and influenced the fishing admirals to hear complaints only when the plaintiffs were their own friends. Sometimes too, the fishing admirals were the masters of ships which had come indirectly to Newfoundland, a practice forbidden by the Act of 1699. None of the admirals kept journals as they were supposed to do, and fear of the displeasure of the commodore was the only thing that caused them to observe any of their legal obligations. Had it not been for the annual appearance of the commodore, the fishing admirals would have

left behind at Placentia, Sept. 23, 1724, C.O. 194: 7, fol. 264; William Keen to the Board of Trade, Oct. 30, 1719, C.O. 194: 6, no. 72; Richard Wheelock to Richard West, Apr. 28, 1722, C.O. 195: 7, pp. 80-81; Popple to the same, Jan. 25, 1723, C.O. 195: 7, p. 86.

262 BRITISH FISHERY AT NEWFOUNDLAND

been more arbitrary and selfish than was actually the case. Some of the more arrogant defied the commodore's authority and deliberately disobeyed his orders. Disregard of the Newfoundland Act and the disastrous effect of the general lawlessness of the fishery influenced the government to strengthen the administration of Newfoundland and led eventually to the establishment of civil government in 1729.[18]

The question of regulation was considered by the authorities in 1715, as the result of the receipt of memorials and petitions from persons interested in Newfoundland. The extensive report of Commodore Kempthorn in 1715, pointing out that the system of control prescribed by the Act of 1699 had broken down, determined the crown to call upon the Board of Trade for recommendations in order that the question might be presented to parliament for consideration and action. The board, therefore, undertook one of its usual investigations, calling upon the Admiralty and other government departments as well as persons interested in the fishery for their opinions. A number of proposals were submitted by persons concerned in the fishery or acquainted with Newfoundland, who in general either advocated amendment of the Act of

[18] Heads of inquiry, clauses 9, 33, and additional instructions, for the commodore, clause 33, Mar. 8, 1715, C.O. 195: 6, pp. 32 ff.; heads of inquiry, clauses 9, 30, additional instructions for the commodore, clause 32, 1717, C.O. 195: 6, pp. 376-391; answers to inquiries, 1715, 1717, 1720, C.O. 194: 6, nos. 10 (i), 50 (i); C.O. 194: 7, fols. 3-10; Lord Vere Beauclerk to the Board of Trade, Oct. 7, 1728, C.O. 194: 8, O 45; Board of Trade representation to the king, Mar. 2, 1716, C.O. 195: 6, pp. 242-261; another, Dec. 19, 1718, C.O. 195: 6, pp. 414-464; heads of a bill for remedying the abuses in the Newfoundland trade, Dec. 24, 1718, C.O. 195: 6, pp. 465-484; Board of Trade representation to the king, Dec. 20, 1728, C.O. 195: 7, pp. 158-174. See also Popple to [Ogle], June 3, 1719, C.O. 195: 6, pp. 503-504; the same to [Burchett], Apr. 6, 1720, C.O. 195: 7, pp. 11-35. See also the following: Capt. Taverner's remarks, received Mar. 19, 1715, C.O. 194: 5, no. 35; Kempthorn's answers to inquiries, 1715, C.O. 194: 6, no. 10 (i); the same to Burchett [Oct. 12, 1715], C.O. 194: 5, no. 99 (ii).

1699 or the adoption of new legislation intended either to strengthen the fishery regulations or to curb the illegal and colonial trade. The suggestions varied in extent, character, and severity, but in general provided for the establishment of a system of fines and penalties, and included provisions intended to bring the regulation of the fishery into closer harmony with mercantilist practices. The restriction of the fishery to ships from Great Britain and the Channel Islands and the prohibition of indirect trade between the West Indies and Newfoundland was proposed in order to restrain the Irish and New Englanders in the interests of British commercial supremacy. Proposals were also made for granting the fishing ships from England precedence over the inhabitants and by-boatkeepers, and for restricting the latter to the use of fishing conveniences in the inshore zone. It was also suggested that a local magistracy be established to prevent the crime and disorder which were so prevalent during the winter; although nothing was proposed which would interfere with the authority of the fishing admirals or curb their excessive zeal, lest the fishery be injured thereby. Those desirous of promoting the welfare of the West Country industry also sought to obtain legal authority for the reëstablishment of the apprenticeship system and the compulsory return of fishermen and sailors to England at the end of the fishing voyage. Valuable information and suggestions were made by such conscientious officers as Commodore Kempthorn, who was on the Newfoundland station in 1715, Captain William Taverner, the royal surveyor, Colonel Roope, the royal engineer at St. John's, and persons acquainted with the trade such as Captain Caleb Wade, Archibald Cumming, and James Campbell.[19]

[19] Official action during the investigation of 1715 is to be found in the correspondence of the various administrative units concerned. James Stan-

On the basis of the suggestion submitted, the Board of Trade was able to report to the king on March 2, 1716, that the Act of 1699 had proved unsatisfactory because of its failure to provide penalities for violations of the regulations. It recommended that supplementary legislation be asked of parliament in order to strengthen the rules, to provide more careful supervision of the fishery and settlement throughout the entire year, and to insure a more strict control over general trade. The board pointed out that the commodore was helpless to enforce the Act of 1699, and that the situation could only be remedied by conferring additional powers upon him by act of parliament. The board further recommended the appointment of harbor magistrates, who should be empowered to adjust differences arising during the winter during the absence of the commodore and fishing admirals. They also proposed a system of licenses for public houses, and the prohibition of retail sales of liquor on Sundays or at any time during the fishing season. Debts were not to be esteemed good unless settled before the commodore, the

hope, Secretary of State to the Board of Trade, Mar. 7, Apr. 2, Oct. 13, 1715, C.O. 194:5, nos. 78, 84, 99; Admiralty to Stanhope, Oct. 12, 1715, C.O. 194:5, no. 99 (i); Kempthorn to Burchett [Oct. 12, 1715], C.O. 194:5, no. 99 (ii); Board of Trade to Stanhope, Oct. 20, Nov. 11, 1715, C.O. 195:6, pp. 144-145, 149; the same to Burchett, Nov. 24, 1715, C.O. 195:6, p. 150; Popple to the same, Nov. 29, 1715, C.O. 195:6, p. 162; Burchett to Popple, Nov. 24, 1715, C.O. 194:5, no. 101; Circular letter from Popple to the mayors of the outports, Dec. 3, 1715, C.O. 195:6, pp. 154-155. Communications were received during 1714 and 1715 from persons interested. Besides the reports and memorials of Archibald Cumming and James Campbell which have been frequently cited, see: Capt. Taverner's proposed amendments to the Act of Parliament, Mar. 19, 1714, C.O. 194:23, no. 14 (i); some amendments to the Act of Parliament desired by the traders, received from Col. Roope, Apr. 21, 1714, C.O. 194:5, no. 42; remedies proposed by Capt. Caleb Wade [Mar. 23, 1715], C.O. 194:5, no. 82 (i); representation from the merchants of Poole to the Board of Trade, received Mar. 17, 1716, C.O. 194:5, no. 78 (i); another from Bideford to the same, Mar. 31, 1715, C.O. 194:5, no. 83.

fishing admirals, or the harbor magistrates. To prevent illegal trade, the board proposed that the shipping to Newfoundland be restricted to vessels which had victualled and cleared from ports in Great Britain and the Channel Islands. In order to prevent the traffic in labor to New England, and to curtail the trading activities of the colonials, the board also recommended the stiffening of the provisions of the Act of 1699 relative to the employment of green men, proposed that factors be prohibited from residing in Newfoundland unless they came directly from Great Britain and represented British mercantile houses, and advised that more severe restrictions be placed upon New England shipmasters. In general, the board felt that these proposals should be incorporated in an act of parliament to supplement the Newfoundland Act. It recommended also that the old act be extended to cover the south coast recently acquired from France, and that steps should be taken to check French trading in that region. The western adventurers, formerly so alarmed and concerned whenever the question of regulations for the fishery was raised, took no active part in the investigation. With the recommendations suggested by the board, however, it may well be imagined that the West Country was content. It is significant, however, that the report was not approved by the king in council and the proposed legislation never recommended to parliament.[20]

The conservative attitude of the Board of Trade which is so apparent in its report of 1716, was again manifested two years later when Governor Richard Phillipps of

[20] Other letters were received from the West Country early in 1716. Mayor of Dartmouth to the Board of Trade, Jan. 9, 1716, C.O. 194: 6, no. 4; mayor of Fowey to the same, Jan. 19, 1716, C.O. 194: 6, no. 8; mayor of Exeter to the same, Jan. 25, 1716, C.O. 195: 6, p. 181; the report of the Board of Trade to the king, Mar. 2, 1716, C.O. 195: 6, pp. 242-261. Neither the *Acts, Priv. Coun., Col.,* nor the *Journals* of either house of parliament contain anything relative to this report or to the proposed legislation.

Nova Scotia and Placentia proposed that as governor of the latter place he should receive the same regulatory powers as were vested in the commodore. The board was unfavorable to this proposal, pointing out that the Newfoundland fishery was regulated by statute, and that it would be inadvisable to attempt any alteration in the administration of the fishery without mature consideration. Indeed, the board was so definitely committed to the idea of a ship fishery governed by the Act of 1699 that it actually suggested the transference of the Newfoundlanders to Nova Scotia, alleging that the abuses current in the industry arose largely from the trade of the inhabitants with New England.[21] In the two years which elapsed between their report of 1716 and their reply to Phillipps in 1718, the Board of Trade showed itself still favorable to the western adventurers.

Meanwhile the need of supplementary legislation was emphasized as a result of the reports received from Commodore Hagar in 1716 and Commodore Passenger in 1717. Both complained that the fishing admirals exceeded their authority and disregarded the Act of 1699. They also pointed out the need, particularly during the winter, of government for the island, since there was no law and order there except for the short time the king's ships were on that station. The independent character of the people made them disinclined to obey the law unless it were supported by the presence of superior authority. Even when they observed the orders of the commodore during his stay, they disregarded them immediately after

[21] Memorial of Col. Richard Phillipps, Gov. of Nova Scotia and Placentia [May 30, 1718], C.O. 218:1, pp. 343-350. In May 1720, Phillipps invited the Newfoundlanders to settle in Nova Scotia. See his letter to Secretary Craggs, May 26, 1720, in the *Governor's Letter Book, Annapolis, 1719-1742*, Nova Scotia *Archives*, II, 60-61. See also the Board of Trade representation to the king on Phillipps' memorial, May 20, 1718, C.O. 194:23, nos. 30, 30 (i).

his departure. Passenger felt that no resident of Newfoundland was fit to govern the island, "for this set of people that live here are those that can't live in Great Britain or any where else, but in a place without government," and he thought that the only solution of the problem would be the appointment of a royal governor.[22] As usual, all concerned with the management or regulation of the fishery were agreed that something must be done, but again no one was willing to adopt any of the recommendations made.

In December, 1718, the Board of Trade submitted to the king an extensive and important report on the state of Newfoundland. This paper contains the most exhaustive summary of the whole question that the author has encountered among the records relating to the fishery. The board, citing earlier documents, reviewed the entire history of the fishery and plantation since 1615. Though it continued to favor the western adventurers, it pointed out that the fishery could not prosper under the existing statutory regulations. It also called attention to the many alleged irregularities which had been introduced in the preceding forty years, and attributed the decline of the fishery to the presence of the inhabitants and byboatkeepers. It held the New England traders responsible for the reduction in the number of seamen trained in the fishery and for the decay of British trade. After discussing the ways in which the operation and regulation of the industry had proved ineffective, and after presenting statistics to demonstrate the decline of the West Country participation since 1615, the board took a position unequivocally in favor of the West of England, saying that "this important fishery at Newfoundland can never be revived or restored to its former flourishing state and

[22] Answers to inquiries, 1716 and 1717, C.O. 194: 6, nos. 29 (i), 50 (i).

condition until it be again wholly carried on by fishing ships, according to its ancient custom, and regulated by laws agreeable thereunto."[23] In order to accomplish this, it recommended that the Newfoundlanders be removed to Nova Scotia or some other plantation, thus eliminating the principal obstruction to the fishery, the element whom it considered most responsible for irregularities and disorders. None of these changes could be effected except by act of parliament, and the board, therefore, submitted a draft bill to the Privy Council for such an act as it desired.

The proposed legislation was intended to replace the Act of 1699. It imposed severe restrictions upon those engaged in the fishery and trade, and provided the long-desired penalties for violation of the rules. Strangely enough the bill did not propose the removal of the Newfoundlanders, but incorporated a number of provisions offering direct encouragement to the West Country and distinctly curtailing the activities of the foreign and colonial traders. In general, the tone of the proposed legislation was decidedly reactionary, harking back to the days previous to the passage of the Newfoundland Act of 1699, taking into account no changes since that time. In spite of the work of the Board of Trade in preparing the representation and draft bill, neither received any attention from either the crown or parliament.[24] It is uncertain just what inspired the great representation of 1718, as no pressure was being exerted either by the West Country fishing interests or by the Newfoundland planters. Prob-

[23] Board of Trade representation to the king, Dec. 19, 1718, C.O. 195: 6, pp. 416-464; *Acts, Priv. Coun., Col., The Unbound Papers*, pp. 109-119.

[24] Heads of a bill for remedying the abuses in the Newfoundland trade, submitted by the Board of Trade to the king, Dec. 24, 1718, C.O. 195: 6, pp. 465-484. This bill was never introduced into parliament by the ministry. Stock, *Proceedings and Debates*, III, covering the session of 1718-1719, contains no mention of it.

ably the application of Governor Richard Phillipps for increased power at Placentia, coupled with the recent suggestions of the commodores that additional supervision was needed, caused the Board of Trade to take the opportunity to present to the higher authorities a recapitulation of the doctrines of the western adventurers. The fact that none of the recommendations of the board, made since 1702, had received any official notice, indicates that the Lords Commissioners for Trade and Plantations had far less influence in determining matters of commercial and colonial policy than has sometimes been assumed.

While the Board of Trade was showing such surprising zeal for the West Country interests in its report of 1718, there was a striking amount of apathy on the part of the western adventurers themselves relative to improvements in the regulations. After 1715 or 1716 there are scarcely any protests or memorials from the West of England, until the establishment of civil government in 1729 aroused the region to some of its old time fury. In these years the regulatory system was in almost complete collapse, and was beyond hope of improvement even by such legislation as the Board of Trade proposed in 1718. It seems incredible in view of the reports of the commodores, the petitions and memorials of persons interested in the fishery and plantation, and the recommendations of the Board of Trade, that the British government should have taken no action to better the fearful state of affairs which existed in Newfoundland. The answer probably is to be found in the fear of the home officials that the creation of a civil government for such a community as Newfoundland would cause expense to the crown. During the same period the nearby province of Nova Scotia also was ignored at Whitehall for the same reason. Perhaps fear that neither of these two regions could be held successfully in another war with France may have been a mo-

270 BRITISH FISHERY AT NEWFOUNDLAND

tivating force in determining British official policy. Whatever the reason, the fact remains that nothing was done until 1728 when the crown finally took Newfoundland's problems into serious consideration for the first time in many long years. During the period from 1718 to 1728, the island fishery suffered from neglect and maladministration by the authorities in England, from the indifference of the commodores, and from the continual bickering of the planters, fishermen, and colonial traders.

In the years just previous to the establishment of civil government utter demoralization existed in the fishery. Again and again the more conscientious commodores pointed out the need of establishing some sort of local judicature to cope with the disorder which was rapidly verging upon anarchy. The situation grew steadily worse after 1720, the commodores became indifferent, careless, and actually neglectful of their duties. The answers to heads of inquiry during the period from 1722 to 1727, during the administration of Commodore Edward Bowler, became stereotyped and worthless. In many cases they were copied almost *verbatim* from previous reports, and the commodore, following the line of least resistance, obtained his information from such untrustworthy authorities as the fishing admirals. His inefficient and unoriginal reports compare unfavorably with the highly efficient and carefully conducted investigation of Lord Vere Beauclerk who went out as commodore in 1728.[25]

During the period from 1713 to 1729 a great many

[25] Capt. Scott to the Board of Trade, Nov. 16, 1718, C.O. 194:6, no. 59; Ogle to Popple, Oct. 13, 1719, C.O. 194:6, no. 71; Percy's answers to inquiries, 1720, C.O. 194:7, fols. 3-10; Bowler's answers to inquiries, 1722, 1723, C.O. 194:7, fols. 99-172, 205-214; the same for 1725, 1726, and 1727 inclosed in his letters to the Board of Trade, Oct. 10, 1725, Oct. 13, 1726, Oct. 16, 1727, C.O. 194:8; St. Lo's answers to inquiries relating to Placentia, inclosed in his letter to the same, Sept. 30, 1727, C.O. 194:8; Lord Vere Beauclerk to the same, Oct. 27, 1728, C.O. 194:8, O 48.

changes took place in the fishery, trade, and settlement. In the first years following the peace with France conditions were deplorable. The western adventurers had abandoned fishing to a considerable extent and were concentrating on trade, with the result that the fishery was managed increasingly by the Newfoundlanders and by-boatkeepers, although its regulation was still in the hands of the West Country shipmasters and fishing admirals. For several years all concerned had suffered owing to the failure of the fish and owing to the closing of the Spanish market.[26] The development by the Newfoundlanders of the offshore fishery on the banks and the gradual accumulation of wealth locally caused a change in the character of the industry. The western adventurers, following the inhabitants onto the banks, soon recovered a fair share of the trade, but never became as dominant as before the War of the Spanish Succession. General trading raised questions which were not covered by the Newfoundland Act, while defects in that law gradually caused it to be disregarded; its violation being assisted by the indifference of the fishing admirals and commodores, as well as by the failure of the British government to provide machinery for its enforcement. Newfoundland as a produc-

[26] The Spanish Basques sought to obtain recognition of their fishing rights at Newfoundland after the conclusion of peace in 1713. They claimed the right by virtue of their early activities there and also by right of article XV of the Anglo-Spanish Treaty concluded at Utrecht in 1713. The British steadfastly denied these pretensions although the Spanish government raised the issue intermittently. See the following references: C.O. 194, 5, nos. 18, 18 (i), 78 (i), 79; C.O. 194: 6, nos. 10 (i), 66, 67, 69; C.O. 194: 23, nos. 8, 9, 31, 31 (i); C.O. 195: 5, pp. 287-288; C.O. 195: 6, pp. 53, 510; Admiralty 2: 48, pp. 256-257; S.P., Dom., Entry Books, 248; S.P., Foreign, Spain, 114. For a detailed discussion of this question see Vera L. Brown, ''Spanish Claims to a Share in the Newfoundland Fishery,'' *Annual Report* of the Canadian Historical Association (1925), pp. 64-82. See also Harold A. Innis, ''The Rise and Fall of the Spanish Fishery at Newfoundland,'' *Transactions* of the Royal Society of Canada, XXV, section II (1931), 51-70, *passim*.

ing area was still important, but its potentialities as a region where British, Irish, and colonial traders might make sales were becoming more and more significant. By 1729 we begin to discern the rudiments of a colonial society, which was to grow slowly for many years before reaching maturity. Committed to the old West Country conception of the fishery the Board of Trade and other British officials were slow to appreciate the changed conditions, and it was not until 1729, after thirty years of uncertainty and disorder since the passage of the Newfoundland Act that the home government finally became convinced that a government more in keeping with the new social and economic order in Newfoundland was necessary. Henceforward the island and its fishery were to be administered in somewhat closer conformity to the standards set for the rest of the British colonial world.

CHAPTER IX

CIVIL GOVERNMENT, 1729-1763

NEWFOUNDLAND entered upon a period of development in 1729 which contrasts strikingly with its early history. Although the codfishery remained the basis of its economic life, political matters received major consideration for the first time. Just as the center of economic control had been gradually transferred from the West of England to the island itself, so the center of political influence was shifted from that part of England to St. John's.

It was not until the winter of 1729 that the British government devoted serious thought to Newfoundland for the first time in many years, and paid attention to the question of its future government. Official recognition of the serious situation had been brought home to the ministry by the very compelling report of Lord Vere Beauclerk in 1728, and the government took action because of the personal interest of the Duke of Newcastle. Beauclerk had voiced his alarm over the general disregard of the Act of 1699, particularly with respect to the detrimental effect lawlessness was having upon the nursery for seamen. He also was exercised over the interference of Samuel Gledhill, lieutenant governor of Placentia, in the affairs of the fishery and settlement. Gledhill refused to acknowledge the superior authority of the commodore on the ground that he was responsible only to Governor Phillipps of Nova Scotia for his conduct.[1] The conflict be-

[1] Lord Vere Beauclerk to the Board of Trade, Oct. 7, 27, 1728, C.O. 194: 8, O 45, O 48; the same to Burchett, Aug. 19, 1728, C.O. 194: 8.

274 BRITISH FISHERY AT NEWFOUNDLAND

tween the military and naval officers unquestionably had a great deal to do with bringing about the reorganization of the entire administration of Newfoundland.

The cabinet asked the Board of Trade to consider the question seriously and to propose such measures as would reduce disorder and confusion in the fishery and improve such an important branch of British commerce.[2] In reply the board maintained its customary attitude of conservatism, and expressed the opinion that no radical changes could be made without the consent of parliament. It proposed that such a bill be introduced into the current session, but suggested that some of the abuses might be corrected by direct exercise of the royal prerogative, since the garrison was already under the king's command and the Newfoundland Act was inoperative in the winter. Specifically the Board of Trade proposed that the commodore be commissioned as commander in chief of all military and naval forces, and that he also be granted civil powers with authority to create a local magistracy to function during the nine months when the fishing ships were not there. Furthermore, the board also proposed that the titles to fishing grounds and real property at Placentia should be cleared. Conservative to the last, however, the board again proposed the removal of the Newfoundlanders to Nova Scotia.[3]

The cabinet considered the future government in March and April, and reached a final decision in May, 1729. Besides the Board of Trade, Beauclerk and Phillipps were also consulted. Lord Vere was chosen as first royal governor, but was disqualified because of his seat in parliament. Eventually it was decided that he should re-

[2] The Duke of Newcastle to the Board of Trade, Oct. 18, 1728, C.O. 194: 8.
[3] Board of Trade report to the king, Dec. 20, 1728, C.O. 195: 7, pp. 158-178.

CIVIL GOVERNMENT, 1729-1763

turn to Newfoundland as commodore, but that the governorship should be conferred upon someone "skilled in the laws," who was to be an annual appointee residing only during the fishing season. The new government was to be very rudimentary in form, and the powers conferred upon the governor were, therefore, few and simple. Placentia was removed from the jurisdiction of the governor of Nova Scotia and placed under the control of the new governor. The Board of Trade inserted very full directions in the draft instructions for the enforcement of the acts of trade, but the cabinet struck these out, being unwilling to expect too much from the new governor until a vice-admiralty court could be established. The new form of administration was approved by order in council, May 22, 1729, and Captain Henry Osborn of H.M.S. *Squirrel* was duly appointed to the governorship.[4] These simple changes were the foundation of civil government.

Besides being commander in chief of the military forces, Osborn's commission authorized him to perform a number of civil functions. He was directed to administer the oaths of supremacy and allegiance, and empowered to appoint justices of the peace and other necessary officers. He was also to order the magistrates to hold general quarter sessions in accordance with British usage, but he was carefully enjoined not to violate the Act of 1699 and to avoid conflicts in jurisdiction with the fishing admirals and the commodore. Furthermore, he was also empowered to establish court houses and erect prisons.[5]

---

[4] Report of the committee of council on the Board of Trade report, Apr. 19, 1729, Additional MSS., 33,028, fols. 192-194. For the consideration and reconsideration of both the report of the board and that of the committee see *Acts, Priv. Coun., Col.*, 1720-1745, pp. 215-221. The recommendations were made effective by order in council, May 22, 1729, *ibid.*, 1720-1745, pp. 221-222.

[5] Commission to Henry Osborn, May 14, 1729, appointing him governor and commander in chief of Newfoundland and Placentia, C.O. 195: 7, pp.

These directions show that the crown was not bent upon erecting an elaborate governmental machine for Newfoundland, and that it was anxious, not only to avoid adverse criticism in the West Country, but also to prevent a clash with parliament over the question of conflicting prerogatives. There is no indication that the new administration was to be the first step in a process of introducing local self-government. On the contrary, action was delayed until anarchy virtually existed. The only type of government possible was an autocratic one, with no checks on gubernatorial authority nearer than Whitehall.

During the first two years of the new government, administrative authority was divided. Lord Vere Beauclerk acted as commodore of the convoy and administered the fishery regulations as prescribed by the Act of 1699, while Governor Henry Osborn, his naval subordinate, performed the civil functions. The duality resulted from Beauclerk's membership in parliament, but it served a practical purpose in making the change in government appear less radical.[6] Osborn's instructions contained am-

183-192. This commission was the basis for those issued to the succeeding governors down to 1750, when changes relating to criminal matters were introduced in the commission to Francis William Drake. Drake's commission in turn became the standard form for his successors until 1763, when the form was altered owing to the inclusion of Labrador, Anticosti, and the Magdalen Islands in the government of Newfoundland. Throughout the period from 1729 to 1763, one is struck with the mechanical uniformity of the royal orders to the governors. A comparison of the available commissions and instructions, clause by clause, indicates very little variation over considerable periods of time.

[6] Instructions to Beauclerk, June 3, 1729, C.O. 195: 7, pp. 205-234; instructions to Gov. Osborn, same date, C.O. 195: 7, pp. 193-204; order in council approving the instructions, May 22, 1729, *Acts, Priv. Coun., Col.*, 1720-1745, §168. Hereafter the heads of inquiry were considered as royal instructions, and were submitted to the king in council, and subsequently issued under the signet and sign manual, instead of going directly from the Board of Trade to the Admiralty. After 1730, those issued to the commodore were merged with the royal instructions issued to the governor. Contrary to

CIVIL GOVERNMENT, 1729-1763 277

plified directions as to his procedure in the administration of justice, the enforcement of the Navigation Act of 1663, and the survey of the coasts. Beauclerk was the real governor, for Osborn's instructions contained a clause directing him to follow the written orders of Lord Vere, provided the commodore's proposals were agreeable to law.[7] It is significant that it was the commodore who took out eleven sets of Shaw's *Practical Justice of the Peace,* thirteen copies of the Newfoundland Act, and a bundle of acts relating to trade and navigation for distribution to the new local magistrates.[8] Although Osborn carried out the formal orders in accordance with his commission and instructions, Lord Vere Beauclerk deserves the real credit as the founder of civil government in Newfoundland.

During the summer of 1729 Osborn was occupied principally in straightening out the difficulties involving the garrison at Placentia and in establishing the judicial system. He visited many of the outports as well as St. John's and divided the island into six districts wherein he appointed justices of the peace and constables. He was scarcely a governor "skilled in the laws," for he was even doubtful whether his commissions to the magistrates had been issued in proper legal form. He was not sanguine of the success of the experiment in local government, for he knew the past history of Newfoundland, and

the usual practice in the case of colonial governors, the charges incident to the passage of Osborn's commission and instructions through the seals were not paid by the appointee but were defrayed from the Treasury, *Cal. Treas. Books and Paps.,* 1729-1730, p. 85. This practice was continued throughout the period down to 1763.

[7] Osborn's instructions, June 3, 1729, C.O. 195: 7, pp. 193-204.

[8] Instructions to Lord Vere Beauclerk, commander of the convoy, June 3, 1729, C.O. 195: 7, pp. 205-234. See especially clauses 47, 48, 49, and the note following clause 50. Osborn's instructions, same date, C.O. 195: 7, pp. 193-204. See especially marginal note 2, Burchett to Popple, Apr. 3, 24, 1730, C.O. 194: 8.

had little confidence in the abilities of his appointees whom he felt were lacking in self-confidence and legal experience.[9] The success of the venture was also threatened almost immediately by a clash with the western adventurers. When Osborn, supported by Beauclerk, sought to levy a tax in the judicial district of St. John's to provide money to build a jail, his tax collectors made the mistake of trying to obtain contributions from the English fishermen. Both Beauclerk and the governor were doubtful of the legality of the tax, but had no other funds at their disposal for such purposes as the home government had provided none. In spite of the fact that the tax was levied in violation of the Act of 1699, they allowed it, and then appealed to the home authorities for a definition upon their return to England.[10] The whole incident was unfortunate as it aroused the ire of the West Countrymen just at a time when organized opposition was most inconvenient.

Both Beauclerk and Osborn desired a more clear definition of the governor's powers. Specifically they wanted to know whether those who refused to contribute to the jail fund could be punished; whether the justices of the peace could decide questions of property during the absence of the fishing admirals; whether the magistrates were restricted in their functions to the maintenance of peace and order; and whether the governor was empowered to levy any tax proposed by the magistrates for the construction or repair of churches and other public works.

[9] Osborn to Newcastle, Oct. 14, 1729, C.O. 194: 8, O 51. Included in this letter is a copy of his commission appointing justices of the peace. A table of judicial districts is in C.O. 194: 8, O 92.

[10] Warrant to the justices of the peace at St. John's, Aug. 23, 1729; copy of a presentment from these magistrates to Osborn, Aug. 25, 1729; Osborn to the justices of the peace, same date; his order to the inhabitants of St. John's, Oct. 5, 1729; his warrant to the tax collectors, Oct. 11, 1729, all inclosed in his letter to Newcastle, Oct. 14, 1729, C.O. 194: 8, O 51.

The Board of Trade considered these questions sufficiently important to consult lawyers, and submitted Osborn's questions to Philip Yorke, attorney general, and to Francis Fane, its own legal advisor.[11] Fane submitted his opinion first. He felt that Osborn's action had been inexpedient, but that the tax for the jail might stand because no legal complaint had been made against it. In future, however, money for such purposes should be raised in conformity with the powers granted to justices of the peace in England. Persons who impeded the magistrates and constables in the exercise of their duties could be indicted and punished by fine or imprisonment. In criminal matters the magistrates possessed the same powers as their prototypes in England, but they had no right to decide questions of property, and neither the governor nor the magistrates could raise money for repairing the church or for constructing public works. Yorke partly agreed and partly disagreed with Fane. He stated that the magistrates had no authority to raise money for the construction of the jail by levying a tax on fish or fishing boats, as this was contrary to the Act of 1699, but he agreed with Fane that the justices of the peace could punish those who held them in contempt. With respect to the right of the governor to levy taxes on his authority alone, Yorke could give no definite answer as he had seen neither Osborn's instructions, nor the magistrates' commissions, but he presumed "that no power is comprised in those instructions of imposing taxes in general without the consent of some assembly of the people." On the basis of these opinions the Board of Trade told Osborn that all his actions had been justified legally, but cautioned him to levy an assessment for the construction of a jail in ac-

[11] Queries from Beauclerk and Osborn [Apr. 30, 1730], C.O. 194: 8. These questions were submitted to Philip Yorke and Francis Fane by the Board of Trade in April and May, 1730, C.O. 195: 7, pp. 246, 247.

cordance with English law, that is, by making the levy in money, although he might accept fish in payment. The point raised by Yorke with respect to general taxation only with the consent of an assembly was an interesting one.[12] Unfortunately, the governors were never empowered to call such a body together, and the financial resources of the local government were, therefore, decidedly limited for many years. Assessments for specific public works projects were all that were legally permissible.

As might have been expected, the establishment of the new government was not effected without a struggle with the western adventurers. The West Country fishermen immediately became jealous of their favored position under the Act of 1699 and flouted the authority of the governor and of the local magistrates whom they considered contemptible. The West of England fishermen had defied Osborn's tax collectors at St. John's in 1729, and during the fishing season of 1730 there were a number of clashes between the local magistrates and the fishing admirals. At Placentia the latter assumed judicial powers and imprisoned one man in the fort, alleging that the justices of the peace were to concern themselves only with small matters "such as the church." The fishing admirals questioned the governor's authority to issue commissions, asserting that he derived his powers only from the Privy Council, while their powers were conferred upon them by parliament. The local magistrates complained of these arbitrary actions and resented the slurs of the fishermen who maintained that they were only "winter justices." In this last respect the fishing admirals were quite correct

[12] Francis Fane to the Board of Trade, Apr. 26, 1730, C.O. 194: 8; Philip Yorke, attorney general, to the same, Apr. 27, 1730, C.O. 194: 8; Chalmers, *Opinions of Eminent Lawyers*, II, 233-235; Board of Trade to Osborn, May 12, 1730, C.O. 195: 7, pp. 248-249; Board of Trade representation to the king, June 12, 1730, C.O. 195: 7, pp. 251-253.

CIVIL GOVERNMENT, 1729-1763

since it had originally been intended that the justices of the peace should function only during that season. However, owing to the utter ineptitude and indifference of most of the fishing admirals, and owing to the necessity of organizing the new government during the time when the governor was present, it had been found impossible to abide by the original intention of the home authorities, and the local magistrates had begun to exercise their authority during the fishing season. Difficulties similar to those at Placentia were encountered elsewhere because of this change in plans. The justices at St. John's were very uncertain of their authority, and asked the governor for a clearer definition of their powers. They understood that they had jurisdiction in all criminal matters, and claimed that the fishing admirals were limited to the adjudication of disputes relating specifically to the fishery. However, the fishing admirals had not only ordered the constables to put criminal sentences into execution, but had also required the magistrates to assist them in punishing offenders against the fishing rules, without bringing the culprits before the justices of the peace. The magistrates felt that this attitude on the part of the fishing admirals could lead only to confusion and uncertainty in the public mind as to the proper place to apply for justice, "since we are but too sensible the commanders of fishing ships are now and will be constantly endeavoring to oppose all authority but their own."[13] Inexperienced and lacking in self-confidence, and accustomed to being bullied by the fishing admirals, the new magistrates had a difficult time during the first few years in asserting their authority and

[13] Justices of the peace at Placentia to Osborn, May 16, 1730, inclosed in his letter to the Board of Trade, Sept. 8, 1730, C.O. 194: 9; copy of the petition of the justices of the peace of St. John's to Osborn, inclosed in his letter to Newcastle, Sept. 25, 1730, C.O. 194: 9.

282 BRITISH FISHERY AT NEWFOUNDLAND

relegating the fishing admirals to the position of administrators of the Act of 1699.

The administration of justice was further hampered by the absence of some of the magistrates in Europe, with the result that sometimes the judicial districts were left without the legal number of justices. To insure the continual presence of a quorum, Osborn removed the limitations originally placed upon their jurisdictions, and authorized any magistrate to act in any part of the island, so that those of one district might ask assistance from others. The ridicule heaped upon his appointees annoyed Governor Osborn greatly, and Lord Vere Beauclerk felt that no satisfactory government could be obtained until the question of the jurisdiction of the governor, justices of the peace, and fishing admirals had been definitely settled.[14] Although disappointed by the ineffectiveness of the measures which he had instituted to bring law and order to Newfoundland, the loyal manner in which Osborn supported his appointees unquestionably had a great deal to do with the permanent success of the entire venture in civil government.

The clashes between the fishing admirals and the justices of the peace came about through their concomitant jurisdiction during the fishing season. Originally, Beauclerk had recommended that they function only during the winter, but subsequently the question was raised as to whether their duties could be interrupted once they had been commissioned. After witnessing the difficulties which Osborn had encountered in 1730 in the matter of jurisdiction, Beauclerk recommended that the succeeding

[14] Copy of Osborn's order to the justices of the peace authorizing them to act anywhere in the island, Aug. 12, 1730, inclosed in his letter to the Board of Trade, Sept. 8, 1730, C.O. 194: 9. See also his letter to Newcastle, Sept. 25, 1730, C.O. 194: 9; and another to the Board of Trade, Nov. 24, 1730, C.O. 194: 24, fols. 48-50.

CIVIL GOVERNMENT, 1729-1763

governor should be given clear and positive instructions to define the respective spheres of action of the two groups of officials, and that the magistrates then holding office should be regarded as temporary appointees until the crown could devise a more effectual method of governing Newfoundland. Beauclerk's recommendations were received with attention at Whitehall, and the matter of defining the jurisdictions of the magistrates and fishing admirals was undertaken by the Board of Trade in the autumn of 1730, final action being deferred until Osborn's return. Meanwhile, the board proceeded to consult Yorke and Fane as to the respective powers of the officers in the fishery and settlement. The attorney general stated that the authority of the fishing admirals was confined to disputes concerning property in fishing rooms and other fishing conveniences in accordance with the Act of 1699, while the magistrates were to concern themselves with breaches of the peace and other criminal matters. Yorke did not consider the governor's power to create justices of the peace as contrary to the act of parliament, and asserted that there was no legal interference between the powers of the fishing admirals under the act and those conferred upon the magistrates by their commission from the governor. A similar, but more detailed opinion was presented by Fane.[15] With these two legal advisors supporting it, the crown became more sure of its ground, and an attempt was made in the summer of 1731 to settle the controversy once and for all.

Captain George Clinton, R.N., later to become royal

[15] Beauclerk to Popple, Sept. 26, 1730, C.O. 194: 9; Newcastle to the Board of Trade, Nov. 26, 1730, C.O. 194: 9; Board of Trade to Newcastle, same date, C.O. 195:7, pp. 253-254; report of Philip Yorke, attorney general, to the king, Dec. 29, 1730, Chalmers, *Opinions of Eminent Lawyers*, II, 235-238. The Board of Trade did not receive the opinions until three months later, Newcastle to the Board of Trade, Mar. 24, 1731, C.O. 194: 9; Francis Fane to the Board of Trade, Mar. 24, 1731, C.O. 194: 9.

governor of New York, was appointed commodore and governor of Newfoundland in 1731. He was the first of a long line of royal officials at Newfoundland to hold both the positions of commodore and governor. He was ordered to inquire into the conduct of the local magistrates and other civil officers, and although he went out under the usual heads of inquiry, these heads were incorporated for the first time into the royal instructions. Owing to a delay in passing Clinton's commission through the seals, the Admiralty decided to send him to Newfoundland equipped only with the customary heads of inquiry for the commodore, and Osborn who accompanied the convoy, continued to function as governor during the early part of the fishing season of 1731, until June 30, when Clinton took over the administration.[16] Thus, in spite of the intentions of the authorities at Whitehall, duality of function continued for part of a third year.

Clinton found that the ill-feeling between the magistrates and the fishing admirals was as strong as ever. The justices continued to complain of the disrespect shown them by the English shipmasters, while the fishing admirals persisted in claiming superior authority by virtue of the Act of 1699, some threatening to oppose any form of government which was not established by parliament. Osborn supported the magistrates during the early part of the season by issuing an order reproving the fishing admirals and other shipmasters for their "scandalous and seditious reflections upon His Majesty's authority and power," and called upon them to obey the law and uphold the justices of the peace in their execution of it

[16] Clinton's commission and instructions are to be found in series C.O. 5, 192, pp. 545-552, 559-597. Both were approved by orders in council, May 11, 1731, C.O. 194: 9. Slight changes were made in the instructions. Osborn continued as governor during the early summer of 1731. Osborn to Popple June 28, 1731, C.O. 194: 9; the same to Newcastle, same date, C.O. 194: 24 fol. 56.

CIVIL GOVERNMENT, 1729-1763 285

General disorder continued, offenders escaped punishment, and a woman and four children were murdered. Clinton had to reprimand one of the fishing admirals at Mosquito Cove for assuming powers not granted to him by the Act of 1699 and to publish an order denying the admiral's claims to authority, taking this opportunity to rebuke all the fishing admirals for assuming unwarranted powers. However, he did not feel oversanguine of his success in governing "this set of people."[17] Probably the principal reason for the arrogant attitude of the West Country fishing admirals and shipmasters was their appreciation of the growing power of parliament and the contemporary decline in the constitutional prestige of the Privy Council. Because of their residence in England the West Country fishermen were in a better position to understand the political changes which were taking place in Great Britain at this time than were the Newfoundlanders. The extent of the growth of parliamentary power was not generally comprehended in North America until after 1763.

There was, consequently, a constitutional point involved in the disputes between the justices of the peace and the fishing admirals, which made the solution of their difficulties a much more delicate matter than its actual importance seemed to imply. This is illustrated in the difficulty of administering the criminal provisions of the Newfoundland Act. The statute required that persons accused of capital crimes and felonies committed in New-

[17] Osborn to Popple, June 28, 1731, inclosing copy of a petition from the justices of the peace at St. John's, June 10, 1731, and copy of his orders to the fishing admirals, June 12, 1731; Clinton to Popple, July 29, 1731, inclosing a copy of the order of "Admiral" Joel Davis at Musketto (Mosquito Cove), May 31, 1731; copy of Clinton's order to the fishing admirals recalling power granted by Davis to John Goss for the apprehension of John Jones, C.O. 194: 9. See also Clinton to [Newcastle], same date, C.O. 194: 24, fol. 59.

foundland should be tried in England. Consequently, the magistrates, though empowered to examine and try offenders before a jury, were not permitted to give a judgment in such cases. Persons found guilty by a jury had to be taken to England for sentence, and the accused, witnesses, and trial record had to be sent home. This procedure entailed the loss of a great deal of time and money. Sometimes witnesses lost a year from business, and the cost of transporting the accused amounted to £50 or more in each case. As no government funds were available to pay these costs and as most of the accused had no property worth seizing, the local magistrates usually had to pay the expenses out of their own pockets or else apportion them among the residents of their judicial districts.[18] This dilemma was not easily surmounted, because for many years the crown either hesitated to ask parliament to alter the Act of 1699 or dared not defy it openly by instituting some other form of judicial procedure.

Osborn had lacked money for contingent expenses and Clinton found himself no better off. None was available for the transportation of felons to England nor were there sufficient funds collected to pay for the recently completed jail at St. John's. The local magistrates met the deficit out of their own pockets when some of the inhabitants refused to pay the assessment levied by Osborn in 1729. In order to recoup their losses the justices were obliged to place a tax on the wages of laborers employed by the planters, setting it at 3*d.* per pound. This time they were wise enough not to attempt to levy on the West Country fishermen, but even so they encountered resistance from both employers and workers who were supported in their defiance by the fishing admirals and other English shipmasters. The magistrates looked to Clinton

[18] Representation of the justices of the peace at St. John's to Clinton, Aug. 20, 1731, C.O. 194: 9.

CIVIL GOVERNMENT, 1729-1763   287

to support them in "so easy and reasonable tax," but he declined to approve of their action on the ground that the assessment was illegal, being contrary to the orders issued by Governor Osborn. Clinton maintained that the original tax would have been quite sufficient to cover the construction of the jail if it had been properly collected, and ordered the magistrates to procure the outstanding £50 of arrears from the delinquents. The governor's authority was sufficiently respected to insure final collection of the entire amount.[19] Thus, the justices of the peace, like their rivals the fishing admirals, would have exceeded their authority at every opportunity if they had not been held in check by the governors.

Clinton was discouraged at the progress made during his brief administration. At the close of the fishing season of 1731, he felt that it was hopeless to govern Newfoundland under the type of administration prescribed by the home government. He complained bitterly of the indifference and opposition of the fishing admirals, having experienced great difficulty in getting them to perform their legal duties. He found that the admirals had a considerable following among the "ignorant," whom they were able to persuade to disobey the governor's orders by emphasizing the rights granted by parliament to the western adventurers.[20] Whatever his later record as governor of New York, Clinton proved himself a fairly able and efficient administrator during the short time he was at Newfoundland, and the ultimate success of the new government was to a considerable extent due to his support of the local magistrates, and to his insistence upon respect

[19] Representation of the justices at St. John's, Aug. 20, 1731; Clinton's orders to the collectors for the prison, Aug. 27, 1731; his public notice relative to the tax, same date; his order to the justices of Conception Bay, same date; his order to Lieut. Richard Hughes, same date, all found in C.O. 194: 9.

[20] Clinton to [Newcastle], Oct. 1, 1731, C.O. 194: 9.

for the gubernatorial office. With Osborn and his immediate successor Falkingham, Clinton stands out as one of the more able of the early governors.

Upon receipt of Clinton's pessimistic report, Newcastle asked the Board of Trade to reconsider the question of the disputed powers of the magistrates and fishing admirals. Following its old time procedure, the board called upon the West Country fishing interests to submit their opinions. But it received only a few replies from the West of England, although these replies were couched in terms distinctly hostile to the summer activities of the justices of the peace, and were filled with statements supporting the claims of the fishing admirals to complete jurisdiction during the fishing season. The merchants of Poole set forth the ridiculous claim that the admirals had always managed the fishery without complaint, and alleged that the new magistrates had usurped the powers of the admirals. They also asserted that the magistrates, some of whom were New Englanders, "and men of little worth and less reputation" had acted arbitrarily toward the English fishing masters and their employees, holding the fishermen from the Channel Islands responsible for much of the disorder and maladministration that occurred. The western adventurers maintained that the powers granted to the fishing admirals by the Act of 1699 were paramount to those of the magistrates, and besought the British government to forbid the justices from exercising their powers during the fishing season, demanding that the commodore himself should assist in enforcing the statute. The investigation produced no definite results. The Board of Trade received scant response from the West Country, and the proceeding was so desultory that Newcastle had to ask the board for its report. The indifference of most of the West Country ports to the request for information is significant of the changes which had

CIVIL GOVERNMENT, 1729-1763 289

taken place in the industry. Only three memorials and one deposition were received from the many ports of the southwestern counties, and these were only obtained after considerable delay.[21]

Although pressed to report, the Board of Trade did not submit any reply until April, 1732. The representation then sent expressed the opinion that the local magistrates had not interfered with the fishing admirals, though both sets of officers had been guilty of irregularities. The board recommended that the next governor inquire into the veracity of the statements made by the West Country merchants, and that he be authorized to remove any magistrates found acting in an unwarranted manner; that the governor be directed to prohibit the magistrates from interfering in any way with the fishing admirals and make such careful observations of conditions in Newfoundland as to enable the board to offer further suggestions after the governor's return to England.[22] In view of the favoritism shown by the board to the western adventurers in past years, it is astonishing to find it questioning in any way the truth of the information received from the West Country, and speaks well for the independence of the board at a time when it has been charged with an indiffer-

[21] Newcastle to the Board of Trade, Nov. 23, 1731; petition of the mayor, etc., of Poole to the same, Feb. 1, 1732; deposition of Peter Shank of Poole, Feb. 11, 1732, petition of the mayor, etc., of Dartmouth to the Board of Trade [Feb. 8, 1732]; petition of the merchants, etc., of Bristol to the same [Mar. 28, 1732]; Newcastle to the same, Feb. 4, 1732, all found in C.O. 194: 9.

[22] When Falkingham's appointment was under consideration an effort was made to enlarge his instructions so that the rivalry of the fishing admirals and the justices of the peace might be cleared up. See the Board of Trade representations to the king, Apr. 6, May 5, 1732, C.O. 195: 7, pp. 263-266, 267-268. His commission was the same as those of his predecessors, May 18, 1732, C.O. 195: 7, pp. 269-277. His instructions contained the changes recommended by the Board of Trade, in article 65, C.O. 195: 7, pp. 263-266. Both commission and instructions were approved by order in council, May 18, 1732, C.O. 194: 9.

ence to its duties. As in this case the initiative was taken by the ministry, it may be that the board did not dare show too much partiality to the interests of the southwestern counties.

By 1732 the situation in Newfoundland had improved slightly. Governor Edward Falkingham found that the various judicial districts were generally well regulated by the magistrates, and that the fishing admirals were the only responsible persons who were remiss in their duties. No complaints were received by the governor of irregularities in the garrison at Placentia, and Falkingham's courts of inquiry revealed that most of the West Country complaints against the magistrates were unsubstantiated, only one case of misbehavior in office being found.[23] With their failure to prove that the magistrates had discriminated against the English fishing interests, the West Country merchants gave up their struggle to maintain their old privileges. After 1732 they made no serious attempt to regain control of the machinery established by the Act of 1699, and although they continued to fish off the island for many years, ceased to play any important part in its politics.

With the exception of Governor Fitzroy Henry Lee, who served well from 1735 through 1737, Governor Falkingham was the last of the early governors to show any ability. For the most part, the succeeding executives were careless and negligent in their duties. Robert McCarty, Lord Muskerry, who followed Falkingham in 1733 and 1734, was an extremely careless and superficial observer of conditions at Newfoundland. Even Lee, who was wide awake enough to appreciate the necessity of extending the governor's judicial powers, and who appreciated the

[23] Falkingham's answers to articles of instructions, articles 12, 27, 28, 65, inclosed in his letter to the Board of Trade, Oct. 4, 1732, C.O. 194: 9, O 147. See his letter to the same, July 30, 1732, C.O. 194: 24, fol. 63.

CIVIL GOVERNMENT, 1729-1763 291

need of a vice-admiralty court to cope with illegal trade, seems to have enjoyed a colorless administration. Captain Philip Vanbrugh, who went out in 1738, and Captain Henry Medley who followed in 1739 and 1740 do not appear to have been very conscientious in the discharge of their gubernatorial duties. Vanbrugh filled his answers to articles of instructions with a number of sweeping generalizations, which were copied almost *verbatim* by Medley in the following year. Sometimes the governors apparently did not bother to submit any reports at all to the home government. At least a search of the "Newfoundland Papers" in the Public Record Office does not reveal them, although there is a good deal of correspondence from earlier executives.[24] In general, the governors tended to regard their civil responsibilities as secondary to their naval duties. They received no additional salary for ruling the island, and their professional training inclined them to neglect or to slight that with which they were unfamiliar.

As before, the most pressing problem confronting the governors was the strengthening of the judicial system in order to control crime, disorder, and illegal trade. Governor Lee proposed its extension to permit the trial of capital crimes and felonies in Newfoundland, and suggested the establishment of a court of vice-admiralty to cope with illegal trade. Although steps were immediately

[24] For citations of the commissions and instructions of the Newfoundland governors, 1729-1763 see Charles M. Andrews, "List of Commissions, Instructions and Additional Instructions to the Royal Governors and others in America," Appendix C, *Annual Report* of the American Historical Association for the Year 1911, I, 393-528. See especially pp. 477-482. There is much less correspondence from the governors after 1732, and a number of the answers to articles of instructions are missing from the records. Even those which are available are brief and unsatisfactory. See Lord Muskerry's answers for 1733, 1734, C.O. 194: 9, O 142, O 155; Lee's for 1736, C.O. 194: 10, P 7; Medley's for 1739, C.O. 194: 10, P 33.

taken to erect a vice-admiralty court, difficulties of a legal nature were encountered when the crown sought to extend criminal jurisdiction. The Board of Trade, which was favorable to an extension of the governor's powers in this direction, received advice from the law officers of the crown that this could be done by exercise of the royal prerogative; but its recommendation to this effect was not acceptable to the higher officials of the British government, who felt that the fifteenth clause of the Newfoundland Act prevented the king from exercising his power contrary to a parliamentary measure. Here again we have evidence of the cautiousness of the crown in seeking to avoid any action which might involve it in a controversy with parliament. The governors of Newfoundland, therefore, were obliged to remain content with the limited judicial powers at their disposal, and repeated, as did Governor Thomas Smith in 1741, the old complaint of insufficient authority.[25]

One of the ablest administrators ever sent to Newfoundland during the first half of the eighteenth century was Captain John Byng, R.N., who was appointed in 1742. Whatever his future record as a naval officer, he was extremely conscientious in suppressing illegal trade. Byng found that by the time he became governor the local magistrates had learned to carry on their duties in a satisfactory manner, and that the people were submitting

[25] Lee's answers to articles of instructions, 1736, C.O. 194: 10, P 7; Board of Trade representations to the king, Apr. 10, 23, 1738, C.O. 195: 7, pp. 410-416, 417-427; Draft instructions for Vanbrugh, 1738, C.O. 195: 7, pp. 428-477, especially articles 67, 68. With the exception of an addition authorizing the governor to investigate the method of curing fish, the other recommended alterations, which were concerned with judicial administration, were disapproved in the order in council, May 25, 1738, C.O. 194: 10. Difficulties encountered in the administration of justice were reported by Gov. Thomas Smith in his answers to articles of instructions for 1741, C.O. 194: 11, P 51; and in his letter to the Board of Trade, Dec. 19, 1741, C.O. 194: 11.

CIVIL GOVERNMENT, 1729-1763   293

peaceably to their rule. He found too, that the justices had ceased to interfere in the fishery and were no longer accused of oppressing the fishermen. "Were it not [he reported] for three or four particular Trading Men at St. John's, who have opposed all Act of Government, there never would be any reason of Complaints." This disaffected group gave Byng some trouble.[26] His administration marks a bright spot in the development of civil government, because for some time thereafter the island suffered from neglect occasioned by the war with France and the consequent military and naval operations in northeastern America. Captain Thomas Smith, who succeeded Byng in 1743, was occupied chiefly with military matters pertaining to the defense of St. John's.[27]

The War of the Austrian Succession checked the attempt to place the civil government of Newfoundland on a secure foundation. In 1744 owing to the impending hostilities the British authorities proposed to omit commissioning the commodore as royal governor and to send him out with only the old heads of inquiry. Captain Charles Hardy, R.N., however, was commissioned governor, but he returned no answers to the articles of instruction and it is doubtful if he even visited the island. In the following year Captain Richard Edwards, R.N., was appointed governor, but again we find no record of his administration. By 1746 naval duties were so pressing that no governor

[26] Byng's commission and instructions were similar to those given to Medley and Smith in 1739 and 1741. Draft commission for Byng, Mar. 2, 1742, C.O. 195: 8, pp. 2-8; draft instructions, same date, C.O. 195: 8, pp. 9-43; approved by order in council, Mar. 16, 1742, *Acts, Priv. Coun., Col.*, 1720-1745, p. 818. See Byng to the Board of Trade, Feb. 22, 1743, C.O. 194: 11, P 60; his answers to articles of instructions for 1742, C.O. 194: 11, P 62.

[27] Smith was reappointed in 1743, his commission and instructions being in the usual form. Board of Trade representations to the king, Mar. 11, 15, 1743, C.O. 195: 8, pp. 47-48, 49; approved by order in council, Mar. 23, 1743, *Acts, Priv. Coun., Col.*, 1720-1745, p. 818.

was commissioned, the care of the fishery being entrusted to the commander of the British fleet at Cape Breton.[28] Instead of royal instructions, heads of inquiry were transmitted through the Admiralty, and Admiral Charles Watson detailed Captain James Douglas to investigate conditions. Acting in the capacity of commodore, Douglas visited the island and sent back fairly full answers to inquiries, but as he had no authority to supervise the civil administration he made no attempt to investigate it.[29] Neither commodore nor governor went out in 1747, and during the entire war period Newfoundland's civil government was very much neglected, although its military and naval defenses were kept in good condition. However, neglect of the civil administration had a demoralizing effect, which temporarily upset the stability attained before the war began.

As during the four fishing seasons from 1744 through 1747 there was no royal governor at Newfoundland the rudimentary civil establishment inevitably suffered from neglect. Vacancies among the justices of the peace due to death or removal from the island remained unfilled, and those magistrates who remained, unchecked by the presence of a representative of the crown, became arrogant

[28] Hardy was appointed governor in 1744. His commission and instructions were similar to Byng's. Board of Trade representations to the king, May 24, 29, 1744, C.O. 195: 8, pp. 51-53, 54. Edwards' commission was similar. Board of Trade representation to the king, Jan. 22, 1745, C.O. 195: 8, p. 56; approved by order in council, Feb. 7, 1745, C.O. 194: 12. The instructions were in the usual form except that article 62, directing the governor to enforce article XIII of the Treaty of Utrecht, was omitted because of the war with France. Draft instructions, Feb. 6, 1745, C.O. 195: 8, pp. 59-96. In the following year the government was entrusted to the commander in chief of the British naval forces at Cape Breton, and the actual supervision of Newfoundland was detailed to Capt. Douglas, who was not commissioned. Corbett to Hill, Mar. 1, 1746, C.O. 194: 12; Hill to Corbett, Mar. 14, 1746, C.O. 195, 8, p. 97; Douglas' answers to heads of inquiry, 1746, C.O. 194: 12, Q 18.

[29] Douglas' answers to inquiries for 1746, C.O. 194: 12, Q 18.

CIVIL GOVERNMENT, 1729-1763   295

and oppressive. Otho Hamilton, lieutenant governor of Placentia, reported in the autumn of 1747 that there was only one justice of the peace left at Placentia, and he refused to perform his duties. Hamilton acknowledged that he had been obliged to exercise civil powers himself in order to keep the peace, admitting that he had usurped authority, because he had no legal means of calling the local magistrates to account for dereliction of duty and no right to fill vacancies. He requested that he be granted civil powers, or at least be advised as to how he could preserve order at Placentia. His request was not granted, but his complaint caused the British government to discontinue the practice of entrusting the supervision of the fishery to the naval officer commanding at Cape Breton, which had been done in 1746 and 1747, and to resume the practice of sending out duly commissioned royal governors.[30] In spite of the four seasons of neglect, the local administration survived very well and a complete collapse, which might have been expected a few years before, did not occur.

From 1748 to the beginning of the Seven Years' War, the administration continued to improve slowly and to enlarge its functions steadily. Admiral Charles Watson, commander of the British fleet at Louisbourg, was commissioned governor in 1748. Several changes were made in his instructions, among them the introduction of a clause which required him to report upon the number of Irish Catholics in Newfoundland and to describe their activities. Disaffection in Ireland was rife during the years

[30] Lieut. Gov. Otho Hamilton to the Board of Trade, Oct. 31, 1747, C.O. 194:12; Board of Trade to [the Duke of Bedford, Secretary of State], Feb. 5, 1748, C.O. 195:8, p. 127; orders in council, Feb. 10, 23, 1748, *Acts, Priv. Coun., Col.*, 1745-1766, p. 55; Board of Trade to Bedford, Apr. 14, 1748, C.O. 195:8, pp. 128-130; the same to the committee of council, May 24, 1748, C.O. 195:8, pp. 179-181; the same to Hamilton, May 24, 1748, C.O. 195:8, pp. 181-182.

296 BRITISH FISHERY AT NEWFOUNDLAND

immediately following the Stuart Rebellion of 1745, and echoes of the troubles were heard in Newfoundland whither there had been a considerable immigration of Irishmen for some years. Another clause directed the governor to ascertain what justices of the peace were functioning, to make removals, and to fill vacancies.[31] No drastic official changes were made in the powers of the governor, but innovations were gradually effected in the method of handling the political and judicial problems of the community. Under Watson's rule the reconstruction of the weakened magisterial system was begun. Except for the years of the war there appears a marked increase after 1740 in local administrative acts.[32] At the same time the royal instructions of the governors became longer and more detailed, although the home government ceased to require replies to all clauses of the instructions, seeking answers rather to those sections only which directly concerned the Act of 1699.[33] The working of the administrative routine shows that the local government had at least become well established and was performing its functions in a confident manner.

Captain George Bridges Rodney, R.N., later one of Britain's great admirals, became governor of Newfoundland in 1749.[34] He was troubled with the difficulty of administering criminal justice, because of the limitations placed upon the judicial functions of the local govern-

[31] Watson's commission was in the usual form. Board of Trade representation to the king, Apr. 21, 1748, C.O. 195: 8, p. 133. Changes were made in articles 58, 63, 66 of the instructions. Draft instructions, Apr. 27, 1748, C.O. 195: 8, p. 133. The commission and instructions were approved by order in council, May 5, 1748, C.O. 194: 12. See his answers to articles of instructions, 1748, C.O. 194: 12, Q 27.

[32] Newfoundland Records, St. John's. A description of this material will be found in the Bibliographical Note, pp. 340-341.

[33] Watson's answers to articles of instructions, 1748, C.O. 194: 12, Q 27.

[34] Rodney's commission is similar to those of his predecessors. Board of

ment by the Act of 1699. His protests, unlike those of some of his predecessors, were heeded and the British authorities took immediate steps to remove the handicap before his successor, Captain Francis William Drake, took office in 1750. As previously, when this question had arisen, the government hesitated to risk infringing the Newfoundland Act, but on seeking the legal advice of Sir Dudley Ryder, the attorney general, it decided to act. Ryder held that the crown's right to erect courts of justice was not restricted by the Act of 1699. Consequently, in 1750 the governor was granted power to establish courts superior to those of the local magistrates for the trial of all offenses except treason. The crown had previously intended in 1738 to authorize the creation of higher courts, and Governor Vanbrugh's draft commission had contained a provision to this effect, even including jurisdiction over cases of treason, but fear of arousing the hostility of parliament probably prevented the government from acting at that time, and Vanbrugh went out without the desired powers. Now with the approval of Sir Dudley Ryder, Governor Drake was authorized by his commission to appoint judges and commissioners of oyer and terminer to hear and determine all criminal cases except treason, to continue the appointment of justices of the peace, and to name other officers necessary for the administration of justice. Furthermore, he was given the

Trade representation to the king, Apr. 14, 1749, C.O. 195: 8, p. 118. Only a slight change was made in his instructions, article 64 relating to the enforcement of article XIII of the Treaty of Utrecht being included because of the conclusion of the war with France in 1748. The only articles of the instructions which were submitted to the king in council were those from 63 to 69 inclusive. Board of Trade representation to the king, Apr. 20, 1749, C.O. 195: 8, pp. 187-188; draft instructions, same date, C.O. 195: 8, pp. 188-195. Both commission and instructions were approved by order in council, May 2, 1749, *Acts, Priv. Coun., Col.*, 1745-1766, p. 777. Rodney's original commission is in Admiralty 5: bundle 33.

298 BRITISH FISHERY AT NEWFOUNDLAND

right to pardon offenders and to remit offenses, fines, and forfeitures, except in cases of wilful murder. In the latter he might grant reprieve if the circumstances warranted. His instructions required that he report fully to the secretary of state and to the Board of Trade whenever the sentence involved the "loss of life or limb," and to submit full transcripts of the court records in order that capital sentences might be allowed or disallowed by the crown. It was necessary, therefore, for the governor to grant reprieves until a final decision had been made by the home authorities. The governor was also instructed not to permit the commissioners of oyer and terminer to hold more than one assize a year and then only when the governor was in residence. Thus, Governor Drake went to Newfoundland in 1750 armed with far greater power than any of his predecessors wherewith to lay the foundation of a more complete judicial system for the island.[35]

From one summer's experience Governor Drake found that the new powers were very effective in checking crime. The establishment of the higher criminal court proved generally satisfactory, although the governor discovered that the limitation placed upon his power to exe-

[35] Rodney to Bedford, Dec. 28, 1749, C.O. 194: 25, fol. 13; report of Sir Dudley Ryder, Mar. 27, 1750, C.O. 194: 12; Chalmers, *Opinions of Eminent Lawyers*, II, 240-241; draft commission for Philip Vanbrugh, Apr. 13, 1738, C.O. 195: 7, pp. 417-427; draft instructions for Vanbrugh, same date, C.O. 195: 7, pp. 428-477, especially article 67. Originally a commission and instructions similar to those given to Rodney were prepared for Drake, but a new commission granting the added judicial powers, and new instructions as to how to exercise them, were drawn up and submitted. Except for these changes the instructions were in the usual form. Board of Trade representations to the king, Jan. 29, Feb. 6, Apr. 6, 1750, C.O. 195: 8, pp. 196-197, 200, 208-209; draft commission, Apr. 6, 1750, C.O. 195: 8, 210-219; draft instructions, same date, C.O. 195: 8, pp. 220-263. See clause 63 of the instructions. Strangely enough his instructions retained the provision that the trial of persons accused of robbery, felony, and murder should take place in England, C.O. 195: 8, pp. 220-263, article 2. See also Board of Trade to Bedford, Apr. 2, 1750, C.O. 194: 12.

CIVIL GOVERNMENT, 1729-1763 299

cute for capital offenses and his obligation to reprieve the condemned awaiting the royal pleasure were detrimental to the efficiency of the judicial system. The condemned would have to wait from one fishing season to the next before knowing their fate, with the result that after the departure of the commodore their friends would be strongly tempted to attempt a jail delivery. There was no prison in the island strong enough to hold prisoners during the winter and such deliveries had taken place in the past and were certain to take place in the future. Therefore, Drake asked that the governor's powers be extended to permit the immediate execution of culprits whenever necessary. The Board of Trade referred this request to Sir Dudley Ryder, who saw no objection to the immediate execution of criminals, provided the governor's authority did not extend to cases of treason or to such as involved the officers of his own ships or the officers and men belonging to the merchant marine at Newfoundland. With these exceptions, the crown granted the governor authority to order the immediate execution of criminals in Newfoundland.[36] Drake used restraint in the exercise of this new power when he returned to the island in 1751. During his residence there two men were sentenced to death for serious crimes and several others were committed to prison at a court of oyer and terminer held at St. John's. Drake was doubtful of the actual guilt of the entire group of men and granted reprieves in order that their sentences might be reviewed in England. Later they were all pardoned, but some of them languished in prison until 1756, when

[36] Drake to the Board of Trade, Dec. 26, 1750, C.O. 194:12; Hill to Ryder, May 8, 1751, C.O. 195:8, pp. 263-265; Ryder to the Board of Trade, May 16, 1751, C.O. 194:12; Chalmers, *Opinions of Eminent Lawyers*, II, 241-243. A number of changes were made in Drake's instructions for 1751, relative to the enforcement of the Act of 1699. Draft instructions, May 17, 1751, C.O. 195:8, pp. 278-320; approved by order in council, June 4, 1751, *Acts, Priv. Coun., Col.*, 1745-1766, pp. 117-119.

Governor Richard Dorrill procured their release.[37] Once armed with the power to condemn to death and to execute immediately, Governor Drake was not as ready to use his authority, as one would have expected from his request to the home government in 1750, but probably when faced with an actual case, he was either too sympathetic or too unfamiliar with criminal law to take a definite stand.[38]

The Irish trouble which had been simmering for some years in Newfoundland, boiled over during the administrations of Bonfoy and Dorrill. After the murder of William Keen, justice of the peace at St. John's, which occurred at the hands of some Irishmen in 1754, a wave of intolerance and persecution spread over the island. Trouble had been brewing for a long time, but Keen's murder brought things to a climax. For many years there had been a steady flow of immigrants to Newfoundland from the southern counties of Ireland because of the promise of employment in the fishery. The presence of considerable numbers of Irish Catholics had been remarked as early as 1720 by Commodore Percy, many of

[37] Drake to the Board of Trade, Nov. 22, 1751, Mar. 24, 1753, C.O. 194: 23; John Pownall to Gov. Hugh Bonfoy, May 31, 1753, C.O. 195: 8, p. 325; Gov. Richard Dorrill to [the Board of Trade], Jan. 22, 1756, C.O. 194: 13; Board of Trade to Henry Fox, Secretary of State, 1756, C.O. 195: 8, pp. 336-337.

[38] The governors who succeeded Drake performed their duties in a conventional manner. Draft commission and instructions for Bonfoy were submitted by the Board of Trade, and approved by the king in council, May 21, 1753, *Acts, Priv. Coun., Col.*, 1745-1766, p. 778. Dorrill's draft commission was submitted by the Board of Trade to the Lords Justices, May 30, 1755, C.O. 195: 8, pp. 333-334; approved by the Lords Justices in council, June 3, 1755, *Acts, Priv. Coun., Col.*, 1745-1766, p. 778. His draft instructions were submitted later and were similar to those given Bonfoy in 1753. Draft instructions, June 5, 1755, C.O. 195: 8, p. 335; approved by order of the Lords Justices in council, June 10, 1755, *Acts, Priv. Coun., Col.*, 1745-1766, p. 778. Only one change had been made for many years, that directing the royal governors to correspond directly with the secretary of state in accordance with the order in council, Mar. 11, 1752, *ibid.*, 1745-1766, pp. 156-157.

CIVIL GOVERNMENT, 1729-1763   301

whom were brought out in fishing ships from Bristol, Bideford, and Barnstaple, which were accustomed to call at Irish ports on the outward voyage for provisions and labor. Without any intention of being facetious, the captains of these vessels employed Irishmen to complete the quota of green men required by the Act of 1699. Usually a good many remained in Newfoundland at the end of the fishing season, and by 1729 the number of those resident in the island had considerably increased. In addition to the Irishmen who came as members of fishing crews, others arrived as passengers to seek employment from the planters and byboatkeepers. Some of the shipmasters deplored the presence of so large a number of these people, but they continued to carry them because their competitors did likewise. The early commodores and the later governors, as well as some of the merchants and shipmasters, held the Irish responsible for a good deal of the disorder committed in the winter, considering them undesirable because of their religion and dangerous to British interests because of their political disaffection. By 1731 a majority of the male population of the island was Irish Catholic, and the Protestant minority feared and despised them. They were considered poor sailors and fishermen, and too indolent to work their return passage home. Generally they were considered as contributing little to the improvement of British shipping.[39]

Following "the '45" the potential danger from a large group of disaffected subjects in Newfoundland became a reality. From that time forward the British government paid more attention to their activities than had been the

[39] Bonfoy to the Board of Trade, Oct. 13, 1754, C.O. 194: 13; his order to the justices of the peace to erect gallows, Greenwich Hospital, Miscellanea, Various, 121: p. 9. Cf. Commodore Percy's answers to inquiries, 1720, C.O. 194: 7, fols. 3-10; Beauclerk to the Board of Trade, Oct. 7, 1728, C.O. 194: 9, O 45; his answers to inquiries, 1730, C.O. 194: 9, O 42; Clinton to [Newcastle], Oct. 1, 1731, C.O. 194: 9.

case in earlier years. In 1746, Commodore Douglas made mention of them in his report, and in 1748, Governor Watson was directed to ascertain the exact number of those among them who were capable of bearing arms, as they were said to outnumber the Protestants. Watson discovered that there was a considerable number in Newfoundland, but that they were not as numerous as circumstances made it appear. They outbalanced the Protestants in St. John's and the southern outports, but comprised only about a fifth of the population in the northern harbors, and the governor estimated that they formed about a quarter of the total population, although immigration from Ireland was proportionately larger than that from England. Both Watson and Rodney, who succeeded him, felt that the situation was alarming owing to the disaffection of the Irish and their unwillingness to take the oath of allegiance to the Hanoverian dynasty.[40] The nearby presence of the French at Cape Breton and on the northern and western shores of Newfoundland undoubtedly increased the anxiety of the British, whom past experience had taught that France might seek to profit by the presence of co-religionists so near at hand.

The murder of William Keen in 1754 caused the whole Irish Catholic community to suffer for the crimes of a few lawless men. Inspired by fanatical hatred, Governor Bonfoy issued orders restricting their activities, and popular feeling was so strongly against them that many fled from the island. The climax of hysterical feeling was reached during the administration of Governor Dorrill in 1755. He ordered the arrest of a Roman Catholic priest for say-

[40] Capt. Douglas' answers to inquiries, 1746, C.O. 194: 12, Q 18; Board of Trade to Bedford, Apr. 14, 1748, C.O. 195: 8, pp. 128-130; Board of Trade representation to the king, Apr. 27, 1746, C.O. 195: 8, pp. 135-137; Watson's answers to articles of instructions, 1748, C.O. 194: 12, Q 27; Rodney's answers, 1749, C.O. 194: 12, Q 43.

ing mass at Harbor Grace, and commanded all Roman Catholics to return to Ireland at the end of the fishing season, except those who had his special permission to remain. In addition to deportation, the governor ordered the local magistrates at St. John's to prohibit Irishmen or persons employing them to retail liquor. Violation of the order was to mean the destruction of houses built by Roman Catholics and the seizure of their lands.[41] Dorrill's act was dictated by the intolerance of the Protestant community, but its arbitrariness shows the danger of entrusting too much civil power to an officer of the navy. The home government remained silent, and it may be assumed that Whitehall did not disapprove of his measures.

The Irish lived miserably, their wages were seldom paid, and they were continually exploited by their employers. A large number were stranded in the island and in order to exist through the long winters they were obliged to work for enough to eat. The traders supplied them with large quantities of rum and charged them exorbitant prices for both food and drink, so that they were perpetually in debt and their families in a distressed condition. Governor Edwards, who succeeded Dorrill in 1757, felt that their presence was a menace to the trade and fishery and although he kept a watchful eye on them he made no effort to enforce Dorrill's orders.[42] From the fact that Edwards mentions a large number of Irishmen still living in Newfoundland, it is evident that Dorrill's order of deportation was not obeyed. During the remain-

[41] Michael Gill, magistrate at St. John's, to Bonfoy, Nov. 22, 1754, inclosed in Bonfoy to the Board of Trade, Feb. 21, 1755, C.O. 194: 23; Dorrill's order to George Garland, justice of the peace at Harbor Grace, Aug. 15, 1755; his order requiring all Irish Catholics to return to Ireland, Sept. 22, 1755; his order to the justices of the peace at St. John's, prohibiting Roman Catholics from retailing spirits, etc., Oct. 17, 1755, Greenwich Hospital, Miscellanea, Various, 121: pp. 10, 14.

[42] Edwards to the Board of Trade, Oct. 29, 1757, C.O. 194: 13.

der of the period of the war with France, the Irish question did not again become acute.

Captain Richard Edwards, R.N., who had been governor in 1745, was reappointed in 1757, and served through 1759. He was also governor again in 1779. Edwards was followed in 1760 by Captain James Webb, R.N., who was reappointed in 1761, but died before he returned to Newfoundland. His place was taken by Admiral Thomas Graves, who was the last of the governors appointed according to the practices which had been followed since 1729.[43] In 1763, the government of Newfoundland was reorganized, Labrador, Anticosti, and the Magdalen Islands being included in the jurisdiction of its governor. The close of the Seven Years' War introduced new factors, and the administration was altered to meet the new requirements. Thereafter, Newfoundland entered upon a new phase of its history which does not concern us, the institutions which had developed previous to 1763 underwent considerable changes after that date, and the social life of the inhabitants approached more closely that of the colonists on the neighboring continent.

During the years from 1729 to 1763, in spite of the very

[43] The commissions and instructions of the later governors such as Edwards, Webb, and Graves are in the usual form, except that the article of instructions relative to the enforcement of the Treaty of Utrecht was omitted on account of the war. Board of Trade representations to the king, Mar. 15, 22, 1757, C.O. 195: 8, pp. 338, 339; Edwards' commission and instructions were approved by orders in council, Mar. 26, 1757, *Acts, Priv. Coun., Col.*, 1745-1766, p. 778. Webb's commission and instructions were similar to Edwards', Board of Trade representation to the king, May 8, 1760, C.O. 195: 8, pp. 347-348; both approved by orders in council, May 13, 1760, *Acts, Priv. Coun., Col.*, 1745-1766, p. 778. Webb was reappointed in 1761, but died and Graves became governor. Draft commission for Graves, May 19, 1761, C.O. 195: 9, pp. 80-81; draft instructions, same date, C.O. 195: 9, pp. 95-148. The instructions omit directions for making inquiries regarding the French inhabitants at Placentia. Both commission and instructions are the same as those intended for Webb. Approved by orders in council, May 19, 26, 1761, *Acts, Priv. Coun., Col.*, 1745-1766, p. 778.

CIVIL GOVERNMENT, 1729-1763

considerable powers enjoyed by the governor, his subordinates, the lieutenant governors, never were permitted to participate actively in civil affairs. When civil government was first established in 1729, the commander of the fort and garrison at Placentia bore the courtesy title of lieutenant governor and for purposes of military administration was subordinate to the governor of Nova Scotia. Colonel John Moody had been the first incumbent at Placentia and he was succeeded by Samuel Gledhill and Otho Hamilton. In 1743 a garrison was established at St. John's under the command of Captain John Bradstreet, who was followed by Captain Christian Aldridge. All these military commanders received 10s. per day in addition to their perquisites as commanders of companies of foot.[44] Their status was comparable with that of similar officers in Nova Scotia before 1749, and the close connection between Placentia and Nova Scotia before 1729 set the style for Newfoundland.

Although these officers were the only representatives of the British government to be continuously resident in the island, they possessed no civil power. Indeed, the need for local government had arisen partly through the continual interference of Lieutenant Governor Gledhill with the fishery and trade at Placentia. Consequently, there was no desire on the part of the authorities to permit these officers to take an active part in any but strictly military affairs, although occasionally the governors ordered them to perform specified civil duties. Osborn, for example, commanded Gledhill to stop illegal trade at Placentia and to prevent the emigration of fishermen to New England, but this order was issued not for the purpose of conferring administrative power upon the lieutenant governor but rather to cause him to desist from participation in these activities. In 1749, Bradstreet at St. John's, was

[44] Clinton to the king, no date, State Paps., Dom., Military: bundle 48.

entrusted with the duty of proclaiming the peace between Great Britain and France as no governor had visited the island to perform that function since the end of the war.[45] Generally, the lieutenant governors were distrusted and disliked by both the governors and the local magistrates.

The reasons for the distrust and dislike of the military commanders were not only the customary jealousy between the army and the navy and between the civil and the military authorities, but also the ambitions of the lieutenant governors to participate in non-professional activities. Gledhill, like Moody before him, had made himself the dictator and magnate of Placentia. Otho Hamilton at Placentia was anxious to obtain civil authority, and Christian Aldridge at St. John's quarreled with the local magistrates. Governor Thomas Smith made the mistake of appointing some of the officers of the garrison at Placentia justices of the peace, but Governor Charles Watson later revoked these commissions and conferred the magisterial authority upon persons whom Hamilton considered partial to the Irish Catholics. The lieutenant governor was anxious to have the justices under his control because he was at the time engaged in a dispute with some Irishmen relative to land which he claimed belonged to his predecessor, Gledhill. Captain Aldridge at St. John's got into a dispute with the local magistrates regarding their respective disciplinary control over the soldiers. He complained that William Keen frequently put the soldiers in jail without notifying the commandant of his action. Governor Bonfoy was ordered to investigate the charges against Keen and to remove him from office if necessary, but the murder of the magistrate intervened to prevent the settlement of the dispute, and no ac-

[45] Osborn's instructions to Gledhill, Sept. 6, 1729, C.O. 194: 8; Clinton's instructions to the same, 1731, C.O. 194: 9; John Bradstreet to Bedford, June 10, 1749, C.O. 194: 25, fol. 12.

CIVIL GOVERNMENT, 1729-1763

tion was ever taken.[46] Civil authority was never conferred upon the lieutenant governors, and throughout the entire period from 1729 to 1763 they retained only their purely military functions.

Whenever the royal governors required the assistance of subordinates to execute their orders, they either relied upon the magistrates or else deputized officers of the navy to act in a civil capacity in the remoter outports. These deputies, or surrogates as they were usually called, were temporarily commissioned to administer oaths to the magistrates in such of the outports as the governor was unable to visit, and to undertake other administrative duties.[47] The jealousies of the military, naval, and civil officers prevented full coöperation among them and contributed not a little to the continuance of disorder and uncertainty even after the new government had been in operation for some years.

The development of government in Newfoundland had its origin in the supervisory powers of the commander of the naval convoy. In the seventeenth century the commodore had been sent out as a protector, supervisor, and investigator of the fishery and shipping. His authority over the settlement was only incidental. During the wars with Louis XIV, his powers were temporarily extended to make him commander in chief of the military forces, but until the passage of the Newfoundland Act of 1699, all his authority was derived from the crown through exercise of the royal prerogative. The Act of 1699 recog-

---

[46] Otho Hamilton to the Board of Trade, May 11, 1751, C.O. 194: 23; [Christian] Aldridge to Bonfoy, Dec. 5, 1753, C.O. 194: 23; Pownall to Bonfoy, June 5, 1754, C.O. 195: 8, pp. 327-328.

[47] Rodney to Capt. John Knight, Aug. 16, 1749, Newfoundland Records, St. John's, I, 21; the same to the same, same date, *ibid.*, I, 22; Rodney to Capt. Francis W. Drake, Aug. 20, 1749, *ibid.*, I, 33; Gov. Drake to Capt. Rodney, Aug. 3, 1751, *ibid.*, I, 217; the same to Capt. Knight, Aug. 13, 1751, *ibid.*, I, 229.

nized his functions and conferred upon him responsibility for the enforcement of the fishery regulations prescribed in that statute. Unfortunately, his powers were ill-defined and he had no authority to correct many of the more glaring abuses committed in the fishery and settlement. When it became increasingly evident that the Act of 1699 was ineffective, the commodore had to assume many of the duties which should have been performed by the fishing admirals. After 1729, the commanders of the convoy still continued to exercise their statutory duties, but they also, as governors, derived additional authority from the crown. Henceforward "the commodore-governor" had a dual position as administrator of both the fishery and the settlement, to which may be added a third function, that of commander in chief of the military forces, while at the same time he continued to perform his regular duties as an officer of the Royal Navy. Altogether, a great deal of authority was concentrated in the hands of one individual. Naval discipline tended naturally to make the governor arbitrary and impatient in his dealings with the Newfoundlanders, and the jealousy existing between the two armed services caused him to look upon the poorly paid, ill-clothed garrisons with scorn. At no time were any checks imposed upon his authority except such as were specifically included in his commission and instructions or in the Newfoundland Act. The only real curb on arbitrary government was the short period in each year when the governor was at Newfoundland. No matter how severe he might be, his subordinates were inclined to perform their duties less zealously once the convoy had put to sea. Moreover, there were many communities which neither the governor nor his surrogates found time to visit, and in those places the existence of a civil government must have meant little to the lawless elements in the population. The enthusiasm of the governor for bet-

CIVIL GOVERNMENT, 1729-1763

ter civil administration was tempered somewhat by the fact that he received no additional salary for performing these functions. In spite of the potentially autocratic form which the government took there is no evidence to show that the island suffered often from gubernatorial dictatorship. In fact, owing to the two wars with France in the middle of the eighteenth century, it was neglected a great deal of the time because of the more pressing naval duties imposed upon its governors. Many of the incumbents were careless and neglectful of their civil duties, and during the Seven Years' War they returned almost no reports to the British government relative to the condition of the fishery and settlement.

In no other British North American dominions did the power of the governor evolve in such a peculiar manner, for here he was unlimited in his authority, having no council or assembly to check him as was the case elsewhere. In view of the popular belief, that wherever British subjects settle, there the seeds of representative institutions are immediately sown, it is curious to notice that in the long history of the British occupation of Newfoundland there is not a single instance of a popular demand for participation in the government until long after the American Revolution. While the neighboring province of Nova Scotia was seething with indignation because of the long delay in calling an assembly during the administration of Governor Charles Lawrence, there was no corresponding demand in Newfoundland. The island was too isolated, its society too primitive, and its concentration upon the fishing industry too intense to produce the atmosphere conducive to the development of popular institutions. In the eighteenth century the government of Newfoundland was quite as anomalous as was its economic position in the British commercial and colonial system.

CHAPTER X

THE END OF AN ERA: RECAPITULATION

THE condition of the trade and fishery from 1729 to 1763 was generally prosperous, in spite of the interruptions caused by the two wars with France, because there were few poor fishing seasons. In 1728, Commodore Beauclerk reported that fish were much more plentiful than had been the case for fifteen years, and they continued to appear in abundance during the ensuing period. The bank fishery proved particularly good and contributed a great deal to the prosperity of the industry. The inshore fishery maintained a lead in production but continued to suffer from congestion in some of the old harbors. By 1731, it was so overdeveloped and overcrowded that Governor Clinton feared lest hard times might ruin some of the young fishermen who had built new stages and flakes. Although retaining the lead in the quantity of fish produced, the inshore fishery in the old fishing region was really less prosperous than it had been before, but this decline was offset by the gradual expansion of the Newfoundlanders into the harbors lying west of Cape Race and north of Cape Bonavista and by the rapid development of the bank fishery. The inhabitants and byboatkeepers produced the greater part of the fish caught inshore, while the English fishing ships tended more and more after 1729 to confine their operations to the banks.[1]

[1] Beauclerk to Burchett, Aug. 19, 1728, C.O. 194:8; Board of Trade representation to the king, Nov. 20, 1728, C.O. 195:7, pp. 158-174; Clinton

END OF AN ERA: RECAPITULATION 311

Statistics gathered during the eighteenth century relating to the fishing industry cannot be relied on, but some idea of conditions in the industry may be obtained by examining the surveys made by persons especially interested in Newfoundland. The governors were accustomed to submit statistical tables with their answers to articles of instruction, which purport to furnish the home authorities with accurate data as to the shipping, population, trade, and production of fish and train oil. But in many cases only a part of the fishery is covered and in others the figures are confused and the totals incorrectly determined. In 1740, Captain William Taverner, formerly surveyor of Newfoundland, submitted estimates to the crown covering the years 1736 to 1739 inclusive. According to his survey there were about 1,100 fishing ships and boats employed on the eastern and southern coasts, which produced about 335,000 quintals per year, while the British ships which fished on the banks caught about 45,000 quintals more. Taverner estimated the total annual British catch at about 380,000 quintals, besides which there was a yearly output of about 1,900 tuns of train oil. The total value of the codfish, at 10s. per quintal, was about £190,000; to which was added another £29,000 derived from train oil, and from seal and whale oil which were beginning to take places among the exports of Newfoundland. Taverner placed the total annual production at over £227,000, and estimated that the fishery and carrying trade employed each year about 8,000 people and about

to Popple, July 29, 1731, C.O. 194: 9. The annual schemes of the fishery covering the years 1729 to 1757 and accompanying the governors' answers to articles of instructions, contain a great amount of statistical information, which, unfortunately, is of doubtful accuracy. The schemes were often hastily compiled, and in most cases do not cover all the harbors fished by the British. They are useful as indices of trade conditions only in a very general way, and should be used with extreme caution. They are to be found in C.O. 194: 9-13, *passim.*

21,500 tons of shipping. Moreover, he considered that the annual importations of fishing equipment, clothing, provisions, sailcloth, cordage, ironwork, and other necessary fittings and supplies amounted to about £80,000.[2] A similar survey made by an unidentified person, covering the years 1738 to 1741 inclusive, placed the number of fishing boats at 1,310, of which 870 were to be found in the old inshore fishing region, 80 were employed north of Cape Bonavista, and there were 360 more west of Cape Race. Our anonymous statistician estimated the annual catch at 450,000 quintals, which at 10s. per quintal was worth £225,000. During these years the average annual production of train oil was 2,650 tuns, valued at £53,000. The industry employed about 8,000 people and 25,000 tons of shipping in fishing and carrying the product to market. At that time the Newfoundland fishery compared favorably with that of New England, which yielded only about 70,000 or 80,000 quintals in its combined summer and winter seasons.[3] In 1741 Governor Thomas Smith furnished William Pitt with an estimate of the gross value of the Newfoundland trade, placing the total production from all sources at £233,000. In 1743, Captain John Masters considered the total trade as worth £300,000 annually including freights. He stated that the industry employed 10,000 British subjects in all its branches and more than 26,000 tons of shipping in the carrying trade alone.[4] Al-

[2] Taverner's scheme of the fishery, 1736-1739 [Feb. 14, 1740], C.O. 194: 10. This is about as reliable as the governors' schemes mentioned in note 1.

[3] Account of the colony and fishery of Newfoundland and the present state thereof, 1744, Additional MSS., 13,972, fols. 1-7.

[4] Figures furnished by Capt. Smith to William Pitt, on a sheet with figures relative to French sugar production, no date, Chatham MSS., bundle 81; copy of the memorial of Capt. [John] Masters to the Admiralty, no date, transmitted by Christopher Kilby, agent for Massachusetts Bay, to the Board of Trade in his letter of Sept. 6, 1743, C.O. 194: 12.

END OF AN ERA: RECAPITULATION 313

though none of these estimates agree in detail, it will be seen that the Newfoundland trade was in a fairly good condition and constituted an important branch of the nation's commerce.

From 1743 until 1763 conditions were more uncertain because of the two wars with France. Owing to its proximity to the scene of naval operations the fishery on the southern coast suffered from neglect. During the period of the War of the Austrian Succession no man-of-war visited Placentia for five years. The nearest point at which ships fishing on that coast could receive convoy protection was at St. John's, but in order to reach that harbor it was necessary for the vessels to beat around Cape Race, a dangerous voyage late in the season, promising delays which sometimes resulted in their missing the convoy altogether. Faced with this handicap only three fishing ships appeared at Placentia in 1746. During the interval between the close of the War of the Austrian Succession and the outbreak of the Seven Years' War there was some improvement, so that by 1753 the fishery was again considered to be in a prosperous condition. In that year 1,675 ships and shallops were engaged in the fishery and carrying trade, with a total tonnage of nearly 57,000; while over 15,000 men were reported as being employed in the various branches of the industry. The value of the entire trade of Newfoundland, including fish, oil, skins, other produce, and freight revenues, averaged over £470,000 annually.[5] Not only was the codfishery in good condition by the beginning of the Seven Years' War, but it is worth noting that seal skins and seal and whale oil

---

[5] Capt. Joseph Gledhill to the Board of Trade, Oct. 9, 1746, C.O. 194: 12; petition of Frederick Calvert, Lord Baltimore, to the king in council, no date, referred to the Board of Trade by the committee of council, July 27, 1753, C.O. 194: 13. The petition was referred previously to the committee by the Privy Council, July 19, 1753, *Acts, Priv. Coun., Col.*, 1745-1766, p. 223.

were now becoming items of some importance in the export trade.

Owing to the incompleteness of the records for the period from 1729 to 1763 it is difficult to obtain accurate data with respect to the carrying trade. In general, however, fully rigged ships, snows, brigs, brigantines, schooners, sloops, and sometimes galleys, pinks, and ketches made market voyages. None of the merchantmen was very large, the biggest fully rigged ship having a burden of 350 tons, the average for all classes being in the neighborhood of seventy-five tons, and the smallest sloops and schooners being of only about twenty tons. The larger vessels were usually well armed, some with twenty or more guns, but the smaller carried only about six or seven cannon. About two-thirds of the tonnage was registered in the British Isles, but probably about eighty per cent of the vessels were built in the plantations, the remainder being about equally divided between British-built vessels and foreign ships made free by condemnation in prize courts. The West of England still provided much of the shipping, but the Newfoundland trade had become concentrated in a few ports such as Bristol, Dartmouth, Teignmouth, Topsham, and Poole. London continued to be well represented, but there were about as many Irish vessels as came from the Thames, while the Island of Jersey had more in the trade than either Bristol, the leading West Country port, or London. Boston was the chief colonial port in the carrying trade to southwestern Europe, although other New England towns, as well as New York and Philadelphia, also participated. A few vessels were built in Newfoundland, and probably about a tenth of the merchant fleet was owned there. St John's led in number, but there were carriers from such outports as Placentia, Trepassey, Ferryland, Bay of Bulls, Harbor Grace, Trinity and Twillingate. Most of the

END OF AN ERA: RECAPITULATION

British, Irish and Channel Island shipping was of greater burden than the Newfoundland and colonial, for the Old World operators relied upon fully rigged ships, snows, brigs, and brigantines to do most of their work, and employed relatively few schooners and sloops. Although the colonials still continued to use the two latter types to a considerable degree, they also used brigs and brigantines and a few snows and fully rigged ships.

By the middle of the eighteenth century the carrying trade had more or less fixed routes. The British vessels usually called at ports in the continental colonies first, then proceeded to Newfoundland for fish, and afterward set their course for southern Europe. The colonial ships usually called at Halifax or Canso, Nova Scotia, for fish before proceeding to Newfoundland and thence to market. Vessels which sailed for Bilbao and other northern ports in Spain such as Coruña usually terminated their voyage at Oporto or Lisbon in Portugal. Those which made for the Strait of Gibraltar sometimes called at Lisbon first, then at Cadiz or Vigo, and entered the Mediterranean to terminate their voyage at Alicante, Spain, or Leghorn, Italy. Others made direct voyages to northern Spain or Portugal and then returned to England. A few called at the Azores and Madeira *en route* or made the latter place their destination. From these foreign ports the British "sackers" returned home laden with southern European products, and the colonial vessels carried similar cargoes to the plantations. By the middle of the eighteenth century the Newfoundland trade had ceased to be an uncertain venture and had become a well established business.

The great duel with France between 1755 and 1763 naturally affected the fishing industry adversely. The West of England, as in the earlier wars, was particularly hard hit. In 1758 the western adventurers complained of the loss of their ships to the enemy, claiming that the convoy system was unsatisfactory, that the war had increased insurance rates, and that their sailors were constantly being impressed. They said that they were not only faced with a loss of trade, but that their financial resources also had been put to a severe strain in order to meet the extraordinary situation created by the war. The merchants of Bideford reported that they had suffered also because they had been obliged to make money advances to the families of seamen engaged for the Newfoundland voyage who were subsequently impressed. Bideford, which formerly fitted out twenty-five ships annually and employed about 1,000 men in the fishery, found that its trade had sunk "to the lowest ebb." Complaints of the same tenor were received also by the Board of Trade from Barnstaple, Poole, Exeter, Topsham, Teignmouth, and Dartmouth. Henry Case, mayor of Bristol, said that there never could be too many fishermen at Newfoundland, but that owing to impressment of their men and insufficient naval protection the Bristol merchants were seriously handicapped. Inadequate naval escort had resulted in the capture of several vessels by the French in 1757, and the imprisonment of their crews in France, where many men had died before they could be exchanged. The Bristol merchants, too, deplored the loss of a considerable amount of money through making advances to seamen for clothing to be worn on the Newfoundland voyage, only to have these men fall into the hands of the impressment officers. Others reported that impressment and capture had made labor for the fishery so scarce and so expensive that shipowners and byboatkeepers were faced with great financial burdens if the war continued.[6] These complaints are not unlike those

[6] Mayor of Bideford to the Board of Trade, Nov. 24, 1758, petition of the mayor, etc., of Barnstaple to the same, Dec. 20, 1758, petition of the mer-

END OF AN ERA: RECAPITULATION 317

presented to the Board of Trade during the War of the Spanish Succession, but the memorials and petitions of the period from 1755 to 1763 are less numerous and more uniform in their style and content than were those submitted during the earlier war. Evidently by this time there were fewer firms and fewer ports engaged in the trade than before, although the total amount of fish made by the West Country people remained about the same as in the old days.

While there was an increase in the productivity of the fishery during the period from 1729 to 1763, except when the wars occasioned interruptions, there is a good deal of evidence to show that the quality of Newfoundland fish had declined. British consuls and merchants residing in Spain, Portugal, and Italy complained for some years that the consumption of British-caught Newfoundland fish had fallen off considerably owing to improper curing, and might result in the eventual loss of the entire trade were not the methods of curing fish improved. The Board of Trade instructed Lord Vere Beauclerk in 1729 to warn the fishing admirals, shipmasters, inhabitants, and by-boatkeepers to pay particular attention to the curing process. Similar instructions were given to the royal governors in later years. But, very little attention was paid to these warnings.

The alleged reason for the decline in quality was the increasing use of bank fish which was considered inferior to that caught inshore. Decay was more rapid because the bank fish lay on board the fishing ship or shallop for a longer time than that which was caught near the coast

chants of Poole to the same, Dec. 26, 1758, petition of the mayor of Exeter and the merchants of Exeter, Topsham, and Teignmouth to the same [1758], petition of the merchants, etc., of Dartmouth to the same, Nov. 16, 1758, Henry Case, mayor of Bristol, to the same, Feb. 22, 1759, all found in C.O. 194: 14.

and brought ashore almost immediately. Moreover, no attempt was made to cull the fish properly with the result that a good deal of badly cured fish was sent to market. Governor Lee made a careful inquiry in 1736, and attributed the decline in quality not to carelessness in curing but to the competition of the sack ships for ladings. The masters of these carriers had developed great rivalry among themselves as to which one of them would be the first to arrive at his foreign destination. In their anxiety to sail ahead of their rivals, the masters frequently accepted fish before it was properly cured, with the result that it spoiled on the voyage. Although the trans-Atlantic races may have been good for seamanship, many merchants and shipowners lost heavily thereby. The quality of British-caught fish declined, though the price remained about the same, and the French were able to undersell their rivals in the Italian market by about a dollar a quintal. Lee's report led to the inclusion of an additional clause in the instructions to the governor, requesting him to offer suggestions for putting a stop to the rivalry between the sack ships. No very practical results were obtained; and most of the governors attributed the bad condition of the fish to warm weather or rains during the drying season, while others thought the decline in quality was due to the practice of the buyer culling the fish. Nothing was done to remedy the defects in the curing process, and at the close of the Seven Years' War the British trade to southern Europe was suffering considerably from poor quality and high prices.[7]

[7] Beauclerk's instructions, 1729, C.O. 195: 7, pp. 205-234. See especially, article 46. His answers to inquiries, 1729, C.O. 194: 8, O 63; the same for 1730, C.O. 194: 9, O 92; Falkingham's answers, 1732, C.O. 194: 9, O 147; McCarty's [Lord Muskerry's] answers, 1733, C.O. 194: 9, O 142; the same for 1734, C.O. 194: 9B, O 155; Lee's answers for 1736, C.O. 194: 10, P 7; Vanbrugh's instructions, 1738, C.O. 195: 7, pp. 428-477, article 68; his answers, 1738, C.O. 194: 10, P 24; Medley's answers, 1740, C.O. 194: 10, P 33;

END OF AN ERA: RECAPITULATION

The harmful effects of selling fish of poor quality at high prices may be seen in the reports received from British commercial representatives abroad during the years following the end of the war in 1763. The British factory at Lisbon wrote the Board of Trade in 1765 that the trade in dried codfish was not as prosperous as formerly. At one time 80,000 quintals had been brought into Lisbon from Newfoundland and New England, but by 1765 the quantity had fallen to about 60,000. The British merchants in Portugal attributed the decline partly to an increase in the import duty and partly to the high prices demanded by New England and Newfoundland for many years. Robert Forrester, deputy consul at Barcelona, stated that Newfoundland cod was suffering in the Spanish market from competition with Norwegian and Russian fish, and that for several years European fish had been preferred in Catalonia because of its larger size, though he did not consider Newfoundland inshore cod inferior. If the bank, Nova Scotia, Laurentian, and Labrador cod were cured as well as the inshore cod, he was sure they would bring a better price and be preferred to the European varieties.[8] Evidently, the small bank fish did not lend itself so well to curing as the larger inshore variety, and the process of salting and drying the fish had not kept pace with the expansion of the fishery into regions where the smaller variety abounded.

Although it is difficult to ascertain the true extent and worth of the trade and fishery because of inaccuracy and conflict in the statistics, nevertheless, it is evident that several important changes were taking place in the eco-

Smith's answers, 1741, C.O. 194: 11, P 51; Byng's answers, 1742, C.O. 194: 11, P 62; Watson's answers, 1748, C.O. 194: 12, Q 12.

[8] Memorial of the British factory at Lisbon, relative to British trade there, 1765, C.O. 388, 96; Robert Forrester, deputy consul at Barcelona, to the Board of Trade, Feb. 15, 1765, C.O. 388: 95.

nomic organization of Newfoundland. It is apparent that by the middle of the eighteenth century people were becoming interested to some extent in other activities than codfishing. Though the salmon fishery never became a rival of the older industry, it became of sufficient importance to employ a number of inhabitants in the regions north of Cape Bonavista and west of Cape Race. The same may be said of sealing, later a most important Newfoundland industry, which had its beginnings in the period from 1729 to 1763. Small shipments of seal oil were made during these years, but the exploitation of this branch of maritime enterprise did not become of real importance until after the Seven Years' War. Also some fur trading and trapping were carried on in the northern and western parts of the island, but the fur trade never became as important an item as codfishing, seal hunting, and salmon fishing.[9] The existence of these industries was significant of that broadening of life in Newfoundland which occurred in the eighteenth century.

The most important development during the period was the extension of the fishery of the northern regions to the coasts of Labrador. This northward movement was the direct result of the gradual expansion which had been taking place in Newfoundland ever since 1713. Between 1729 and 1763 the "Schemes of the Fishery" which each governor submitted with his annual report contained the names of an increasing number of harbors along the coasts of Newfoundland. The removal of the French from the south coast eventually resulted in immigration into the fishing harbors of that region, and although the French still continued to fish north of Cape Bonavista after 1713, nevertheless, the Newfoundlanders had set up

[9] The salmon fisheries and the fur trade are referred to regularly in the answers to articles of instructions, and figures relating to fur, salmon, and seal are given in the annual schemes of the fishery.

END OF AN ERA: RECAPITULATION   321

salmon fishing establishments there, and were gradually becoming familiar with the northern region. During the Seven Years' War fishing ships penetrated the northerly regions, and visited the coasts of Labrador north of the Strait of Belle Isle, reporting the existence of many good harbors and a plentiful supply of cod. Interest in Labrador was aroused, and the possibilities of commercial exploration and colonization there were pointed out to the British government by merchants in England and the American colonies, who sought to obtain exclusive privileges of trade. Nothing came of these proposals, partly because they were made before the end of the war with France, but principally because the Hudson's Bay Company claimed exclusive rights to the region.[10] The actual settlement and exploitation of Labrador occurred after 1763, but it is significant that interest in the region was first awakened by the Newfoundlanders.

During the early years of the new civil government the governors paid little attention to the suppression of illegal trade or to the prevention of emigration to New England. At first they were too busy with local questions and too preoccupied with the problem of conflicting jurisdiction between the justices of the peace and the fishing admirals to devote much energy to the enforcement of the acts of trade. Although required by their instructions to enforce the Navigation Act of 1663, they paid little attention to this obligation. Lord Vere Beauclerk, though not actually governor, seized a ship at Petty Harbor in 1729

[10] Henry Case, mayor of Bristol, to the Board of Trade, Feb. 22, 1759, C.O. 194:14; Claudius Amyand, secretary to the Lords Justices, to the same, May 11, 1752, C.O. 388:45; petition of John Thomlinson, John Hanbury, Samuel Touchett, and Henry Thrale inclosed with Amyand's letter, above; Board of Trade report to the Lords Justices, July 23, 1752, C.O. 5:6, fols. 38-58; memorial of the governor, etc., of the Hudson's Bay Company, July 20, 1752, C.O. 388:45; order in council, Dec. 20, 1752, *Acts, Priv. Coun., Col.*, 1745-1766, §189.

for bringing in French brandy, although he was unable to convict the master. He believed that illegal importations were seldom made from foreign countries, and he could find no evidence of any indirect trade in plantation goods through Newfoundland. Falkingham agreed with Beauclerk, and considered the market unsuited for the extensive consumption of European goods, and assumed rather naïvely that fear of the law deterred traders from engaging in smuggling. Needless to say, he obtained no convictions while governor in 1732. His successor, Robert McCarty, Lord Muskerry, also denied the existence of illegal trade in European goods. It is to be feared that none of these early governors were particularly observing. They were at Newfoundland for such a short time and were so busy with other things that they had no time to undertake any serious investigation. Moreover, the new government lacked the very essential customs machinery for curbing illegal trade, having no naval office where vessels might be required to enter and clear, no representatives of the customs to levy duties upon imports and exports, and no vice-admiralty court in which cases could be tried when offenders were apprehended.[11]

After 1735 the government was well enough established for those in authority to pay some attention to illicit trade. Governor Lee was more concerned about it than any of his predecessors, recognizing the fact that the navigation acts had been scarcely observed at all. He knew that large quantities of foreign wines and brandies were imported, and in order to prevent this he ordered the officers of the navy under his command to exercise the utmost vigilance and to seize offending vessels and pro-

[11] Beauclerk's answers to inquiries, 1729, C.O. 194: 8, O 63; Falkingham's answers to articles of instruction, 1732, C.O. 194: 9, O 147; McCarty's answers for 1733, C.O. 194: 9, O 142; the same for 1734, C.O. 194: 9B, O 155.

END OF AN ERA: RECAPITULATION   323

hibited commodities when they could. But in spite of his anxiety to enforce the Act of 1663, he found it very difficult to obtain real proof of violations of the law, even though he was perfectly aware of what was going on and knew that European goods were regularly consumed. During the three years of his administration he was able to seize but two vessels, and in one case the ship was discharged by the vice-admiralty court in Boston. In spite of his efforts, Lee was never able to obtain any reliable information regarding the traffic in European goods from Newfoundland to the continental colonies. He was sufficiently wide awake to realize that smuggling was very extensive, and that it could only be stopped by erecting a vice-admiralty court at St. John's to confiscate unlawful goods and condemn them on the spot, as the shipmasters would be unwilling to take risks with a court so near at hand.[12] His investigation was the first conscientious effort any governor had made to ascertain the extent of illicit trade and to propose a practical method of stopping it. His recommendation bore fruit, for the home government authorized the erection of such a court and the appointment of a preventive officer representing the customs.

The character of the foreign imports during this period did not differ greatly from that of earlier times. Generally, wines and brandies formed a large part, but Governor Vanbrugh reported that ships from Jersey brought foreign-made fishing tackle to the northern outports. By 1760 the local government was making a concerted drive on illegal trade. Inspired by the St. John's merchants, Governor Webb ordered his officers to search and seize all vessels and houses in the outports suspected of containing contraband goods, and to bring the vessels and goods to St. John's for condemnation. Although the war with France offered an excuse for this action, the real reason

[12] Lee's answers to articles of instructions, 1736, C.O. 194: 10, P 7.

for the anxiety of the St. John's merchants was their desire to control the trade of the entire island.[13] The patriotic zeal of the St. John's business men is amusing in view of the notorious violations of the acts of trade committed in that port in times past. Altogether, smuggling was not curtailed very greatly. It was curbed at St. John's, where the courts were established and where the governor remained when at Newfoundland, but it flourished for many years in the outports.

During the period from 1729 to 1763 the usual complaints were made against the unfair and unethical trading methods of the Irish and New Englanders. These old offenders had been joined by a third group of miscreants, the Channel Islanders. All were looked upon with disfavor by the western adventurers who were now quite as much concerned in general trade as they were in fishing. The West Countrymen alleged that trade from outlying British possessions such as the Channel Islands, Ireland, and New England, was carried on in violation of the Newfoundland Act, which they asserted confined the trade and fishery to ships of England, Wales, and Berwick-on-Tweed. This complaint elicited no favorable response at Whitehall, the government always maintaining that all vessels of British registry belonged constructively to the realm of England.

Actually the plantation trade was comparatively small during the period from 1729 to 1763, although it probably comprised a considerable proportion of the total import trade. Its character after 1729 was similar to that of the preceding period, consisting of provisions, live stock, and

[13] Vanbrugh's answers to articles of instructions, 1738, C.O. 194: 10, P 24; order of Gov. James Webb to Lieut. Brown, June 29, 1760, Newfoundland Records, St. John's, III, 74. For the fishing and trading activities of the Channel Islanders, see H. W. Le Messurier, ''The Early Relations between Newfoundland and the Channel Islands,'' *American Geographical Review* (1916), I, 449-457.

END OF AN ERA: RECAPITULATION 325

enumerated commodities, most of which were consumed in the island. The plantation traffic was estimated as being worth from £10,000 to £16,000 annually. Traders from the colonies continued to sell for hard money or bills of exchange, or to barter their goods for refuse fish which they carried to the West Indies and Madeira.[14] The Newfoundlanders remained dependent upon the neighboring plantations for provisions, especially in times of war. There is little evidence, however, to show that the transshipment of plantation commodities to foreign countries was as frequent as had been the case in the early days of colonial trade, for New England had now developed her own carrying trade to the point where her ships were making direct and frequent voyages abroad. It was unnecessary for the colonial traders to use Newfoundland any longer as a base for the exchange of unlawful goods.

While there may not have been as much smuggling at Newfoundland as in times past, the home authorities and the governors continued to complain of the constant emigration of able fishermen and sailors to the continental colonies. The governors continued to follow the old practice, originated in the days of the commodores, of demanding bonds from colonial shipmasters not to take more men away than they had brought to Newfoundland, but in spite of this futile effort to curb it, the traffic went on as extensively as ever. As a matter of fact it was too well organized to be easily suppressed. Many of the more important merchants of the island were concerned in the business and took great care to see that the passenger traffic was carried on from the outports where the governor and his subordinates were not likely to interfere. The English shipmasters and byboatkeepers also con-

[14] Mayor of Poole to the Board of Trade, Feb. 12, 1729, C.O. 194: 8; also the answers to articles of instructions, 1729-1748, cited in note 7, C.O. 194: 8-12, *passim*.

tinued to do their part to encourage emigration by permitting their men to remain in Newfoundland at the end of the fishing season. Probably several hundred men a year deserted the Newfoundland fishery for the more congenial climate and living conditions of New England, and thus placed themselves out of reach of the press gangs.[15] This phase of the activities of the Newfoundlanders and their allies in New England, which had always been regarded with concern by the authorities, became of increasing importance during the wars with France.

Once the system of local magistrates was well established and functioning smoothly, the British authorities took steps to provide other forms of administrative machinery, particularly such offices as were necessary for the suppression of illegal trade. A beginning was made as a result of Governor Lee's investigation of smuggling. A court of vice-admiralty was erected in 1736 or 1737, and William Keen, senior, one of the prominent businessmen and a magistrate of St. John's, was appointed the first judge. The court does not appear to have been particularly active in the prosecution of cases until the advent of Governor Byng in 1742, probably because of the absence of officers responsible for the prosecution of smuggling cases.[16] The only representative of the commissioners of the customs was a preventive officer. In 1737, John Lewes was appointed to the post which had been vacant since Archibald Cumming's retirement or death some years be-

[15] The answers to articles of instructions, 1729-1748, C.O. 194: 8-12, *passim*.

[16] William Brown, deputy register of the Admiralty Court, to Thomas Corbett, secretary to the Admiralty, May 1, 1736, Admiralty 1: 3673; record of Byng's appearance in the vice-admiralty court at St. John's, Aug. 1, 1742, Adm. 1: 3878; other cases and a considerable amount of correspondence between William Keen, judge of the vice-admiralty court, and the Admiralty officials at home, 1742-1746, are found in Adm. 1: 3738, 3881. Cf. Wood c. Grove, Jan. 26, 1758, Admiralty Court, Instance and Prize Libels, &c., File 113, no. 194.

END OF AN ERA: RECAPITULATION 327

fore.[17] The preventive officer remained the sole representative of the customs service until a customs house was established at St. John's in 1767. Meanwhile some attempt was made to keep track of the entry and clearance of vessels. Governor Thomas Smith appointed the first naval officer and deputies in 1741. The identity of the first incumbent is unknown, but Governor Byng appointed William Keen, junior, son of the judge of the vice-admiralty court, to the office in 1742. No deputies are mentioned in Byng's report to the home government, but subsequently one was appointed for Bay of Bulls and vicinity. William Wigmore succeeded Keen in 1752, and was followed in turn by Nicholas Gill in 1760. At first the merchants and shipmasters resented the erection of a naval office, and many sea-captains refused to pay any attention to it, or to acknowledge the authority of the naval officer. Several merchants informed Governor Byng that they would never allow their shipmasters to report their cargoes.[18] In spite of the resentment in mercantile and shipping circles, the new officers—the preventive officer, naval officer, and judge of the vice-admiralty court—were effective in reducing the amount of illegal trade. Both governors Watson and Rodney reported that by 1748 and 1749 there was no prohibited trade carried on, although the war with France probably accounted for the decline.[19]

[17] John Lawes was deputed preventive officer, Aug. 30, 1737, *Cal. Treas. Books and Paps.*, 1735-1738, p. 336.

[18] Smith to the Board of Trade, Dec. 19, 1741, C.O. 194: 11; commission appointing Nathaniel Brooks naval officer of Bay of Bulls, etc., no date, Newfoundland Records, St. John's, I, 35; Drake's commission to William Wigmore, appointing him naval officer of Newfoundland, Aug. 5, 1752, *ibid.*, I, 325; Webb's commission to Nicholas Gill, July 3, 1760, *ibid.*, III, 76; order of Graves to Gill, Sept. 10, 1761, *ibid.*, 130; Byng to the Board of Trade, Feb. 22, 1743, C.O. 194: 11.

[19] Douglas' answers to heads of inquiry, 1746, C.O. 194: 12, Q 18; Watson's answers to articles of instructions, 1748, C.O. 194: 12, Q 27; Rodney's answers, 1749, C.O. 194: 12, Q 13.

At any rate smuggling appears to have been much less important in the closing years of this epoch than at any previous time in Newfoundland's history.

Strange as it may seem, the official records of Newfoundland from 1729 to 1763 contain very few references to French fishing activities at Newfoundland. In 1728, a group of renegade Frenchmen from Cape Breton was reported as settling in the vicinity of Port aux Basques on the southern coast in violation of the Treaty of Utrecht, and subsequently Captain Taverner sent word that the French were planning to establish a settlement on "a little island near the north end of Gaspey" which may have been St. Paul Island, the Magdalen Islands, or Anticosti in the Gulf of St. Lawrence. During times of peace the governors were directed by their instructions to enforce the provisions of the Treaty of Utrecht which applied to Newfoundland, but there is little evidence to show that they paid much attention to the performance of this part of their duties.[20] As a matter of fact, the removal of the French from Placentia and the southern coast had succeeded in making the British at Newfoundland less aware of the competition of their rivals and less afraid of the dangers which they faced from French military and naval aggression. That these dangers were still existent was manifested during the War of the Austrian Succession and the Seven Years' War. Finally convinced that the struggle with France in North America was to be

[20] Extract of a letter from France relating to the design of the French to settle a little island near the north end of Gaspey [Cape Breton Island?], in Newfoundland, and to establish a fishery there [Aug. 14, 1729], inclosed in Taverner to Popple, Nov. 20, 1729, C.O. 194: 8. For further information concerning French activities see Beauclerk's answers to inquiries, 1729, C.O. 194: 8; Taverner to the Board of Trade [Feb. 2, 1734], Feb. 12, 1734, C.O. 194: 9; Board of Trade to Newcastle, Apr. 24, 1734, C.O. 195: 7, pp. 340-342, the same to Lee, May 26, 1736, C.O. 195: 7, pp. 403-404. See also the royal instructions to the governors, *passim*.

END OF AN ERA: RECAPITULATION

a bloody one, the British government set about to improve the defenses of the island. During the Seven Years' War the French attacked and held St. John's for some time until dislodged by a joint military and naval attack under the command of General Amherst and Admiral Thomas Graves in 1762. During this war attempts were made to put the defenses in order. The fortifications at Placentia were rehabilitated, and those at St. John's were virtually reconstructed and the place garrisoned for the first time since the War of the Spanish Succession, except for a brief time in 1743 when Captain Thomas Smith landed a body of marines there to defend the port against an impending French attack.[21] For the most part, however, there is little reference to French fishing interests and an overwhelming amount of material relating to the defense of Newfoundland, for at last the British government considered the island worth keeping.

Newfoundland was worth holding because of the importance of its fishery as a training school for seamen. During the period from 1729 to 1763 the old economic theories regarding the fishing industry and its place in the nation's commercial system were reiterated over and over again, but owing to the wars with France there was a tendency to pay less attention to the place of the fishery

[21] There are many references to military affairs, particularly for the period from 1740 to 1763. The royal instructions contain much concerning the maintenance of the forts and garrisons. During most of the period a large part of the correspondence relates to the condition of the troops, the progress made in strengthening the forts at St. John's and Placentia, and the building of additional defenses at Ferryland, etc. The French attack of 1762, the surrender of the garrison at St. John's, and the expedition from New York and Louisbourg to recover the island, receive considerable attention. There are abundant references to military matters in the following: C.O. 194: 11-15, 23-26, *passim;* C.O. 195: 8, 9; C.O. 323: 32; W.O. 1; the Loudoun Papers in the Huntington Library, LO 465, 510, 2133, 4028, 4200, 4459; the general manuscripts of that library also contain a few references to military affairs in Newfoundland, HM 2092 A & B, 23,352.

in the balance of trade and to emphasize increasingly its function as a contributor to the naval strength of Great Britain. Some of the contemporary pamphleteers and propagandists considered the Newfoundland trade as one of the most valuable commercial ventures in which the nation was engaged. John Ashley and George Heathcote, two of the most important writers of the time, estimated the returns which Great Britain received both directly and indirectly from the North American fisheries at about £260,000 per year, of which sum Newfoundland contributed about two-thirds. Both of these men devoted considerable space in their pamphlets to the trade and fishery and both emphasized the importance of Newfoundland as a training school for seamen and a contributor to the maintenance of a balance of power favorable to Great Britain. Heathcote provided elaborate tables to show the respective naval strength of Great Britain and France, and pointed out that it was French preponderance in the North American fisheries which gave them the advantage in the struggle for naval and colonial supremacy.[22] In general, however, the politico-economic writings of this period lack the vividness and forcefulness of those of the late seventeenth century and of the time of the War of the Spanish Succession. Indeed, in reading them one is impressed with their lack of originality and depth of thought. Mercantilist doctrine was already becoming con-

[22] Anonymous, *The Importance of the Sugar Colonies to Great Britain Stated* (London, 1731), pp. 21-22; anonymous, *Observation on the Case of the Northern Colonies* (London, 1731), p. 5; Henry Robinson, *England's Safety in Trade's Insurance* (London, 1741), p. 13; anonymous, *Some Considerations on the British Fisheries* (Dublin, 1750), p. 4; anonymous, *A Letter from a Merchant of the City of London, to the R[igh]t Hon[oura]ble W[illiam] P[itt] Esquire, upon the Affairs and Commerce of North America, and the West Indies* (London, 2d ed., 1757), pp. 23-24, 38-39; John Ashley, *Memoirs Concerning Trade* (London, 1740), I, 18-19; George Heathcote, *Letter on Trade* (London, 3d ed., 1762), pp. 40-42.

END OF AN ERA: RECAPITULATION 331

ventional and was on the way to being discarded altogether in the years following the close of the Seven Years' War.

The social condition of the Newfoundlanders was bettered somewhat by the establishment of civil government, although many of the old evils continued to beset them. A considerable group of influential merchants had grown up in St. John's. These men had accumulated a good deal of property and dominated the business of the entire island, but the condition of the poor was no better than before. Unquestionably, life and property were safer after the erection of a local government, but it is probable that a majority of the inhabitants derived little benefit, as they continued to be the victims of the merchants, tavern-keepers, and well-to-do planters, and spent much of their time in idleness and debauchery. The question of enforced indebtedness continued to plague the community and no efforts were made by the civil authorities to relieve the poverty-stricken fishermen and planters. There was some improvement, however, for the western adventurers made little trouble for the Newfoundlanders and in general the community was much more orderly than had been the case before 1729. The fact that the settlement weathered the difficulties of the two wars with France, 1743-1748, and 1755-1763, demonstrates that conditions were much better than had been the case during the earlier wars at the turn of the century. Indeed, after the stirring days of continuous quarrels between the western adventurers and the Newfoundlanders over the proper method of regulating the fishery, one feels that the period from 1729 to 1763 was somewhat of an anticlimax in the history of Newfoundland.[23] With the exception of the

[23] Answers to articles of instructions, 1729-1749, C.O. 194: 8, O 63; 9, O 92, O 142, O 147; 9B, O 155; 10, P 7, P 24, P 33; 11, P 51, P 62; 12, Q 12, Q 13, Q 18. These are filled with information relative to the condition

brief struggle of the West Country fishing admirals to maintain their authority during the early years of the new government, there is only one other incident during the period which is reminiscent of old times. That was the attempt of Lord Baltimore to reassert his claim to Avalon in 1753. The government, however, refused to consider his plea, and the only significance which the incident has, was the revival for a short time of the memory of the early struggles between the proprietors and the western adventurers. Feudal control of Newfoundland, like West Country dominance in the fishery, was finally a thing of the past.[24]

The fishery of Newfoundland occupies a unique position in the British commercial system of the seventeenth and eighteenth centuries, because it was never classed as a part of the plantation system which prevailed elsewhere in British North America and the West Indies. Although the trade in dried codfish was always considered a fairly important branch of the nation's commerce, it was looked upon as performing a supplementary function, benefiting Great Britain only indirectly. In this respect it resembled the African trade, regarded as a contributor to national prosperity, but in an accessory capacity only. The only branch of the plantation trade which was at all like it was the traffic in rice from the southern colonies, which though rice was never removed from the "enumerated" list, was granted a measure of relief when the discovery was made that the sole market for rice was in southern

of the inhabitants, but it is widely scattered and difficult to extract. After 1749 the governors did not submit answers to articles of instructions, and it becomes increasingly difficult to discover the social and economic condition of Newfoundland to the close of the Seven Years' War.

[24] Petition of Frederick Calvert, Lord Baltimore, to the king in council. Referred to the Board of Trade by the committee of council, July 27, 1753, C.O. 194:13. See also the reference to committee by the order in council, July 19, 1753, *Acts, Priv. Coun., Col.*, 1745-1766, p. 223.

END OF AN ERA: RECAPITULATION 333

Europe. The climate and soil of Newfoundland, the inaccessibility of the hinterland, and the paucity of important fur-bearing animals ruled out agriculture or activities along the lines followed by the Hudson's Bay Company. The abundance of cod determined from the first that the island's economic role should be as a base for the prosecution of the fishery. However, the absence of a market for this product in Great Britain, which only accepted train oil from Newfoundland, made the trade of the island a branch of the nation's *foreign* commerce. By pushing the sales of dried fish in the Roman Catholic countries of southwestern Europe, Great Britain was assisted in maintaining a favorable balance of trade with Spain, Portugal, and Italy. This function was recognized by both the merchants and the government, and most persons acquainted with the Newfoundland fishery acknowledged that it performed a very different duty from that of the plantations, with which it had little in common except a geographical location in North America.

The natural limitations of the island for settlement were understood and their significance appreciated at a very early date. Although we cannot admire the tactics employed by the West Country fishing interests and must acknowledge the narrowness of their commercial theories, nevertheless, we must admit that their conception of the trade and fishery was perfectly logical in view of the current ideas concerning the function performed by Newfoundland. Maintenance of the western adventurers became a cardinal principle of British policy, not because the authorities admired their methods more than those of others, but rather because they feared that the erection of colonial government and any encouragement to the Newfoundlanders would result in making the island a second New England. It must be remembered that the British government had always before its eyes the horrible example of New England and the other "bread" colonies of the north, which were included in the plantation system, but whose products were never suited to a regulatory system designed to promote Virginia tobacco and Barbados sugar. The desire to prevent such a situation in Newfoundland became a vital part of British commercial policy from 1675 onward. It explains the constant support given to the western adventurers, the persistent refusal of the crown to recognize the proprietary claimants, to establish civil government under royal auspices, or to apply the acts of trade at Newfoundland. The successive failures of the western charters of 1661 and 1676 and the Newfoundland Act of 1699, and the utter collapse of the fishery regulations in the period from 1713 to 1729, owing to the economic changes which took place, finally caused the government reluctantly to accept the inevitable and provide for some small measure of civil control in the period from 1729 to 1763. The delay of the crown to recognize the existence of the Newfoundlanders was not dictated by heartless indifference, but was based upon the firm conviction that the best interests of the nation could be served by maintaining the western adventurers in their privileged position.

If mercantilism implies a rigid control over the commerce of outlying possessions in the interests of both the mother country and the colony, it never found application in Newfoundland. Technically, a "free fishery" always existed there, wherein all British shipping was permitted to participate. The western adventurers, opponents of proprietary control, sought to convert the industry into an exclusive monopoly for themselves, but the competition of the byboatkeepers and planters prevented them from assuming exclusive privileges. In spite of the persistence of the Newfoundland settlers and their mercantilist allies in Great Britain, the government was never

END OF AN ERA: RECAPITULATION   335

willing to include the island in the commercial system which operated in other parts of British North America. The laxity of the Act of 1699, the unwillingness of the crown for many years to make any serious attempt to apply the acts of trade there in spite of the tremendous amount of illegal trade that existed, lead to the conclusion that the British government did not desire to apply mercantilist practices in Newfoundland.

During the years when Great Britain was building up her navy, the function of the fishery as a training school for seamen was considered as being quite as important as the trade. Many of the regulations, included either in the charters of 1661 and 1676 or in the Act of 1699, were intended to encourage the training of sailors. Opposition to schemes which promised to develop a local industry in Newfoundland was natural because favoring the West Country implied the continuance of a nursery for seamen which could be more readily supervised by the Royal Navy and its men made more accessible in time of war. During the years when Anglo-French rivalry became more acute, this feature received increasing emphasis. The correspondence of the officers charged with the supervision of the fishery, both at home and at Newfoundland, shows constant anxiety because of the ineffectiveness of the regulations in maintaining or improving this important naval interest. There were unfortunately no more effective legal means of enforcing the rules in this respect than in any other, and consequently the training of seamen labored under as many handicaps as did the purely commercial function of the fishery. In the field of national economy, Newfoundland played only an indirect part in contributing to the wealth of Great Britain, but in the field of international politics the training of seamen was vital to the nation, and in that particular the fishery participated directly in building and maintaining British naval power. The direct contribution of the trade and fishery of Newfoundland to the maintenance of the *balance of power* is of as much if not more significance than its direct assistance to the maintenance of the *balance of trade*. The history of Newfoundland, so different from that of the adjacent continental and West Indian colonies, cannot be fully explained for the period from 1660 to 1763 on any other basis.

Fundamentally, Newfoundland's insular situation and peculiar natural environment determined from the first the course of its economic and political development. During the seventeenth and eighteenth centuries Great Britain's "oldest colony" was never a colony at all judged by contemporary definition, but was always regarded as an overseas fishing industry to which were attached for better or worse a number of permanent settlements, the inhabitants of which proved a constant embarrassment to those who sought to maintain the West Country policy and to preserve the nursery for seamen. The history of Newfoundland to 1763, at least, must be approached from an entirely different standpoint, and appraised according to totally different standards from those that are used in interpreting the history of other parts of pre-Revolutionary British America.

BIBLIOGRAPHICAL NOTE

Materials for the history of Newfoundland are abundant, particularly for the period following 1660. Several of the large collections of original documents relating to colonial history contain papers bearing upon the fishery and settlement. A considerable amount of this source material is available either in transcript or in printed form. Direct reference to the originals or to trustworthy transcripts is advisable, as many of the calendared abstracts are too brief, poorly edited, or badly cited to be relied upon with confidence. The principal depositories of original documents, used by the author, are four in number: the collections of State and other official papers in the Public Record Office of Great Britain; the manuscript collections of the British Museum; the Newfoundland Colonial Records at St. John's; and the papers pertaining to English and American colonial history in the Henry E. Huntington Library and Art Gallery, San Marino, California. Transcripts of many of the important Newfoundland manuscripts for the seventeenth and early eighteenth centuries, the originals of which are either in the Public Record Office or the British Museum, were consulted in the Public Archives of Canada at Ottawa. Furthermore, a number of papers of significance pertaining to the period from 1712 to 1763, were transcribed or photostated from the originals in the Public Record Office for the author, and are now deposited in the General Library of New York University, University Heights, New York City.

GUIDES, LISTS, AND OTHER AIDS.

The path of the investigator has been smoothed by the publication of several guides to documentary materials, and lists of manuscripts relating to specific subjects. For the Public Record Office, the most valuable and trustworthy assistance was obtained

338 BRITISH FISHERY AT NEWFOUNDLAND

from Charles M. Andrews, *Guide to the Materials for American History, to 1783, in the Public Record Office of Great Britain*, 2 vols., Washington, 1912, 1914. Volume I, *State Papers*, was the more valuable for the author's purposes, although some Newfoundland papers were located by using Volume II, *Departmental and Miscellaneous Papers*. Location of the commissions and instructions of the royal governors of the island was facilitated by reference to Charles M. Andrews, "List of Commissions, Instructions, and Additional Instructions Issued to the Royal Governors and others in America," Appendix C, *Annual Report of the American Historical Association for the Year 1911*, Washington, 1913. The search for manuscripts in the British Museum was shortened considerably by reference to Charles M. Andrews and Frances G. Davenport, *Guide to the Manuscript Materials for the History of the United States to 1783, in the British Museum, in the Minor London Archives, and in the Libraries of Oxford and Cambridge*, Washington, 1908. The usefulness of this *Guide* has been somewhat impaired by the recent refoliation of many of the manuscripts in the British Museum. The Colonial Records of Newfoundland at St. John's are described in David W. Parker, *Guide to the Materials for United States History in Canadian Archives*, Washington, 1913.

MANUSCRIPT SOURCES IN THE PUBLIC RECORD OFFICE.

By far the most important collections of manuscripts relating to Newfoundland are in the Public Record Office in London. The most pertinent group is the series of Colonial Office Papers, which contains most of the essential material used in the preparation of this work. Chronologically, the first of these great collections is Colonial Office Papers, Class 1 (C.O. 1), extending in the main from 1622 to 1697, but with papers dating back to 1574. A great many of the more important papers have been transcribed by the Canadian Government and were consulted in the Dominion Archives at Ottawa. There are two extensive collections in this group, both of which relate exclusively to Newfoundland: namely, Colonial Office Papers, Class 194 (C.O. 194), and Class 195 (C.O. 195). Class 194 covers the period from 1696 to 1782 and comprises original correspondence with the Board of Trade and the secretary of state. Many of the documents submitted to the board are duplicates, *mutatis mutandis,* of those submitted to the secretary of state. Class 195 consists of a series of entry books covering the period from 1708 to 1769 and later. The early volumes contain out-letters from the Board of Trade and the secretary of state, as well as copies of some of the outgoing correspondence of earlier agencies. The most important documents in the series are the draft commissions and instructions of the royal governors, heads of inquiries for the commodores, and Board of Trade representations, as well as many inter-departmental letters. Some in-letters received by the board and by the secretary of state have been copied into the earlier volumes. Colonial Office Papers, Class 199 (C.O. 199), is concerned also with Newfoundland but most of the papers relate to the period following 1783. Some material is also to be found in other series of Colonial Office Papers, notably Class 5 (C.O. 5), Class 217 (C.O. 217), Classes 323 and 324 (C.O. 323 and C.O. 324), "Plantations General"; and Classes 388 and 390 (C.O. 388, C.O. 390), "Board of Trade, Commercial."

There are a number of documents relating to Newfoundland in the State Papers, Domestic, covering the reigns of Anne, George II, and George III. The State Papers, Domestic Entry Books, Regencies, 1716-1755; and Entry Books, Petitions, 1688-1760, contain some papers of interest. Some information relative to military and naval affairs is to be found in State Papers, Domestic, Military, and in State Papers, Domestic, Entry Books, Naval; while State Papers, Foreign, France, contain a number of valuable manuscripts relative to the Treaty of Utrecht and Anglo-French relations in the fishery.

There is a small but significant amount of Newfoundland material to be found in the Admiralty Papers. Class 1 (Adm. 1), contains a number of papers relative to the establishment and operation of vice-admiralty courts at St. John's; while similar material, as well as some concerning the prize court there is to be found in Class 5 (Adm. 5). Class 7 (Adm. 7), contains the Register of Passes, 1683-1783, and Register of Foreign Passes, 1729-

340 BRITISH FISHERY AT NEWFOUNDLAND

1784. Admiralty Papers, Greenwich Hospital Papers, Miscellanea, Various, vol. 121, is of great importance, but will be discussed below in connection with the Newfoundland Records at St. John's. There are a few references to Newfoundland in the papers of the High Court of Admiralty. The Prize and Libel Files contain decrees and sentences of the Instance Court, a few of which relate to Newfoundland or vessels in the Newfoundland trade, the most important being Calvert's libel against Kirke. Admiralty Court, Miscellanea, has papers relative to the abortive attempt to establish a vice-admiralty court at St. John's in 1708.

The export trade from Great Britain to Newfoundland may be traced by following through the Custom House Papers, Accounts, Ledgers of Imports and Exports, 1697-1780. Other papers of a statistical nature are scattered through the Chatham Papers.

MANUSCRIPT SOURCES IN THE BRITISH MUSEUM.

The Egerton Manuscripts in the British Museum comprise one of the most important sources for the history of Newfoundland to be found outside the Public Record Office. The Lansdowne, Stowe, Harleian, and Sloane Manuscripts also contain a few papers of value, while the Additional Manuscripts, including the Newcastle Papers, have a few significant documents relating to the fishery and settlement. Transcripts of many of these papers are available in the Canadian Archives.

THE COLONIAL RECORDS OF NEWFOUNDLAND.

There are thirty-five entry books covering the period from 1743 to 1830, deposited in the office of the Colonial Secretary in the Court House at St. John's, Newfoundland. The series originally comprised thirty-six volumes, but volume II is missing. Unquestionably, the missing book is volume 121 of the Admiralty Papers, Greenwich Hospital Papers, Miscellanea, Various, as it is similar to the entry-books at St. John's in subject matter, arrangement, and binding. Probably it was carried home by one of the early governors. Including the volume in the Public Record Office, this series contains copies of the governors' commissions, orders, warrants, and commissions issued by the governors to

BIBLIOGRAPHICAL NOTE

their deputies or surrogates, justices of the peace, naval officer and deputies, the sheriff, and military officers. The records of the courts, both civil and criminal, are also found there. Besides material relating to internal affairs, there are entries of correspondence with the governors of Nova Scotia and Massachusetts Bay, and with the French officials at St. Pierre and Miquelon after 1763.

HUNTINGTON LIBRARY MANUSCRIPTS.

There are a number of papers relating to Newfoundland in the seventeenth and eighteenth centuries scattered through the manuscript collections of the Henry E. Huntington Library and Art Gallery, San Marino, California. The papers of William Blathwayt in the Sir Thomas Phillips Collection of American and Colonial Papers contain a number of manuscripts relating to Newfoundland. The manuscripts of the Earl of Bridgewater in the Ellesmere Papers contain numerous notes in the earl's own hand taken down at meetings of the Board of Trade, some of which relate to Newfoundland. The manuscripts of the Earl of Loudoun contain a few letters and papers pertaining to the military situation in Newfoundland during the Seven Years' War. The general collection, styled "Huntington Manuscripts" contains a number of papers bearing either directly or indirectly upon the history of the settlement and fishery.

CALENDARS OF PAPERS IN THE PUBLIC RECORD OFFICE.

The most important series of printed calendars and abstracts of documents for the history of Newfoundland from 1574-1721, is the *Calendar of State Papers, Colonial Series, America and the West Indies,* 1574-1721, 26 vols., London, 1860-1933. This series contains many abstracts, some, all too brief, others, particularly in the more recent volumes, very full. Reference to the original manuscripts in the Colonial Office Papers, Classes 1, 194, 195, etc., or to transcripts has been made whenever possible throughout this work, and citations from the *Calendar* omitted, except when the printed abstract was the only source. Many of the orders in council, committee reports, and representations of the

342   BRITISH FISHERY AT NEWFOUNDLAND

Board of Trade and the earlier agencies are printed in whole or very fully summarized in the *Acts of the Privy Council of England, Colonial Series*, 6 vols., Hereford, London, 1908-1912. Some supplementary material for the sixteenth and early seventeenth centuries was found in the *Calendar of State Papers, Domestic Series*, 1547-1683, 1689-1698, London, 1858-1933; *The Calendar of Treasury Books*, 1660-1700, 14 vols., London, 1904-1933, is useful as is *The Calendar of Treasury Papers*, 1557-1730, 6 vols., London, 1868-1889. For the eighteenth century the *Calendar of Treasury Books and Papers*, 1729-1745, 6 vols., London, 1889-1903, is of great assistance. Some assistance was obtained from the *Journal of the Commissioners for Trade and Plantations*, April 1704 to December 1758, 9 vols., London, 1930-1933. This series is commonly cited as the *Board of Trade Journal*.

OTHER PRINTED OFFICIAL SOURCES.

There are a number of calendars containing important abstracts concerning Newfoundland which have been published from time to time by the Historical Manuscripts Commission. Among them are the *Calendar of the Manuscripts of the Marquis of Salisbury*, 15 vols., London, 1883-1930; the *Calendar of Manuscripts of the Duke of Northumberland*, Appendix to the *Third Report*, London, 1872; the *Manuscripts of the Corporation of Plymouth*, and the *Manuscripts of the Corporation of the Borough of Barnstaple*, Appendices to the *Ninth Report*, London, 1872; and the *Cowper (Coke) Manuscripts*, Appendix to the *Twelfth Report*, London, 1888-1889. Papers in the archives of the House of Lords have been published in several series. *The Calendar of Manuscripts of the House of Lords*, Appendix to the *Third Report*, London, 1872, 1873; *The Manuscripts of the House of Lords, 1690-1691*, Appendix, Part V of the *Thirteenth Report*, London, 1892; *The Manuscripts of House of Lords, 1692-1695*, Appendix to the *Fourteenth Report* of the Historical Manuscripts Commission, London, 1894. These are all useful but supplemental to material found elsewhere. A continuation of the above series, but printed separately, is the *Calendar of the House of Lords Manuscripts*, 8 vols., London, 1900-1923, usually called

BIBLIOGRAPHICAL NOTE 343

the "New Series." Much of the Newfoundland material contained in these volumes is of but incidental value.

TREATIES AND DIPLOMATIC PAPERS.

Texts of various treaties affecting Anglo-French relations at Newfoundland are found in George Chalmers, *A Collection of Treaties between Great Britain and other Powers*, 2 vols., London, 1790. The texts of the Treaty of Utrecht of 1713, the unsuccessful commercial treaty of the same year, the Treaty of Aix-la-Chapelle of 1748, and the Treaty of Paris of 1762 are printed in full. Texts of earlier treaties relating to Newfoundland are to be found in Frances G. Davenport, *European Treaties bearing on the History of the United States and its Dependencies*, 2 vols., Washington, 1917, 1929. Volume I contains the treaties down to 1648, and Volume II those entered into between 1650 and 1697, ending with the Treaty of Ryswyk.

A number of important documents relating to the negotiations of France previous to the conclusion of peace in 1713 are found in L. G. Wickham Legg, *British Diplomatic Instructions, 1689-1789*, Volume II, *France, 1689-1721*, edited by James F. Chance for the Royal Historical Society, London, 1925.

STATUTES, PROCLAMATIONS, AND LEGAL OPINIONS.

Printed texts of the laws relating to Newfoundland and to trade and navigation in general are available in two great collections of British statutes. The first of these is the *Statutes of the Realm* [*containing Charters of Liberties A.D. 1101-1301, and the Statutes from Magna Charta to the End of the Reign of Queen Anne*], 9 vols., London, 1810-1822. Another series is the *Statutes at Large*, 4 vols., London, 1763. The author has used the former series down to 1698, and the latter from 1699 onward. Printed texts of the laws and administrative acts of the Commonwealth and Protectorate are available in C. H. Firth and R. S. Rait, *Acts and Ordinances of the Interregnum, 1642-1660*, 3 vols., London, 1911. This contains some valuable references to the fisheries in general and to Newfoundland in particular. Only two royal proclamations relating to Newfoundland have been printed, and these appear in Clarence S. Brigham, *British Royal Proclama-*

tions relating to America, 1603-1783 (*Transactions and Collections* of the American Antiquarian Society), Worcester, 1911. Legal opinions on trade, fisheries, and civil government are available in George Chalmers, *Opinions of Eminent Lawyers, on Various Points of English Jurisprudence, Chiefly Concerning the Colonies, Fisheries, and Commerce, of Great Britain: Collected and Digested, from the Originals, in the Board of Trade, and other Depositories*, London, 1814.

PUBLICATIONS OF BRITISH NORTH AMERICAN GOVERNMENTS.

The most important collection of printed material relating to Newfoundland aside from the *Calendar of State Papers, Colonial Series* is the series of documents published in connection with the dispute between Canada and Newfoundland over the boundary of Labrador. Besides containing the legal arguments of both parties presented to the Judicial Committee of the Privy Council, the series contains many papers relative to the history of Canada, Newfoundland, and Labrador from the earliest times down to the present. The briefs and supporting documents were collected and edited under the supervision of Messrs Charles Russell & Co., Solicitors for the Dominion of Canada, and Messrs Burn and Berridge, Solicitors for the Colony of Newfoundland. Of particular interest is *Documents Relating to the History of Newfoundland*, being Part IX, Volume IV of the Joint Appendix of *In the Privy Council. In the Matter of the Boundary Between the Dominion of Canada and the Colony of Newfoundland in the Labrador Peninsula. Between the Dominion of Canada of the one part and the Colony of Newfoundland of the other part*, 12 vols. [London, 1928]. Many of the early charters, official papers, legal opinions, etc., relative to the fishery are printed in Part IX, Volume IV. Some have been taken from original sources, others are reprinted from D. W. Prowse, *History of Newfoundland*, the *Calendar of State Papers, Colonial Series*, and the *Acts of the Privy Council, Colonial Series*, but there are a number of papers unavailable elsewhere in print. Most of the material reproduced relates to the period after 1763, and makes the series a valuable source-book for the modern history of British North America.

BIBLIOGRAPHICAL NOTE 345

Two volumes issued by the Public Archives of Canada contain documents relating to the early history of Newfoundland. These have been ably edited by Henry P. Biggar, *The Precursors of Jacques Cartier, 1497-1554; A Collection of Documents relating to the Early History of the Dominion of Canada* (*Publications* of the Canadian Archives, no. 5), Ottawa, 1911; and *The Voyages of Jacques Cartier, Published from the Originals with Translations, Notes, and Appendices* (*Publications* of the Canadian Archives, no. 11), Ottawa, 1924.

PUBLICATIONS OF HISTORICAL SOCIETIES.

Among the publications of historical societies the most useful for the early period are the collected works of Richard Hakluyt and Samuel Purchas, which were published under the auspices of the Hakluyt Society. Richard Hakluyt, *The Principal Navigations Voyages Traffiques and Discoveries of the English Nation Made by Sea or Over-land to the Remote and Farthest Distant Quarters of the Earth at any time within the compass of these 1600 Yeeres*, 12 vols. [for the Hakluyt Society], Glasgow, 1904, and popularly known as the "Extra Series." Volumes VII and VIII contain a good deal of valuable material relating to Newfoundland between 1497 and 1600. Samuel Purchas, *Hakluytus Posthumus, or Purchas His Pilgrimes, Contayning a History of the World in Sea Voyages and Lande Travells by Englishmen and others*, 20 vols. [for the Hakluyt Society], Glasgow, 1906. Volume XIX contains a good deal of information relative to the early attempts of the proprietors to colonize Newfoundland. A number of documents relating to English activities at Newfoundland in the sixteenth century have been printed in Carlos Slafter, *Sir Humfrey Gylberte, and his Enterprise of Colonization in America* (*Publication* of the Prince Society), Boston, 1903. Papers pertaining to illegal trade are to be found in R. N. Toppan, *Edward Randolph* (*Publication* of the Prince Society), 7 vols., Boston, 1898-1909. Volumes II, III, IV, V, and VII contain most of the references to Newfoundland. Similarly, Volume II of the *Andros Tracts* (*Publication* of the Prince Society), 3 vols., Boston, 1868-1874, contains documents relating to the same subject.

CONTEMPORARY BOOKS, PAMPHLETS, AND TRACTS.

There are numerous books, pamphlets, and tracts of the seventeenth and eighteenth centuries relating either directly or indirectly to the political and economic features of the settlement and fishery of Newfoundland. Although not exactly contemporary, John Reeves, *History of the Island of Newfoundland*, London, 1793, presents an eighteenth-century view of Newfoundland's problems. Reeves, who was Chief Justice of the colony, had a much clearer appreciation of its status than most of the later writers whose accuracy is obscured by excessive local patriotism. It is far more authoritative than the later standard histories.

The early colonization projects provoked the appearance of a considerable amount of propaganda intended to arouse interest in the possibilities of investment and settlement. Most of these effusions are florid in style and are typical of the literature issued by the stock-jobber and land promoter of every age. Among the most famous is Lord Bacon's "Essay on Plantations," found in numerous editions of his works, and written to promote the London and Bristol Company. Of more direct interest, Richard Whitbourne, *A Discourse and Discovery of New-Found-Land, with Many Reasons to Prove how Worthy and Beneficiall a Plantation may there be made, etc.*, London, 1620, 2d ed., 1623, which was written to bolster the waning fortunes of the company and probably to influence the crown in its behalf. A similar work, Captain John Mason, *A Briefe Discourse of the New-found-land, with the situation, temperature, and commodities thereof, inciting our Nation to goe forward in that hopfull plantation begunne*, Edinburgh, 1620, reprinted in *Captain John Mason (Publication* of the Prince Society), Boston, 1887, was inspired by the hope of interesting Scotsmen in the colony at the time that Mason procured financial assistance in Edinburgh for the Newfoundland Company. Calvert's Avalon was advertised in *A Letter from Captain Edward Wynne, Governor of the Colony at Ferryland, within the Province of Avalon, in Newfoundland unto . . . Sir George Calvert, Knt., H.M. Principal Secretary of State, July 1622*, London, 1622. The eccentric Sir William Vaughan wrote three interesting and amusing works in verse to give his settlement at Trepassey publicity. His *Cambrensium Caroleia,* London, 1625, wherein he hurls anathema against the West Country fishermen in Latin, has been reproduced in the "American Series of Photostated Reproductions," no. 169, issued by the Massachusetts Historical Society, Boston, 1926. *The Golden Fleece Divided into three Parts Under which are discovered the Errours of Religion, the Vices and Decayse of the Kingdom, and lastly the wayes to get wealth, and to restore Trading so much complayned of. Transported From Cambrioll Colchos, out of the Southermost Part of the Iland, commonly called Newfoundland, by Orpheus Junior, For the generall and perpetuall Good of Great Britaine,* London, 1626, and *The Newlands Cure . . . Published for the Weale of Great Britaine,* London, 1630, reprinted in the *North American Review* (March, 1817), IV, 289-295, both contain propaganda for his colonial enterprise.

There is a great deal of printed material upon matters of political and economic interest during the seventeenth and eighteenth centuries. Most of these books, pamphlets, and tracts are concerned with foreign and colonial policies in general, and with the international rivalries of the period. Some of them contain comments and descriptions of the political and economic functions of the Newfoundland fishery. Among those of importance are: Thomas Jenner, *London's Blame, if not its Shame: Manifested by the Great Neglect of the Fishery,* London, 1651, which contains observations on the British fisheries in general, and mentions the position occupied by the Newfoundland trade at that time; Thomas Mun, *England's Treasure by Forraign Trade, or the Ballance of our Forraign Trade is the Rule of our Treasure,* London, 1694, which presents the contemporary theory of Newfoundland as an indirect contributor to the favorable balance of trade. In contrast, Sir Josiah Child, *A Discourse of the Nature, Use and Advantages of Trade,* London, 1694; and *A New Discourse of Trade,* 3d ed., London, 1718, give the "free trade" point of view. Child, who was a West Country ship-chandler, has the views and prejudices of that region, but presents them most clearly. John Collins, *A Plea for the Bringing in of Irish Cattel,* and *Keeping out of Fish caught by Foreigners* . . ., London, 1680; and *A Discourse of Salt and Fishery*, London, 1682, were written to promote the Royal Fishery Company of which he was an officer, but contain some remarks on Newfoundland. Sir Francis Brewster, *Essays on Trade and Navigation*, London, 1695; and his *New Essays on Trade*, London, 1702, contain comment on the rivalry with France. Joshua Gee, *The Trade and Navigation of Great Britain Considered*, 3d ed., London, 1731, 4th ed., 1738, has a few remarks on Newfoundland, but his comments are not as extensive as those of Child, Brewster, Mun, and others. Daniel Defoe, *Some Further Observations on the Treaty of Navigation and Commerce between Great Britain and France;* . . ., London, 1713; and *A Plan of the English Commerce*, 2d ed. London, 1728, 3d ed., 1737, contain very pertinent observations on Anglo-French rivalry at Newfoundland. William King, *The British Merchant, or Commerce Preserv'd*, 2d ed., London, 1721, contains one of the most important discussions of Newfoundland to be found outside the pages of the above mentioned works. John Ashley, *Memoirs Concerning Trade*, 2 parts, London, 1743, has some worthwhile comments, particularly upon the training of seamen. A criticism of Ashley is found in George Heathcote, *A Letter to the Right Honourable The Lord Mayor, the Worshipful Aldermen, and Common Council; Merchants, Citizens, and Inhabitants of the City of London. From an Old Servant*, 3d ed., London, 1762, commonly called his *Letter on Trade*. There are many others, but much of the material relating to Newfoundland is repeated or even copied by other authors. In general, an examination of the political and economic literature of the seventeenth and eighteenth centuries, convinces one that the Newfoundland fishery and trade were only of incidental importance in the British commercial system.

SECONDARY AUTHORITIES.

There is comparatively little secondary material relating to the history of Newfoundland before 1763 which is at all trustworthy. There are a number of books concerning the international position of the fishery since the American Revolution, but they are useless for the earlier period. There are also a number of local histories purporting to portray the story of the island accurately, but most of them are unreliable from the point of view of the historical scholar. The standard work completely covering the history until nearly the end of the nineteenth century is Judge D. W. Prowse, *History of Newfoundland*, London, 1892, 2d ed., 1893. Judge Prowse wrote an extremely full, though disconnected account. Complete or partial quotations from original sources are included in the appendices at the end of each chapter. Unfortunately, Prowse was not a trained scholar, with the result that his work has a decided bias in favor of the planters and shows extreme prejudice against the western adventurers whom he considers to be responsible for all the ills which have beset his native island. In a number of cases his anxiety to defend the cause of his ancestors has permitted him to quote only those portions of important papers which supported the planters' side of the case. In spite of its local patriotism and insularity it remains unfortunately the best history of Newfoundland written since Reeves' *History of the Island of Newfoundland* appeared in 1793. A briefer and more scholarly work is J. D. Rogers, *Historical Geography of Newfoundland*, Vol. V, part IV of the *Historical Geography of the British Colonies*, edited by Sir Charles Lucas, Oxford, 1911. Rogers' volume has also been published separately. The most recent narrative history of Newfoundland for the period previous to the American Revolution is found in Chapter V of the *Cambridge History of the British Empire*, VI, *Canada and Newfoundland*, Cambridge and New York, 1930. This account is from the pen of Professor A. P. Newton of the University of London, and is excellent, though placing too much reliance upon Prowse.

The most recent work to appear is Dr. Charles B. Judah, Jr., *The North American Fisheries and British Policy to 1713* (*Illinois Studies in the Social Sciences*, XVIII, nos. 3-4), Urbana, Ill., 1933. Dr. Judah emphasizes the rivalry of the English financial groups for control of the American fisheries. He propounds the thesis that the vacillating policy of the government may be accounted for on this ground. With this conclusion the author is in accord, but he feels that Dr. Judah has permitted himself to look upon the question very much in the same light as did Judge Prowse. The diverging lines of development of the Newfoundland and New England fisheries were not altogether historical, but were governed fundamentally by different geographic conditions. He tends to minimize the importance of the fishery as a training school for seamen, which from the author's point of view accounts very greatly for the support given the western adventurers by the crown throughout most of the period previous to 1713. The volume is well documented, and Dr. Judah has done a tremendous amount of research. The book deserves careful consideration by historians of pre-Revolutionary British America.

Two useful works, containing much documentary material are by the late Henry Harrisse. *The Discovery of North America, A Critical, Documentary and Historic Investigation with an Essay on the Early Cartography of the New World, etc.*, London and Paris, 1892, is extremely useful for the sixteenth century discoveries in the vicinity of Newfoundland, but it has been superseded to a very considerable extent by Biggar, *Precursors of Cartier. Le Découverte et Évolution Cartographique de Terre-Neuve et des Pays Circonvoisins, 1497-1501-1769. Essais de Géographie Historique et Documentaire*, Paris, 1901, is valuable with respect to the light it sheds upon the early European fishing activities at Newfoundland and in adjacent regions.

Information relative to the early proprietors may be gleaned from William H. Browne, *George and Cecilius Calvert, Barons Baltimore of Baltimore*, New York, 1890; Henry Kirke, *The First English Conquest of Canada*, London, 1871, 1908; William W. Scott, *The Constitutions of English, Scottish, and Irish Joint-Stock Companies to 1720*, 3 vols., Cambridge, 1910-1912; George P. Insh, *Scottish Colonial Schemes, 1620-1686*, Glasgow, 1922; the *Cambridge History of the British Empire*, I, *The Old Empire to 1783*, Cambridge and New York, 1929. The rivalries of the western adventurers and the planters in the late seventeenth century are discussed uncritically in George Louis Beer, *The Old Colonial System*, Part I, *The Establishment of the System, 1660-1688*, 2 vols., New York, 1912. Chapter I, Volume II contains the account of Newfoundland during the period of the Restoration. Robert Perret, *La Géographie de Terre-Neuve,* Paris, 1913, is one of the most useful works on Newfoundland discovered by the author, who has drawn upon it heavily particularly in the Introduction to this work.

INDEX

Acadia, French fishery at, 21, 56, 61, 172, 182; Alexander attempts to found colony in, 47; cession of, demanded by English, 235; Prior's proposals regarding, 236. See also Nova Scotia.

Administration, of Newfoundland in West of England, 77; transferred to fishing admirals, 214. See also Regulations, fishery.

Admiral, Lord High, reports to, on fishery, 133; not to allow civilians to go to Newfoundland on warships, 175.

Admiral(s), fishing, shipmaster first arriving in harbor becomes, 75; privileges and duties of, 75-76; powerlessness of, 77; more power proposed for, 130-131; notified of suspension of restrictions against planters, 158; advantage of planters and byboatkeepers in disputes with, 206-207; administration of fishery committed to, 214; indifference and incompetence of, 215, 261, 270, 271, 290; distribute beaches, 232; neglect inshore regulations, 250; fail to relieve debtors, 252-253; disregard Act of 1699, 266; governor to avoid conflict with, 275; relations with magistrates, 280-282, 284; jurisdiction of, 282-283; rebuked, 285; support opposition to jail tax, 286; held in check by governor, 287; Board of Trade reviews powers of, 288-289; commodores forced to assume duties of, 308; warned to cure fish properly, 317.

Admiralty, London and Bristol Company apply for assistance from, 41; jurisdiction over offenses at sea, 76; given powers to be exercised by commodores, 132; heads of inquiry transmitted through, 159; to direct commodore to suppress illegal trade, 196-197; to prevent foreigners buying fish at Newfoundland, 197; anxious to strengthen royal administration in Newfoundland, 227; Board of Trade solicits opinion of, 262; give Clinton heads of inquiry for commodore only, 284; transmits heads of inquiry, 294.

Admiralty, High Court of, 41.

Adventurers, western, proposed exclusion of, from carrying trade, 38; rivalry with planters of London and Bristol Company, 39; John Guy the enemy of, 39; protected at Newfoundland by caretakers, 40; dispute of, with planters reviewed, 40-44; struggle of, with Newfoundland Company, 46; seek legislation against proprietary monopolists, 49-52; decline of trade of, 59, 111-112, 118, 120-126, 161-164; indifferent to Dutch competition, 6; attitude toward foreign carriers, 61-66; attitude toward naval defense, 62; Londoners willing to coöperate with, 64; fishing rights of, confirmed, 64, 71-77; deny transporting fish in foreign bottoms, 68; represented on shipping committee, 70; effect of charter of 1634 on, 77;

relations with Kirke group, 79, 80, 82-83, 88, 96-102; oppose impost on foreign buyers, 80; differ with proprietors on conduct of fishery and carrying trade, 86-87; proprietary groups lack complete support of government because of, 87; employ coöperative labor, 87; less conciliatory than Kirke group, 88; maintenance of fleet cuts profits of, 89; localism of, 89, 188; proprietors and London merchants desire contracts with, 89; disadvantages of system of, 89-90; oppose innovations, 90-91; favored by Puritan government, 91; situation of, after 1660, 108-110; losses of, 111; influential at court of Charles II, 112; attitude toward civil government for Newfoundland, 115, 139-140, 202, 217, 280 ff.; forehandedness in obtaining law prohibiting taxation of fish, 116; present views to committee, 117-118; ships of, lose time after arrival in Newfoundland, 118, 122; contest of, with planters and merchants, 126-148; reopen question of fishery regulations, 127-128; proposed charter amendments insure supremacy of, 128-132; attitude toward removal of planters, 133, 161; accept naval supervision of fishery, 134; increased external competition with, 135; increasingly reactionary, 136; financial predicament of, 138; charged with insincerity, 138; uphold charter regulation, 139-140; report of Lords of Trade a victory for, 140-141; fishery left in hands of, 144-145; triumph of, 147, 148; accusations against, 151, 154, 160; dissatisfied with charter, 154-155; actions of representatives of, in investigations of 1677-1681, 156-158, 166-168; shortsightedness of, 160; seek to restrict immigration, 161; planters not desirous of ousting, 166; lose dominance in fishery, 169-170, 245, 271; dependence on navy during wars, 171; close relations with crown, 171, 176-177; attitude toward nursery for seamen, 174-176, 180; cause growth of passenger traffic, 177; unable to meet New England wages, 178; unable to restrain planters, 180; seek support of parliament, 180-181; attitude toward French danger, 182-188; control export trade to Newfoundland, 189-190; attitude toward New England trade, 190-191, 194; declining power of, 209-210; secure passage of Newfoundland Act, 210, 213-214; attitude toward defense of Newfoundland, 212, 217; advocate employment of green men, 223; turn to general shipping, 245, 248, 271; conservatism of, 246; balk colonization policy, 246-247; wage system harmful to, 248; slow to enter bank fishery, 250; indifferent to Board of Trade investigation, 265; board continues favorable to, 266, 267, 268, 269, 272; apathy toward improvements in regulations, 269; admirals cite parliamentary rights of, 287; maintain superior powers of admirals, 288; complain of losses, 315-316; allege trade in violation of Newfoundland Act, 324; make little trouble, 331; logical conception of, regarding trade and fishery, 333; maintenance of, becomes principle of British policy, 333-334; competition prevents monopoly of, 334.

INDEX

Agents. *See* James Campbell, John Downing, William Downing, Thomas Oxford, —— Parrett, John Pollexfen, Benjamin Scutt.

Agriculture, Newfoundland ill-suited to, 5-6; colonization as means of developing, 24.

Aldridge, Capt. Christian, lieutenant governor at St. John's, 305; quarrels with magistrates, 306.

Alexander, Sir William, seeks to colonize Nova Scotia, 46, 61.

Allen, Capt. ——, R. N., reports on illegal trade, 201.

America, Northeastern, discovery of, 19; all English subjects free to fish in, 50; parliamentary legislation concerning, 52; Northeastern, colonization in, result of commercial competition, 61; Northeastern, Kirke and lords apply for grant to fishing area of, 78; Spanish, war to break down monopoly in, 92.

Amherst [Jeffrey], General, in recovery of St. John's, 329.

Andalusia, Irish trade to, 34-35.

Andros, Sir Edmund, instructions to, 199; attempts of, to prevent smuggling opposed, 200.

Anglesey, Lord (Lord Privy Seal), views of, 140.

Anglo-Portuguese expeditions to Newfoundland, 21.

Anglo-Spanish Treaty, 271 n.

Animals in Newfoundland, 6.

Anticosti, included in Newfoundland jurisdiction, 304.

Appalachian physiographic province, 5.

Arctic Current, passage between Flemish Cap and Grand Bank, 8; course and character of, 10-11; temperature of, 11; carrier of field ice and bergs, 12; plankton brought by, 14; Labrador cod and, 16.

Army, Newfoundland fish for, 34.

Artillery, Royal, detachment at Fort William, 211.

Arundell, Earl of, 44.

Ashley, John, opinions of, 330.

Assembly, consent of, necessary for taxation, 280; absence of, in Newfoundland, 309; delay in calling in Nova Scotia, 309.

Attorney General, approves fishery regulations, 72; to report on charges against Kirke, 82-83; to prepare bill confirming charter amendments, 131; ordered to change criminal procedure, 132; opinion of, 279-280, 283, 297, 298.

Austrian Succession, War of, effect on Newfoundland government, 293-294; lack of convoy during, 313; discloses France still as rival in America, 328.

Avalon, charter of. *See* Charter, Avalon.

Avalon, Province of, name of Calvert's grant, 47; Baltimore opposes Kirke patent in, 79; proprietary government in, 106-107; collapse of Baltimore's government in, 107-108; deputy governor of, attempts to enforce Navigation Act, 195; Baltimore reasserts claim to, 332.

Avalon Penninsula, axis of highland in, 3.

Azores, expeditions to Newfoundland from, 20.

Baccalaos, native name for codfish, 19; early name for Newfoundland, 20.

Bait, planters prevent fishermen from hunting seabirds for, 40;

small fish used as, 40, 57; English fishermen may take freely, 129.
Balance of power, Newfoundland contributes to, 330, 336.
Balance of trade, less emphasis on fishery in, during French wars, 329-330; favorable to Great Britain, 333; Newfoundland assists in maintaining, 336.
Ballads, preserve stories of fishermen, 207.
Ballast, thrown into harbors, 165.
Baltimore, Lord, joins opposition to Kirke patent, 79; petitions for restoration of Avalon, 104; legal dispute with Kirke, 104-106; Avalon restored to, 106; establishes proprietary government, 106-107; John Rayner, deputy governor for, 195; reasserts claim to Avalon, 332.
See also Calvert, Sir George.
Banks, extent and character of, 7-9; Newfoundland, course of Arctic Current across, 10; good spawning places for cod, 16; French began fishing on, 21; wet fish caught on, 55; included in grant sought by Kirke, 78; development of fishery on, 249-250, 310.
Barbados, refuse fish carried to, 192, 193.
Barbary pirates, attack Newfoundland shipping, 59; attack English shipping, 61, 62; raid coasts of England, 62; threaten to raid Newfoundland, 62; war with, 169; naval protection against, 180.
Barnstaple, fishing ships from, 55; Newfoundland trade of, under Charles I, 63; representatives of, before Privy Council, 67; agents of, procure charter of 1643, 71-72; mayor of, to enforce rules of charter of 1634, 76; vessels of, retain share system, 248; complaints from, 316.
Barter, practiced by New England traders, 191-193.
Basques, activities of, at Newfoundland, 21, 23, 31; Irish trade with, 34-35; seek recognition of fishing rights in Newfoundland, 271 n.
Bay d'Espoir River, 4.
Bay of Bulls, public houses in, 254; deputy naval officer appointed for, 327.
Bay of Islands, 3.
Bayonne, merchants of, 184.
Bay St. George, 3.
Beaches, fishing admirals have first choice of, 75; proprietors and planters to delay occupation of, 81.
Beauclerk, Commodore Lord Vere, efficiency of, 270; compelling report of, 273; appointed governor, but disqualified, 274; returns as commodore, 275; divides authority with Osborn, 276-277; wants definition of governor's powers, 278; opinion of, on conflicting jurisdictions, 282; recommendations of, 282-283; reports on catch of fish, 310; instructed to warn industry to cure fish properly, 317; seizes ship for smuggling, 321-322; opinion of, on Newfoundland market, 322.
Beef, imported from New England, 192.
Beer, sold at taverns, 83; brewing of, proposed, 90.
Belle Isle, Strait of, divides Newfoundland from mainland, 1; northern end of Long Range, 3; unnamed bank between Cape Race and, 8; Arctic Current in, 10; temperature of water in, 11; ice in, 11; fog in, 18.
Bell Island, iron ore field on, 49.

INDEX 357

Beothuks, or "Red Indians," 6.
Berries, economic value of, 6.
Berry, Capt. Sir John, R. N., commodore of convoy, 149; discovers new fishery regulation to be inoperative, 149-150; investigates charges, 150-152; reports of, 152-154, 156, 158-159, 161-162, 186, 192; inspires planters to defend own interests, 168; corroborated, 186-187; lists English vessels in illegal trade, 196.
Bideford, fishing ships from, at Newfoundland, 55; vessels from, retain share system, 248; complaints from, 316.
Birds, native to Newfoundland, 6; used for bait, 40; sea-, fishermen may hunt, 42.
Biscayan. See Basques.
Board of Trade, attitude of, toward foreign shipping at Newfoundland, 197; proposed creation of, by parliament, 210; concerned over disregard of Act of 1699, 215; recommendations of, to parliament, 215-216; admonishes shipmasters to observe Act of 1699, 216; rejects proposals for civil government, 217; seeks authority for searches of New England vessels, 224; Cumming informant for, 225; must submit annual reports on fishery, 226; attempts to enforce Act of 1699, 226-227; recommendation of, 229; opposes draft treaty, 235-236; suggestions to, for betterment of planters, 246; favors western adventurers, 247; unsuccessful in checking migration, 260; crown asks for recommendations, 262; recommendations of, to crown, 264-265, 267-269; rejects Phillips proposal, 265-266; proposes transfer of Newfoundlanders to Nova Scotia, 266, 268, 274; continues favorable to western adventurers, 266, 267, 268, 269, 272; slow to appreciate changed conditions, 272; proposes measures to prevent disorder, 274; refers Osborn's questions to Yorke and Fane, 279, 283; cautions Osborn, 279-280; undertakes definition of jurisdictions, 283; asked to reconsider question, 288; report of, 289; recommendation of, not accepted, 292; governor to report to, 298; refers Drake's request to Ryder, 299; complaints received by, 316-317, 319; instructs Beauclerk on proper curing of fish, 317.
Boards, imported by New England traders, 192.
Boats, fishing, kept at Newfoundland by fishermen, 40, 58; described, 57; carried from England, 58; shore space allotted in proportion to, 75; English fishermen assured right to construct, 129; planters keep about one quarter of, 163; proposed reduction in number of, 169; employed by adventurers and byboatkeepers, 169; French, described, 187.
Boats' rooms. See Rooms, fishing.
Bolingbroke, negotiates with France, 236.
Bolingbroke ministry, desires concessions from France, 238; uninfluenced by mercantilists, 238, 243, 244; fall of, 239; mercantilists' view of, 241.
Bonavista Bay, 3.
Bonds, required of shipmasters and owners, 130; commodore exacts, from New Englanders, 178, 325.
Bonfoy, Capt. Hugh, R. N., Irish trouble during administration of,

300 ff.; to investigate charges against Keen, 306.
Bordeaux, merchants of, 184.
Boston, relation to bank area, 8; ships at, violate bonds, 198-199; merchants of, oppose attempts to suppress smuggling, 200; revolt at, 200; engages in smuggling tobacco, 201; trade to southwestern Europe from, 314; vice-admiralty court at, 323.
Bottomry, fishing ships mortgaged at, 121-122, 138.
Bowler, Commodore Edward, quoted, 254; inefficiency of, 270.
Bradstreet, Capt. John, lieutenant governor at St. John's, 305; proclaims peace, 305-306.
Brandy, sources of importation of, into Newfoundland, 191, 196; exported to New England, 192-193; smuggling of, 199-200, 321-323; imported into Virginia via Newfoundland, 201.
Bread, baking of, proposed by Kirke group, 90; imported from New England, 192.
Bretons, early fishermen at Newfoundland, 21.
Brewster, Sir Francis, opinion of, 230.
Bridge, Capt. ——, R. N., removes Lloyd and Moody, 212.
Bridgman, Sir Orlando, Baltimore's petition referred to, 105.
Bristol, interest of, in Newfoundland fishery, 36; merchants of, interested in carrying trade, 37; merchants of, interested in London and Bristol Company, 38-39; interest of, in colonization, 47; fishing ships from, at Newfoundland, 55; Newfoundland trade of, under Charles I, 63; fears loss of trade, 111; merchants of, favor civil government, 115; crown against merchants of London and, 155; Newfoundland trade at, 314; complaints from, 316.
British Isles, proposed fishing monopoly in, 78; tonnage registered in, 314.
Browse, Robert, 107.
Bullionism, influence of, on Newfoundland policy, 36.
Burgeo Bank, 8.
Butter, imported from New England, 192.
Byboatkeepers, aid in establishing fishery regulations, 110-111; restriction of transportation of, prohibited, 112; attempt to enforce regulations against, 114; competition of, 119; complaints against, 120-121, 127-128; supported by London merchants, 121; restrictions upon, proposed, 129-130; numbers of, 169, 228; relation of, to nursery for seamen, 171-174; transported as passengers, 173-174; increase of, causes decline in West Country trade, 189; investigation of origin of commodities consumed by, 192; have no legal title to riparian land, 206-207; Act of 1699 places slight restrictions upon, 214; increased expenses of, 247; catch of, 251; a detriment to fishery, 267; greater share in management, 271; Irishmen seek employment from, 301; larger factors in inshore fishery, 310; warned to cure fish properly, 317; encourage emigration, 325-326; competition of, prevents monopoly, 33.
Byboatmen. See Byboatkeepers.
Byng, Capt. John, R. N., administration of, 292-293; quoted, 293; tries to suppress illegal trade, 326, 327.

INDEX 359

Cabot, John and Sebastian, quoted, 19.

Cabot Strait, southern end of Long Range, 3; northern limit of Nova Scotia banks, 8; Arctic Current in, 10; temperature of water in, 11; variability of currents in, 11; ice in, 11; poor spawning place for cod, 16.

Calvert, Cecilius (Lord Baltimore). See Baltimore, Lord.

Calvert, Sir George, influence of, on reversal of royal policy, 45; on committee to consider petition of Newfoundland Company, 45; Newfoundland Company disposes of land to, 46; confirmation of holdings, 47; occupies Avalon, 48; quarrels with London merchants, 48; religion of, 48; leave for Virginia, 48; views of, on legislation concerning America, 52; settlements of, remain, 55; financial difficulties of, 88; appoints governor of Avalon, 104; engages Catholic priest at Ferryland, 207.

Cambrioll Colchos, Vaughan's colony at Trespassey, 46-47.

Campbell, Collin, Jackson quarrels with, 208; deputy agent for prizes, 218.

Canada, absence of Canadian plant life, 5; absence of animals from, 6; French sedentary fishery in, 61; Kirke's project to colonize, 61, 78; illegal trade to, 198-199.

Cape Bonavista, limit of French and English fishing zones, 55; limit of region prohibited to shore settlement, 81; number of French ships fishing north of, 184, 187; ships from St. Malo fish north of, 186; French catches north of, 232; congestion at, 233; Newfoundlanders occupy harbors north of, 310; number of ships north of, 312; salmon fishery north of, 320.

See also Capes Bonavista and Race, region between.

Cape Breton Island, relation of Newfoundland to, 1; distance from Port aux Basques to North Sydney, 2; Indians from, 6; Lord Ochiltree's colony driven from, 61; included in grant sought by Kirke group, 78; trade with, continued, 225; cession of, demanded, 235; proposals on, 235, 236; French activity at, 242; advantages of fishery at, 242; French handicap settlement at, 246; commander of fleet at, supervises fishery, 294, 295; presence of French at, causes anxiety over Irish, 302; Frenchmen from, settle near Port aux Basques, 328.

Cape Chidley, northern limit of banks, 7; bank between Cape Race and, 9.

Cape Cod, southwestern limit of banks, 7.

Cape Norman, distance of, from Cape Ray, 2.

Cape North, Cape Breton Island, distance of, from Cape Ray, 1.

Cape Race, distance of, from Flemish Cap, 7; bank from, to Strait of Belle Isle, 8; bank between, and Cape Chidley, 9; limit of English and French fishing zones, 55; limit of region prohibited to shore settlement, 81; congestion at, 233; Newfoundlanders occupy harbors west of, 310; number of ships engaged west of, 312; salmon fishery west of, 320.

See also Capes Bonavista and Race, region between.

Cape Ray, distance of, from Cape North, 1; distance from, to Cape

Norman, 2; Arctic Current at, 10; ice near, 11-12.
Cape Sable, southern limit of Nova Scotia banks, 8.
Capes Bonavista and Race, region between, foreigners prohibited from fishing in, 129; naval vessels to patrol fishery in, 133; number of ships in, 161-162; congestion in fishery in, 164-165, 168-169.
Cape Spear, distance of, from Cape Anguille, 2.
Cape Verde Islands, date for sailing of vessels to, 130; illegal goods from, 196.
Caplin, diet of, 14; effect of environment on, as food for cod, 16; used as bait, 40, 57.
Carey, Sir Henry. *See* Falkland, Viscount.
Carlisle, Earl of, Lord Hay, 45.
Carrying trade, development of, advocated, 24; problem of foreigners in, 61-66, 67-71, 87, 89, 96-97, 195, 197; depression in, under Charles I, 63-64; proposed improvements in regulation of, 64-65; Privy Council institutes investigation of, 67-68; Londoners seek to build up by control of fisheries, 78-79; Kirke group agrees to enter, 80; Kirke and planters threaten prosperity of, 83; disagreement as to conduct of, 86-87; plan to control offered, 88; as cause of first Dutch war, 92; need of reorganizing, 109-110; decline in, 119, 120-121; planters unable to develop their own, 153; decline in use of fishing ships in, 164; fishermen useless as recruits for, 174; effect of embargoes on, 179; naval protection for, 180; Irish and New England competition in, 190, 193, 194; data on, 311-315.

See also Shipping.
Case, Henry, on handicaps of Bristol merchants, 316.
Cason, Edward, and others seek to engage in bank fishing, 78.
Catching, method of, in bank fishing, 56; equipment for, not to be defaced, 75.
Catholic League, vessels of, seized by English, 32.
Catholic religion handicaps Calvert, 48.
Catholics, Irish, numbers of, in Newfoundland, 108, 295, 301; trouble with, 300 ff.; persecution of, 302-303; miserable condition of, 303; magistrates partial to, 306.
Catholic(s), Roman, Coney considered for governorship, 166; Calvert brings priest to Ferryland, 207; Franciscans at Placentia, 208; priests in Newfoundland, 208.
Census, Berry takes, 149.
Channel, English. *See* English Channel.
Channel Islands, French operate in English zone as inhabitants of, 185; aid in illegal trade, 255; responsibility of fishermen from, for trouble, 288; shipping of, 315; complaint against inhabitants of, 324.
Chaplains, of St. John's garrison, 208.
Charles I, wars of, 59; inadequacy of navy under, 62; adventurers seek confirmation of fishing rights from, 71; favors fishing monopoly, 78; proposes letters patent to Kirke group, 79; promises never to annul Baltimore's patents, 79; guarantees inviolability of charter of 1634, 82.
Charles II, reaction under, 92; heeds

INDEX

merchants, 92-93; acts favorably to Baltimore, 104-108; fishery under, 104-125; grants charter of 1661, 125; adventurers influential at court of, 112; commercial and fishery policies under, 147-148; Hinton's claim to governorship from, 166; pro-French attitude of, 183.

Charlestown, ship of, 198.

Charter, Avalon, rights of fishermen recognized in, 47-48; hampers settlement, 88; restitution of, requested, 104; legal struggle with Kirke over, 104-106; restoration of, 106.

Charter, Gilbert's, 25.

Charter, Kirke's, of 1637, applied for, 78; objections to, 79; reservations in, favorable to West Country, 81-82; violations of, 83; hampers settlement, 88; proprietors willing to accept restrictions in, 88; compared to proposed amendments of 1671, 134.

Charter, Newfoundland Company's, gives no jurisdiction over fishery, 43; no changes in, 44-45; hampers settlement, 88.

Charter, Western, of 1634, applied for, 71; intended as temporary, 72; not on patent rolls, 73; perpetuates regulation of fishery, 74; as fundamental law of Newfoundland, 74; terms of, 74-77; weakness of administration created by, 77; adventurers have to defend, 77; inviolability of, guaranteed, 82; hampers settlement, 88; charter of 1661 almost identical with, 112; provisions of, relative to nursery for seamen, 173.

Charter, Western, of 1661, mentioned, 71; issued, 112; attempts to enforce, 114; barrier to proprietary control or royal government, 125; basis of fishery regulation, 127; additions and amendments to, 128-135; question of continuation of, 136; adventurers favor regulation of fishery in accordance with, 139-140; provisions of, relative to nursery for seamen, 173; failure of, contributes to change of policy, 334; purpose of certain regulations of, 335.

Charter, Western, of 1676, mentioned, 71; approved by Lords of Trade, 140; changes in, relative to criminal jurisdiction, 146; issued, 147; superseded by Newfoundland Act of 1699, 147; becomes effective after Berry's report, 152; adventurers dissatisfied with, 154-155; suspension of clause requiring withdrawal of planters, 156-158, 161; Lords of Trade consider enforcement of, 165-167; validity of, maintained, 166; provisions included in, 171-172; protection of nursery for seamen in, 177; planters question validity of restrictive clauses in, 180; restrictions not enforced, 206; ineffectiveness of, demonstrated, 209; failure of, contributes to change of policy, 334; purpose of certain regulations of, 335.

Charters, early, not invalidated by Western Charter, 156-157; semi-private control of fishery under, terminated in 1699, 204; Newfoundland Act heir to, 214.

Child, Sir Josiah, on decreased Spanish demand for fish, 119.

Church of England, planters seek minister of, 161; beginnings of, at St. John's, 208.

Churches, not active in Newfoundland until late, 207-208.

Civil wars, effect of, on attitude toward naval protection, 102-103.
Clarke, Richard, favors colonization, 27.
Claypoole, John, Baltimore's accusation against, 104-105.
Cleaning of fish, carried on ashore, 40; methods of, 56; labor employed in, 58.
Clement, Alderman, statements of, on Plymouth trade, 63.
Climate, effect of ice on, 11; description of, 17-18; probable effect of, on Calvert's abandonment of Avalon, 48; more favorable to colonization in New England, 54; would prevent overpopulation, 138; unfavorableness emphasized, 143; advantages of, in French area, 169, 232; English to take advantage of, 169; prevents development of agriculture, 331.
Clinton, Capt. George, R. N., commodore and governor of Newfoundland, 283-284; has heads of inquiry for commodore only, 284; rebukes fishing admirals, 285; lacks funds, 286; holds jail tax illegal, 287; discouraged over administration, 287; record in New York, 287; ability of, 288; fears congestion in fishery, 310.
Clothing, planters dependent on outside sources for, 189; exported to New England, 192.
"Clubb-law," recourse to, 253.
Coastline, irregularities of, 2.
Cod, distribution and habits of, 15-17; as food for fishermen, 58; destruction of spawning places of, 249; inshore, yield more oil, 250; as medium of exchange, 250; dried, trade in, declines, 319; competition of European fish with, 319; comparative quality of bank and inshore, 319; determines economic rôle of Newfoundland, 333. See also Fish.
Codfishery. See Fishery.
Cod fry, diet of, 14; taken with seines, 57; destruction of, harmful to fishery, 118, destruction of, reported, 165, 249.
Codliver oil. See Train oil.
Cod roe, temporarily forms part of plankton, 16; food for adults, 16.
Cokayne, William, and others seek monopoly of carrying trade to southwestern Europe, 38.
Colbert, France attempts to control fisheries under, 182.
Colonies, interest in relationship of Newfoundland with, 189; position of Newfoundland in relation to, 202-203.
Colonies, continental, proposed removal of Newfoundland planters to, 133; economic position of Newfoundland differs from, 144; adventurers face competition from, 190; transshipment of goods to and from, at Newfoundland, 194; activity of inhabitants of, in smuggling through Newfoundland, 198; suppression of smuggling in, 198-202; vessels built in, 314; British vessels call at ports of, 315; seek exclusive privileges in Labrador, 321; lack of information as to legal trade of, with Newfoundland, 323.
Colonization. See Settlement.
Commerce. See Trade.
Commercial policy, during Interregnum, 92-104; none during 1660-1692 respecting fishery, 189.
Commercial status, of Newfoundland in seventeenth century, 202-203.
Commission, to investigate charges against Kirke group, 84.

INDEX 363

Commissioners, action of, appointed to investigate charges against Kirke, 98; recommended to govern plantation and fishery, 99-100; effect of use of, on later policy, 102; Baltimore's governor as possible royal, 107; proposed to investigate French activities, 187.

Committee of Council for Trade and Plantations, to hear proposals for Newfoundland government, 116.

Committee of Trade and Plantations. *See* Lords of Trade.

Committee on Shipping, actions of, 70-71.

Commodities, enumerated plantation, fish differ from, 144; exported through Newfoundland, 194, 198, 200-201; brought to Newfoundland illegally, 219; decreased transshipment of, through Newfoundland, 325.

Commodities, foreign, illegally imported through Newfoundland, 194, 200-201, 219; cleared through English customs, 196; seized by customs officials, 199-200; in colonial trade, 257-258.

Commodore(s), instructed to supervise fishery, 103; powers of, 127, 132-133, 226; adventurers accept supervision of fishery by, 134; to furnish information, 136; instructions for, 140; to supervise removal of inhabitants, 145; become supervisors and investigators of fishery, 158-159, 214; heads of inquiry for, 159; not to transport civilians on warships, 175; attitude of, toward emigration to New England, 178; to investigate French activities, 186, 187-188; to investigate origin of commodities, 192; attempt to enforce Navigation Acts, 195-196; to suppress illegal trade, 196-197; report on foreign ships, 197; unable to stop smuggling, 198; report occupants of riparian lands, 207; settle disputes, 207; as military commander in chief, 211; creates militia, 212; on disregard of Act of 1699, 215; commissioned to try piracy and prize cases, 217-218; unable to restrain emigration, 224, 259, 260; proposed enlarged authority of, 226-227, 274; uncertainty of reports of, 256, 257; defied, 260, 262; fishing admirals override, 261; requests and reports of, 269; indifference of, 270, 271; Gledhill refuses to recognize authority of, 273; governor to avoid conflicts with, 275; also governor, 284; proposal to omit commissioning, as governor, 293; not sent out, 294; holds Irish responsible for disorder, 301; government in Newfoundland originates in powers of, 307; powers and duties of, 307-308.

Commons, House of, representation of West Country in, 49-50, 87, 210; debates in, 49-52; committee of, on Newfoundland bill, 213.

Commonwealth, attempts to collect impost from foreigners in Newfoundland, 86.
See also Interregnum.

Company of French Merchants, of Exeter, 84-85; of London, 84-85.

Competition, proposed elimination of, 23; elimination of French, attempted, 32; increase in, 108-109.

Conception Bay, as fishing resort, 3; poor spawning place, 16; iron deposits of, 49; French raid on, 212.

Coney, ——, candidate for governorship, 166.

364 BRITISH FISHERY AT NEWFOUNDLAND

Congestion in fishery, 164-165, 168-169.
Congregationalists, 208.
Constabulary, proposed, 217.
Conveniences, fishing, rights of English fishermen, in, 129; shipmasters favor destruction of, 150-151.
Convoy, provided because of royalist activities, 97; an important contribution of Interregnum, 102-103; Robinson as commander of, 126; regularity of service, 127; influence of commander of, on development of government, 307-308; zeal of governor's subordinates wanes in absence of, 308; lack of, during War of Austrian Succession, 313; inadequacy of, 313.
Cookrooms, maintained, 40; destruction of, 84.
Cornwall, interest of, in fishery, 36, 49, 50; parliamentary representation of, 49-50; powers of vice admiral of, 74, 76.
Cortereal, Gaspar, discoverer of Newfoundland, 20.
Council for Foreign Plantations, on complaints against byboatkeepers, 127-128; recommends additions to charter of 1661, 127-128; proposes increased powers for commodore, 132; arguments presented to, 136; abolished, 140; considers nursery for seamen, 175.
Council of State, committee of, reports of, 99-100.
Council of Trade, on importance of fisheries, 93; established, 103; concerned with fisheries, 112-113; considers Newfoundland affairs, 114.
Court, prize. *See* Prize court.

Courts, proprietary, jurisdiction of, 82.
Courts, vice-admiralty. *See* Vice-admiralty courts.
Courts of law, absence of, 252.
Crews. *See* Seamen.
Crimes, at Newoundland, jurisdiction over, 43, 74, 76, 131-132, 146, 285-286, 292, 298, 299; at sea, jurisdiction over, 76.
Cromwell, Oliver, war with Spain during rule of, 92; fishery under, 92-104; Newfoundland policy of, 93-105; naval defense of fishing and market fleets under, 102-103; complicity of, with Kirke, 104; fishery and commercial policies established under, 147-148.
Crown, title of English, to Newfoundland, 21-22; rejects Whitbourne's recommendations, 42; reverses policy, 45; western adventurers seek confirmation of rights from, 71; grants charter of 1634, 72, 77; assumes jurisdiction over fishery, 82; London merchants and planters seek support of, 114-115; confirms privileges of adventurers, 128-134; reserves right to make further fishery regulations, 129; interests of, appealed to, 137; Newfoundland government a possible financial burden to, 143; supports adventurers, 147, 155; to reconsider Newfoundland question, 154; relies upon official investigators, 158-159; considers civil government, 166; change in attitude of, 167, 168; interest in nursery for seamen, 170-180; support of adventurers wavers, 180-181; unappreciative of French danger, 188-189; has no fixed commercial policy before 1692, 189; acts to prevent illegal trade, 196-

INDEX 365

197; fails to enforce clauses against planters, 206; threatened reversal of policy on civil government, 209; neglects Newfoundland troops, 212; rivalry between parliament and, 213; commodore as representative of, 214; authority of, in fishery after 1699, 217; seeks to regulate fishery, 262; rejects Board of Trade recommendations, 265, 267-268; financial considerations of, 269; takes problem seriously, 270; policy of, regarding government, 276; supported by legal advice, 283; refuses to extend prerogative, 292; right of, to erect courts, 297; jurisdiction over capital offenses, 298, 299; powers of commodore derived from, 307, 308.

Cumming, Archibald, appointed officer, 218, 220-221, 225; makes suggestions, 265.

Curing of fish, carried on ashore, 40, 56; labor employed in, 58; equipment used in, not to be defaced, 75; improvement in methods of, needed, 122; English fishermen may go ashore for, 129; improved French methods of, 185, 232; English methods of, defended, 186; improper methods of, 317, 318, 319.

Currents, ocean, effect of wind on, 10; effect of, on icebergs, 13. *See also* Arctic Current *and* Gulf Stream.

Custom House, in England, 196; at St. John's, 327.

Customs, inadequate machinery of, 322, 326-327; early appointees as officers of, 327.

Customs, Commissioners of, asked for information on Newfoundland trade and fishery, 136; Randolph reports to, on smuggling, 198; instruct Andros on smuggling, 199; classify Newfoundland as foreign country, 202; send Larkin to America, 215; to direct preventive officer, 225; anxious to strengthen royal administration in Newfoundland, 227.

Customs, fishing, recognized in charters, 39, 74, 75-76, 81-82; ignored by Guy, 40; adventurers refuse to abandon, 43.

Customs, officers of, to enforce order against foreign shipping, 65-66; to enforce charter of 1661, 173; seize foreign goods and prosecute shipmasters, 199; keep lists of delinquent shipmasters and byboatkeepers, 226; a check on smuggling, 257.

Customs duties, proposed on foreign shipping at Newfoundland, 69; only taxation permitted upon Newfoundland products, 82; on fish shipped to England, 85-86.

Customs revenue, from Newfoundland trade prior to 1671-1675, 120; value of, from indirect trade, 137; on imports from southern Europe, 139.

Dartmouth, fishing ships from, 55; Newfoundland trade of, 63, 314; representatives of, before Privy Council, 67; agents of, procure charter of 1634, 71-72; mayor of, to enforce rules of charter of 1634, 76; complaints from, 111, 316; loss of shipping in wars, 121.

Davies, Capt. ——, R. N., supports Berry, 152.

De Belleville, etc., petition of, 107 n.

Defense, military, proposed for Newfoundland, 115, 210; adventurers object to proposed, 117;

civil government would provide, 137; would increase capital in fishery, 138; rejected by Lords of Trade, 142; planters renew plea for, 161; orders relating to, 170; French provide, at Placentia, 182; established at St. John's, 209, 211, 293; need of, demonstrated, 210; militia created for, 212; permanence of, 212; adventurers ask for increase of, 217; during War of Spanish Succession, 217; kept in good condition, 294; importance of, 329.

Defense, naval, proposed, 41, 44; opposed, 62; adventurers accept, 134; value of nursery for seamen in, 137; provided in winter by ice, in summer by ships, 139; adequate in summer, 142; of French southcoast fishery, 186; during War of Spanish Succession, 217; kept in good condition, 294.

Delaware Bay, tobacco traffic from, 201.

Delivery of fish, method of making, 66.

Depopulation, forcible, recommended, 140; Lords of Trade approve, 145; impossibility of carrying out, 149-150; consequences of, if forcible, 151; suspension of, 157-158; requirement of, withdrawn, 161; charter clauses providing for, never expunged, 166; proposed in interest of nursery for seamen, 174; proposed for trade advantage of adventurers, 190; Board of Trade recommends, 266, 268, 274.

Deserters, 130.

Desertion, impressment leads to, 179-180; encouraged by foreign shipmasters, 197.

Devon, interest of, in fishery, 36, 49-50; parliamentary representation of, 49-50; assizes of, 70; powers of vice admiral of, 74, 76; justices of peace report destruction of property, 83; advocates relaxed restrictions against foreign carriers, 96.

Dieppe, merchants of, 184.

Discovery, of the Banks, 19-20; of Newfoundland, 20.

Disorder, causes of, 40-41, 245; most frequent in winter, 142; responsibility of factors for, 253; capitalists desire protection against, 255; in fishery, 262; verges on anarchy, 270, 276; effect of, on nursery for seamen, 273; preventive measures needed for, 274; extension of judicial system to control, 291; Irish responsible for, 301; jealousies of officers contribute to, 307.

Dissenters, 208.

Dongan, Gov. Thomas, of New York, complaints of, 199.

Dorrill, Gov. Richard, releases prisoners, 299-300; Irish trouble during administration of, 300 ff.

Dorset, interest of, in fishery, 36, 49-50; parliamentary representation of, 49-50; powers of vice admiral of, 74, 76.

Douglas, Capt. James, as commodore, 294; mentions Irish Catholics, 302.

Downing, John, petition of, as agent for planters, 154, 156-157; succeeded as agent by brother, 165; effect of successful attack on charter by, 167; residence in Newfoundland, 167-168; statement of, on colonial trade at Newfoundland, 192.

Downing, William, agent for planters, 165; advocates stable government, 165; presents case for

INDEX 367

planters, 166; calls attention to Newfoundland, 167.
Drainage, of interior, 4.
Drake, Bernard, goes to Newfoundland, 28; attacks Spanish shipping, 28-29.
Drake, Capt. Francis William, governor of Newfoundland, commission to, 276 n.; enlarged powers of, 297-298; administration of, 298-300.
Drake, Sir Francis, plans to attack Spaniards at Newfoundland, 29.
Drunkenness, prevalence of, 58; forces fishermen to remain in Newfoundland, 177; responsibility for, 193; complaints of, 194; effects of, 208-209, 331.
Dublin, interest of, in Newfoundland, 35.
Dutch. *See* Holland.
Duties, export, at Newfoundland, 86; on Newfoundland fish, 96.

Earl marshal, jurisdiction of, in Newfoundland cases, 74; office of, abolished, 131-132.
Eastern shoals, location and description of, 9.
East Loo, mayor of, 76.
Edinburgh, interest of, in colonization, 47.
Edmunds, Sir Thomas, 45.
Education, backwardness of, 207, 208.
Edwards, Capt. Richard, R. N., governor of Newfoundland, commissioned, 293; alarmed over Irish Catholics, 303; administrations of, 304.
Edward VI, legislation during reign of, 22.
Elizabeth, legislation during reign of, 33 n.

Embargo, effect of, on Newfoundland trade, 179.
Emigration, to New England, continues, 245; forcible collection of debts encourages, 253; of indentured servants, 259-260; lieutenant governor to stop, 305; ineffectual efforts to check, 325-326.
England, distance of, from St. John's, 2 n.; dependent on Iceland for stockfish, 19; early trade and fishery at Newfoundland, 21; title of, to Newfoundland, 21-22, 74; shipping of, 23, 37, 70; benefits to, from colonization of Newfoundland 24-25; sovereignty of, over Newfoundland, 25, 26; fishery of, injured by wars with Spain, 30; divides control of fishery with France, 30-31; decline in consumption of fish in, 33; Dutch in coastal fisheries of, 34; foreign trade in fish of, 35, 36, 37; trial of offenses at Newfoundland in, 43, 74, 285-286; fishing zone of, 55; loses trade to foreigners, 60; Dutch as buyers of fish from, 60, 67, 85; increased rivalry with France in fisheries and trade, 60, 61, 78; suffers from France and Barbary pirates, 61-62; monopoly of train oil in, 64; foreigners excluded from south-coast fishery of, 65; decline of commerce of, 65; administration of fishery located in, 77; fishing rights of subjects of, 81, 129; disturbed political conditions in, 84, 88; export duties charged in, 86; Interregnum and Restoration significant epochs in commerce of, 92; export of fish from, in foreign bottoms, 96; provision trade of, 130; competition of New England with, 142, 190, 191-192, 223, 257-258; Newfound-

land fish cannot be taken to, 144, 333; commercial and fishery policies of, 147-148; war with Barbary pirates, 169; competition in fishery with France, 182, 184-185; products and manufactures of, 189; effect of illegal trade on commercial system of, 194-195; foreign goods for Newfoundland cleared through, 196; relation of Newfoundland to, 202-203; difficulty of obtaining material from, 211; training of sailors vital to, 223; number of ships from, 228, 251; seeks dominance of fisheries and trade, 233-237; dissatisfaction in, over peace treaty, 240-243; scarcity of skilled labor in, 247; laborers of, unwilling to work offshore, 250; precedence proposed for fishing ships from, 263; return of fishermen and sailors to, 263; proposed monopoly of Newfoundland trade, 265; monopoly of, in trade with British possessions, 324; fishery as contributor to naval strength of, 329-330, 335; balance of power favorable to, 330.

England, West of. *See* West Country *and* Adventurers, Western.

English Channel, Newfoundland fleet intercepted in, 29, 61-62.

"Englishman, an," anonymous pamphleteer, 123-125.

Enumerated Commodities. *See* Commodities, enumerated plantation.

Equipment, fishing, appropriation or defacement of, forbidden, 75; dependence upon England for, 189; exported to New England, 192; kind of, on banks, 249; annual importation of, 312.

Europe, plant life similar to, 5; southern, markets for cod in, 34; southwestern, direct voyages from Newfoundland to, 37; control of shipping to, 69; southern, trade within, 139, 169; illegal trade from, 195.

Eustace, Sir Maurice, 106.

Exchange, bills of, 193-194.

Exeter, interest of, in carrying trade, 37; fishing ships from, 55; fears of merchants of, 84-85; help pay expenses of West Country agents, 84-85.

Expedition, military and naval of 1696-1697, 209, 210-211.

Exploits River, 4; home of Beothuks, 6.

Exports, increased by colonization, 24-25; England seeks to increase, to Spain, 36; restricted to English, Irish, or Scottish ships, 65; of fish from West Country, 66; character of, 192.

Factors, hold over catch, 169; exploit planters, 253; New England, 259.

Falkingham, Capt. Edward, R. N., governor of Newfoundland, ability of, 288, 290; administration of, 290; on Newfoundland market, 322.

Falkland, Viscount (Sir Henry Carey) on committee to consider Newfoundland Company petition, 44-45; probable influence of, on reversal of royal policy, 45; Newfoundland Company disposes of land to, 46; advertises for settlers in Ireland, 47; financial difficulties of, 88.

Falmouth, fishing ships from, 55.

Fane, Francis, opinion of, 279, 283.

Farmers, of West Country, interested in fishery, 90.

Fécamp, fish importation declines at, 230.

INDEX

Ferryland, part of Calvert's Avalon, 47; Calvert resides at, 48; representatives of Baltimore at, 106-108; Calvert brings priest to, 207; French attacks on, 210; public houses in, 254; vessels owned at, 314; defenses at, 329 n.

Finch, Sir Heneage, solicitor general, 105; Baltimore's petition referred to, 105; one of committee on Baltimore's second petition, 106; to prepare bill for amending charter, 131; to change judicial procedure, 132.

Fines, West Country mayors may impose on violators of charter, 76; fishing admirals lack right to impose, 76; Act of 1699 provides none, 215, 264.

Fires, in fishing settlements and woods, 58.

First Western Charter. *See* Charter, Western, of 1634.

Fish, habits of, 13; designated by act of parliament, 33; decline of consumption of, in England, 33; sale of, to foreigners, 34; a factor in peace with Spain, 36; cleaning of, 40; stolen by pirates, 41; bill to remove tithes on, 50; seizure of, 59; West Country overstocked with, 65-66; export trade in, 65, 66, 68; failure of, during 1633, 68; transported in foreign bottoms, 70; appropriation of, forbidden, 75; taxation of, 113, 116; decline in Spanish demand for, 119; returns from, 120; enumeration system not applicable to, 144; data on production of, 153, 161-164, 311-313; bought by foreigners at Newfoundland, 185, 197; used in barter, 191-193; New Englanders trade to sack ships, 193; minister's salary paid in, 208; accumulate in Newfoundland during War of Spanish Succession, 229; reasons for decline in quality of, 317-318.

Fish, bank. *See* Fish, wet.

Fish, dry, where caught, 55; methods of making, 55, 56-59; French make, at Newfoundland, 137, 182; French tariff on, 185; admitted to Spain by treaty, 185; sale of, in Portugal and Italy, 185.

Fish, refuse, sold to undiscriminating buyers, 57; carried to Barbados, 192, 193; carried to West Indies and Madeira, 325.

Fish, wet, or cor fish, making of, reported, 23; caught on banks, 55; method of making, 56; as food for fishermen, 58.

Fisheries, ineffective regulations for, 65; proposed monopoly of, 69; importance of, recognized, 93-94.

Fisheries, British, decline of, 33; extension of, to Newfoundland, 36; Dutch hold on, 60; proposed monopoly in, 78; a cause for Dutch war, 92; encouraged by Council of Trade, 103; Council of Trade concerned with, 112-113.

Fisheries, North Atlantic, interest of merchants in, 47; monopoly in, considered, 49-52, 64, 77-79; proposed Dutch activity in, 60; rivalry of England and France in, 61; colonization as protection for, 61; French nursery for seamen projected in, 61; Council of Trade concerned in, 112-113; France seeks to control, 182; expansion of France in, 183.

Fishermen, time lost by, in quarreling, 24; antagonistic to Gilbert and colonization, 26, 27; individual initiative of, 35-36; need for defining and limiting activities of, 40; Guy limits activities of, 40; relations with planters, 40-41, 44, 47, 75, 82-84, 152-153, 158; dispute with London and Bristol Company reviewed, 40-44; rights of, recognized in Avalon charter, 47-48; methods of operation, 55-59; enjoy free market in London, 65; regulated by charter of 1634, 74-76; protected by Kirke charter, 81-82; other occupations of, 120; not to remain after season, 130, 165-167; Berry's charges against, 150-151; proportion of production by, 161-164; lack of pioneering spirit among, 165; number of, in 1684, 169; untrained in higher branches of seamanship, 174; winter in Newfoundland, 177; attracted to New England, 178; unwilling to occupy south coast, 246; wages of, 247.

Fishery, bank, method of conducting, 55, 56; proposal to engage in, 78; development of, 249; catches of, 249-250; develops rapidly, 310.

Fishery, British Newfoundland, beginnings of, 21-22; discussed by Parkhurst, 23-25; as training school for seamen, 33-34, 329-330, 335-336; in formative state before 1610, 35; activity of London and Bristol (Newfoundland) Company in, 36-37, 39, 46; much work of, carried on ashore, 40; contest over monopoly in, 49-53; number of ships in, 63-64, 118-120; investigated by Privy Council, 67-71; regulated by charter of 1634, 71-77; activities of Kirke group in, 77-82, 83, 88; conflicting views of Kirke group and adventurers on, 86-92; under Cromwell and Charles II, 92-125; commissioners appointed to govern, 99-100; condition of, after 1660, 109-112; becomes concern of Council of Trade, 112-113; Privy Council concerned for, 114; inefficient methods in, 118, 122; changes in regulations for (1671), 129-133; Lords of Trade investigate (1675), 136-141; Lords of Trade report on, 141-146; English policy regarding, laid down under Cromwell and Charles II, 147-148; important position of planters in, 153-154; adventurers accused of monopolizing, 154; regulations of 1676 unsatisfactory for, 154-155; Lords of Trade investigate (1677-1689), 159, 160-161, 165-166; commodores to report on, 159; conditions in (1675-1700), 161-165, 168-170; failure of charter regulation for, 180-181; danger of French competition in, 182-189; foreign ships in, 195-197; economic function and importance of, 204-206, 213-214; regulated by Act of 1699, 214-215; effects of War of Spanish Succession on, 227-229; demands for elimination of France from, 229, 233, 234; advantages of French fishery over, 231-233; conditions in (1713-1729), 247-255, 270-272; responsibility for decline of, 267; Board of Trade recommendations for improvement of, 267-269; lack of interest of government in, 269-270; disorder in, 270; oversight of, entrusted to naval officer, 294; prospect of employment in, draws Irish, 300-301; Irish a menace to, 303; powers of commodores in, 308; general prosperity in (1729-1763), 310; statistics on, 311-313; compared with New England, 312; effect of wars with France on, 313,

INDEX

315-317; increase in productivity of, 317; decline in quality of product of, 317-318, 319; competition of European fish with, 319; competition of new industries with, 319-320; extension of, 320, 321; advantage to French of preponderance in, 330; unique position of, in British commercial system, 332-333; economic conditions force change in regulations of, 334; technically always free, 334; contribution of, in maintenance of balance of power, 336.

Fishery, French Newfoundland, dry fish industry of, 56, 182; and settlement of south coast, 137; advantages of, over English, 183, 231-233; military and naval defense of, 183; competition of, 184; fort at Placentia intended to protect, 186, condition of, in 1676, 187; extent of (1699-1713), 231; freedom from disputes in, 232; negotiations respecting, 237; activity in (1729-1763), 328, 329.

Fishery, inshore, English and French engaged in, 55; methods of conducting, 56-59; projected by France in Acadia and Canada, 61; reasons for failure of, 249; suffers from congestion, 310; decline in prosperity of, 310.

Fishery, New England, English attempts to monopolize, 37, 38, 50, 51, 53; never confused with Newfoundland fishery, 53; exports of, 65, 96; colonization of Canada proposed to protect, 69; not mentioned in Act of 1659, 103; competition with Newfoundland fishery, 141, 142; importation of salt to, 195; compared to Newfoundland, 312; product of, in Lisbon, 319.

Fishery, Norman, declines, 230.

Fishery, Nova Scotia, protection of, 69.

Fishery, salmon, 320, 321.

Fishing methods, 56-59, in bank fishery, 249-250.

Flakes, fishing admirals have first choice of, 75.

Flanders, ships of, in carrying trade, 34, 37, 60.

Fleet, fishing, size of, in 1578, 23; cost of maintaining, 89; effect of embargo on, 179; fitted out in England, 189.

See also Ships, Fishing, and Shipping.

Flemish. See Flanders.

Flemish Cap, eastern limit of banks, 7; one of banks, 8; a shoal, 9; icebergs held at, 13.

Flour, imported from New England, 192.

Fog, a menace to navigation, 17-18; cause of, 18.

Foreigners, attempted exclusion of from fishery and carrying trade, 23, 38, 64-68, 129; elimination of, by war and piracy, 28; buy fish in England, 34, 35; danger of Newfoundland Company becoming involved with, 39; warning of encroachments by, 52; gain in fish trade during wars, 60; shipping carries coal and fish, 65; West Country methods of dealing with, 66; proposed tax on, 80, 85, 86, 98; lease best fishing places, 83; attitude of adventurers to, 89; readmitted to carrying trade, 95-96; planters would deal with, if in control of Newfoundland, 139; presence at Newfoundland reported, 164; illegal trade of, 195; provisions of Acts of 1660 and 1663 relating to, 195; attempt to suppress illegal trade of, 196-197; complaints against, 197.

Forests, 5.

Forrester, Robert, on declining market, 319.

Fortifications, proposed, 23; forbidden within six miles of shore, 82; erected at Placentia, 182; erection of, at St. John's, 209, 211; surrender of, to French, 211; extension of, proposed, 217; reconstructed at St. John's, 329; strengthened at Placentia, 329 n.; built at Ferryland, 329 n.

Fortune Bay, as a fishing resort, 3.

Fort William, St. John's, 212.

Fowell, John, views of, on foreign carriers, 96.

Fowey, fishing ships from, 55; mayor of, to enforce charter of 1634, 76.

France, early fishery of, at Newfoundland, 21, 23; Spain's wars with, 30; divides control of fishery with England, 30-32; finds market for fish, 31; a market for English-caught fish, 32; ships of, deal in English-caught fish, 34, 68, 185; shipping of, in direct market voyages, 37; proposed monopoly of trade with, 38; danger of Newfoundland Company becoming involved with, 39; Calvert aids fishermen against, 48; fishing activities of, described, 55-57; effect of war of Charles I with, 59-62; nursery for seamen of, 61, 172; commercial competition results in colonization by, 61; proposed exclusion of, from carrying trade, 68; English monopoly would create difficulties with, 78; difficulties of Kirke with, 84-86; more important competitor than Holland, 86; leanings of Charles II toward, 92-93; Kirke accused of favoring, 98; competition of, 109-111, 115-119, 121-125, 135, 137, 139, 141-142, 156, 182-189; extends operations in Newfoundland, 137; illegal trade with, predicted, 139; changed attitude of England to, after Dutch Wars, 146; settlements of, a probable haven for English planters, 151, 186; fishing zone of, prevents English expansion, 164; planters and fishermen uninterested in ousting, 165; adventurers depend on royal navy because of wars with, 171; need of naval protection against, 180; fortifies Placentia, 182; expands fisheries, 183-184; advantages of, on south coast, 183-184; encourages her fisheries, 184-185; foreign markets of, 185; activities of, reported, 186-187; good voyages made by fishermen of, 187; competition of, intensified after 1677, 188; war with, begins, 188; export trade to Newfoundland from, 190-191; wines of, exported to New England, 192; illegal trade at Newfoundland from, 195-196, 201; less feared than New England, 203; attacks English settlements, 208, 209, 210; military expedition against, 209; effect of war with, 209, 212, 217; defense of Newfoundland against, 212, 217; competition of, causes English reverses, 229; controls fish market, 230; trade with, continues, 255; seeks to profit by presence of Catholics, 302; civil government neglected during wars with, 309; effect of wars with, on fishery, 310, 313, 315, 326, 327; ships seized and crews impressed by, 316;

INDEX

naval importance of fishery during wars with, 329-330; improved conditions during last two wars with, 331.
Franciscans, establish mission at Placentia, 208.
Fuel, wood used as, 40; conservation of firewood, 75; stages and fishing conveniences destroyed for, 151.
Fundy, Bay of, 8.
Fur trade, unimportance of, 6; as chief interest of France in America, 139; forts at Placentia as protection for, 185-186; a factor in development of Newfoundland, 320; lack of, 333.

Gander River, 4.
Garrison, established at St. John's, 209, 329; personnel of, at Fort William, 211-212.
See also Defense, military.
Gee, Joshua, criticizes Treaty of Utrecht, 242.
Geology of Newfoundland, 3, 5.
Gibsone, Col. Sir John, commander of expedition of 1696-1697, 211.
Gilbert, Sir Humphrey, proposes voyage, 22, 25; patent of, 25; abortive expedition of 1578, 25-26, 28; unsuccessful attempt to found colony, 26; loss at sea, 27; effect of his efforts, 35, 54.
Gill, Nicholas, appointed naval officer, 327.
Glaciers, effect of, 5.
Gledhill, Samuel, lieutenant governor of Placentia, interferes with fishery and trade, 273, 305; salary of, 305; as dictator and magnate, 306.
Goffe [William], 104-105.
Gorges, Sir Ferdinando, land policy of, 46; monopoly of, in New England fishery, 51; patent of, 52.
Gould, John, London merchant, view of, on fishery, 121, 122-123; opinion of, on New England trade, 190-191.
Government, civil, West Country opposes, at Newfoundland, 54, 113, 117-118, 190; Kirke charter checks establishment of, 82; need of, 87; proposed by Protectorate, 102; blocked by legislation, 113-114, 125; proposed by merchants and planters, 114-116; proposed by Capt. Robinson, 126-127; Hinton's proposals regarding, 135-137; reasons for and against, 137-140; proposals for, rejected by Lords of Trade, 141-144; planters renew plea for, 161; Downing and Oxford advocate, 165; crown considers establishing, 166, 170, 181, 209, 213; absence of, makes suppression of smuggling difficult, 202; Board of Trade rejects adventurers' proposal for, 217; crown takes steps in direction of, 217-218; advocated, erected at St. John's, 225 n.; established, 262; need for, emphasized, 266-267; protests over establishment of, 269; expense a deterrent in establishment of, 269; demoralization preceding, 270; foundation of, 275; Beauclerk real founder of, 277; success of, due to Osborn, 282; influence of Byng on, 293; suffers from neglect during war, 293, 294; has little meaning for lawless elements between visits of officials, 308; better conditions after establishment of, 331-332.
Governor(s), John Guy as, 39-40; Capt. John Mason as, 44; Kirke goes out as, 82; planters seek appointment of, 166; candidates for, 166; ordered sent to Newfoundland, 170; John Rayner, deputy,

195-196; Richard Phillips as, 265-266, 269, 273, 274; first royal, appointed for Newfoundland, 274-275; authority of, 275-276, 282-283, 308-309; flouted, 280; dual position of commodore-, 284, 308-309; Board of Trade recommendations on, 289; carelessness of, 291; salary of, 291, 305; proposal to extend power of, 291-292; war prevents commissioning of, 293-294; commissioning of, resumed, 295; directed to investigate magistrates, 296; altered instructions to, 295-296; enlarged powers and duties of, 297-298; limitation of, in capital offenses, 298-299; attitude toward Irish, 301, 303; relations with lieutenant governor, 305-306; appoint surrogates for outports, 307; submit statistics, 311-315; warn industry to cure fish properly, 317; opinion of, on decline in quality of fish, 318; enforcement of acts of trade by, 321-323; require bonds from colonial shipmasters as check on emigration, 325.

See also individual names of royal governors.

Governors, lieutenant, administrations of, 304-307.

Grand Bank, location and characteristics of, 8-9; early French fishery on, 21; French vessels on, 231.

Graves, Admiral Thomas, governor of Newfoundland, 304; instructions to, 304 n.; recovers St. John's, 329.

Graydon, Capt. ——, R. N., commodore, reports on disobedience of Act of 1699, 215.

Green Bank, 8.

Greenland Company, monopolizes train oil imports, 64.

Greenland trade, clause in Act of 1699 on, 214.

Green men, quotas of, 130, 224, 226; scarcity of, 247; Irishmen complete quotas of, 301.

Gulf of St. Lawrence. See St. Lawrence, Gulf of.

Gulf Stream, course of, 10; plankton brought by, 14.

Guy, John, settlement made by, 28; as governor of London and Bristol Company plantation, 39; asserts jurisdiction over fishery, 39-41, 42-43; opposition to monopoly of, 50; supports Gorges and Londoners in Commons, 51-52; settlements of, remain, 55; objections to Kirke same as to, 82; Kirke group profits by mistakes of, 88; traditional influence of, 115; begins period of charter regulation, 181; colony of, had Puritan minister, 207.

Hagar, ——, commodore, report of, 266.

Hakluyt, Richard, 22-23.

Hamburg, shipping of, 37; extends trade, 60.

Hamilton, Marquis of, grantee of Newfoundland, 61; applicant for charter, 78.

Hamilton, Otho, lieutenant governor of Placentia, complaint of, 295; salary of, 305; anxious for civil authority, 306.

Hamilton-Kirke group. See Kirke, Sir David.

Hampshire. See Hants.

Hants, jurisdiction of vice admiral of, 74, 76.

Harbor Grace, priest arrested at, 302-303.

Harbors, fishing, choice of fishing rooms, etc., in, 75-76; reserva-

INDEX

tions for Kirke proprietors in, 83; filled in with ballast and refuse, 165.

Hardy, Capt. Charles, R. N., governor of Newfoundland, 293.

Hatsell, Capt. Henry, views on foreign carriers, 96.

Hawkins, Capt. Charles, R. N., commodore, reports of, 178, 193.

Hawkins, Sir John, plans to attack Spaniards at Newfoundland, 29.

Hay, Lord. See Carlisle, Earl of.

Health, conditions of, in fishery, 58-59.

Heathcote, George, opinions of, 330.

Henry of Navarre, 32.

Henry VIII, legislation during reign of, 22, 23.

Herring, diet of, 14; as food for cod, 16; effect of environment on, 16; to be exported only in English ships, 65.

Hill, Charles, on character of Baltimore's officials, 107.

Hill, Capt. William, governor of Avalon, 104.

Hinton, William, represents planters at investigation of 1675, 135-136; supports civil government, 137; claims royal promise of governorship, 166; calls attention to Newfoundland, 167; residence in Newfoundland, 167-168.

Holland, Earl of, applicant for charter, 78.

Holland, competition of, in English domestic fishery, 32-33, 34; shipping of, in direct market voyages, 37, 60; extends trade at England's expense, 60; proposed fishery of, in North America, 60-61; ships of, in English ports to buy herring, 67; exclusion of, from carrying trade considered by Privy Council, 68-69; reëxport trade of, 68; ships built in, employed in Newfoundland fishery, 68; impost advocated as means to eliminate, 85-86; decline of competition of, 86; wars with, 92, 102-103; 118, 121, 126, 134, 135, 141, 145-146, 171, 174; Navigation Act directed against, 95; fear of loss of trade to, 111; seizure of ships from, 195-196; open intercourse at Boston with, 200; illegal trade with, 201, 219, 220; French fishermen emigrate to, 230.

Houblon, James, opinions of, on fishery, 121-122.

Hudson's Bay Company, claims exclusive rights in Labrador, 321; Newfoundland cannot follow methods of, 333.

Hudson Strait, Portuguese whalers in, 20.

Huguenots, ports held by, open to English, 61.

Humber River, course of, 3-4.

Hunting, restricted, 6; for sea-birds used as bait, 40.

Hydrography, of adjacent seas, 10-13.

Hyman, Robert, 47.

Ice, character and movements of field, 11-13; boats of exploring parties crushed by, 57; as a maritime defense of Newfoundland, 139, 142.

Icebergs, character and movements of, 11, 12-13.

Iceland, as source of fish for England, 19; Newfoundland fishery classed with, 22.

Illiteracy, prevalence of, 207.

Immigration, restrictions upon, into Newfoundland, 129-130, 132, 173.

Imports, proposed restriction of, to English ships, 64-65; from Newfoundland described, 192; character of, brought in colonial vessels, 192-194; illegal, 195, 219-221.

Impost, upon foreigners buying fish at Newfoundland proposed, 80, 85, 89, 115; attempts to levy, 85, 86, 98, 182.

Impressment of seamen, under Charles I, 59; proposed, 65; difficulty of obtaining men for, 97; relaxed during Interregnum, 103; seamen trained at fishery available for, 171; unavailability of planters and byboatkeepers for, 173; men remain in Newfoundland to avoid, 174, 179-180, 326; recommendations regarding making men available for, 175; embargoes permit completion of, 179; source of annoyance to fishing industry, 316.

Indebtedness, causes of, 252-253, 254; forced, 331.

Indians, of Newfoundland, described, 6; Portuguese trade with, 20; well treated by adventurers, 74.

Infantry, at Fort William, 211.

Inhabitants. *See* Planters.

Inquiry, heads of, prepared by Lords of Trade, 159; for 1677 emphasize French activities at Newfoundland, 187-188; answers to, become stereotyped, 270; considered as royal instructions, 276 n., 284; Clinton has those of commodore only, 284; proposal to equip commodore with, 293; transmitted through Admiralty, 294.

Interlopers, attempt to exclude, 46; western adventurers as, at New England, 53.

Interregnum, attempts to collect impost from foreigners during, 86; merchants become of political importance during, 92; a period of transition for Newfoundland, 93-94; policy of governments toward Newfoundland, 93-105; legislation during, 93-96; governments of, favor adventurers, 96; governments of, appoint commissioners for Newfoundland, 99, 100; fishery regulations of, 101; Council of Trade established during, 103.

Investigation, conducted by Privy Council into rivalry between planters and fishermen, 41; of 1651, 98-102; of 1668, 117-118; of 1675, 135-147; of 1675-1676, by Lords of Trade, 152-154; of 1677, 155-158, 160-161, 165-166; of 1715, by Board of Trade, 262-265.

Ireland, distance of, from St. John's, 2 n.; proposed invasion of, 30; Newfoundland fish as provisions for, 34; shipping of, in trade to Spain, 34-36; Falkland advertises for settlers in, 47; ships of, may export fish, 65; unrepresented on shipping committee, 70; Newfoundland products exempt from taxation at, 82; export of fish from, 96, 190; provision trade of, 130, 135; trade of, at Newfoundland, 193, 314; immigration from 208, 246, 296, 300, 302; cheap labor from 248, 250; disaffection in, 295-296; fishing ships call at ports of, 300; Catholics ordered to return to, 303; complaints of unfair practices of traders from, 324.

Irish. *See* Catholics, Irish; Ireland.

Iron, production of, proposed, 49, 90.

Italy, as market for fish, 67, 185; ships from, at Newfoundland, 197; ports of call in, for Newfoundland trade, 315; complaints of quality of Newfoundland fish from, 317; British balance of trade with, 333.

INDEX

Jackson, Rev. John, first permanent Anglican priest in Newfoundland, 208; quarrels of, 208, 211.

Jago, John, signs agreement for St. John's government, 255 n.

Jamaica, proposed removal of planters to, 133.

James I, legislation respecting fisheries under, 33; peace negotiations of, 36.

James II, flight of, 167; disturbed conditions during reign of, 188-189; royal policy under, 209.

Jersey, Island of, vessels of, in Newfoundland trade, 314.

Jones, Capt. Daniel, R. N., commodore, charges violation of charter of 1676, 178.

Judges, governor authorized to appoint, 297.

Judicial procedure, Finch asked to propose, 132.

Justice, criminal, limitations on administration of, 296-299.

Justices of the peace, of Devon, complaints of, 83; of southwestern counties given jurisdiction over Newfoundland, 131.
See also Magistrates.

Keen, William, Sr., advocates civil government for Newfoundland, 255; murder of, 300, 302, 306; charges against, 306; first judge of vice-admiralty at St. John's, 326, 327.

Keen, William, Jr., appointed naval officer, 327.

Kempthorn, commodore, opinion of, 252, 253; report of, 262; mentioned, 263.

King (of England). *See* Crown.

King, William, criticizes Treaty of Utrecht, 242.

Kirke, Sir David, influence of Hyman on, 47; grantee of Newfoundland, 61; proposals of, for collection of duties at Newfoundland, 69; and associates apply for grant of northeastern America, 77-80; associated in Canadian enterprise, 78; and associates obtain patent, 81-82; goes to Newfoundland as governor, 82; friction with fishermen, 82, 83; charges against, 83-84; friction with France, 85; plans of group of, 88-90; investigated by Puritan authorities, 96-102; royalist activities of, 97; Baltimore's contest with, over Avalon, 104-108; death of, 106; introduces byboatkeepers into fishery, 110; French refuse to pay "acknowledgement" instituted by, 182.

Kirke, George, asked to assume proprietorship of Avalon, 107.

Kirke, Sir James, seeks confirmation of patent of 1637, 105.

Kirke, Lewis, associated with brother in Canadian and Newfoundland enterprises, 78; seeks confirmation of patent of 1637, 105.

Knoles, desertion of, 26.

Labor, employed afloat, 57; employed ashore, 58; proposed method of payment for, 89-91; quota of green men required for, 130; earnings of, 150; quality of, employed, 169; cheaper in France than in West Country, 184; inefficiency of, in Newfoundland, 211; scarcity of, in West Country, 245, 247; in Newfoundland, 247; traffic in indentured, 259-260; scarcity of, due to impressment and capture, 316.

Labrador, Indians from, 6; Arctic Current on coast of, 10; harbors of, blocked by ice, 12; habits of cod off, 16; Portuguese whalers at, 20; included under Newfoundland jurisdiction, 304; prospects of market for codfishery at, 319; extension of fishery to, 320, 321; awakened interest in, 321.

Labrador Banks, 8.

Labrador Current. *See* Arctic Current.

Land, sale of, 46, 49.

Lands, riparian, rights and titles to, 206-207.

La Poile River, 4.

Larkin, George, reports on disregard of Act of 1699, 215; delivers commissions for trial of pirates, 217; sets up prize court, 217-218.

Laurentian physiographic province, 5.

Laurentian Valley, a submarine channel, 8.

Law merchant, recognition of, for Newfoundland, 77.

Law officers. *See* Attorney and Solicitor general.

Lawrence, Charles, governor of Nova Scotia, 309.

League of Augsburg, War of the, effects of, 209, 210.

Leake, Capt. ——, R. N., commodore, on effects of rum, 194.

Lee, Fitzroy Henry, governor of Newfoundland, administration of 290-291, proposes extension of judicial system, 291; investigates reasons for poor quality of fish, 318; results of report of, 318; efforts of, to prevent illegal trade, 322-323, 326.

Legends, of early fishing voyages, 19-20.

Legislation, early, respecting fisheries and Newfoundland, 22, 33-34; proposed, respecting tithes on fish, 50; proposed, bill for free fishery, 50-52, 71; against train-oil monopoly, proposed, 64; king's power to make, for Newfoundland, doubted, 72; failure of proprietary groups to obtain, 87; during Interregnum, 93-96, 95 n.; effect of, passed in 1663 on history of Newfoundland, 113, 114.

See also Navigation Acts; Newfoundland Act of 1699.

Lent, English law on observance of, 33.

Lewes, John, appointed preventive officer, 326.

Liquor, taverns for, 75, 83; traffic with New England in, 143, 191; forced on planters, 252; evils of traffic in, 253-254; extent of traffic, 254; Irishmen prohibited from retailing, 303.

Lisbon, early expeditions to Newfoundland from, 20.

Littleton, John, appointed by Commonwealth to seize Kirke, 98 n.

Lloyd, Capt., later Major, Thomas, quarrels with Moody, 212; surrenders to French, 212.

Localism, of West Country, 53, 62, 89, 91.

Logwood, illegal export of, 198.

London, English fishermen to enjoy free market in, 65; Dutch ships at, 67; shipping group of, represented at hearings, 67; shipping of, 68, 71; ships of, engage in illegal trade, 196; Newfoundland trade at, 314.

London, Bishop of, interested in St. John's church, 208.

London, merchants of, interest of, in carrying trade, 36, 37, 38; interested in London and Bristol Company, 38; interested in colonization, 47; Calvert quarrels with, 48; monopolistic activities of, 49-51, 64; resentment of, against Dutch, 60; relations with adventurers in fish trade, 64, 68, 70; propose customs duties at Newfoundland, 69; represented on committee concerning shipping, 70; support Kirke's enterprises, 78-79, 81; fear conflict between Kirke and France, 84-85; interested in proprietory schemes, 87; wider vision than adventurers, 89; appeal to spirit of economic nationalism, 91; support planters, 108, 110-111, 121-125; seek civil government in Newfoundland, 114-117, 137; advocate improved production methods, 122-125; fail to oust adventurers from control, 134; adventurers victorious over, 140-141, 147; attitude toward nursery for seamen, 175-176, 188.

London and Bristol Company. See Newfoundland Company.

Long Island, plant life of, 5.

Long Range, backbone of Newfoundland, 3, 4.

Lords, House of, ignores bill for free fishing, 50, 51; indifferent to trade, 87.

Lords of Trade, Hinton's petition referred to, 135-136; investigate Newfoundland question, 140, 152, 155-157, 165-167; recommend additional rules of 1671, 145; program of, ordered put into effect, 146; receive petition from West Country, 155; suspend restrictive clauses of 1676, 157-158; prepare heads of inquiry, 159; favor charter regulation of fishery, 161, 167; seek information on French activities, 184; report on French activities, 185-186; indifferent to French competition, 188; accept views of adventurers on trade competition, 191; direct commodore to stop illegal trade, 196.

Louisbourg, French at, 243, 246; British fleet at, 295; expedition from, 329 n.

Louis XIII, French activities in America under, 61.

Louis XIV, encourages French fishing industry, 184-185; French seek control of American fisheries under, 182; powers of commodores during wars with, 307.

Lyme, mayor of, empowered to enforce charter of 1634, 76; shipping losses of, 121.

McCarty, Capt. Robert (Lord Muskerry), R. N., governor of Newfoundland, lacks ability, 290; denies existence of illegal trade, 322.

Madeira, wines of, imported into Newfoundland, 191; refuse fish carried to, 325.

Magdalen Islands, included under Newfoundland jurisdiction, 304.

Magistrate, chief, proposed for Newfoundland, 217.

Magistrates, local, appointed by royal governor, 275; legal texts supplied to, 277; functions of, 278-279; conflicts of, with fishing admirals, 280-282; governor to investigate, 284; ill-feeling of, toward fishing admirals, 284; constitutional point in dispute, 285-286; difficulties of, over jail tax, 286-287; Board of Trade reconsiders powers of, 288; complaints against, unsubstantiated, 290; cease interference in fishery, 293; become oppressive, 294-295;

Hamilton complains against, 295; governor to investigate, 296; courts superior to those of, provided, 297; ordered to prohibit Irishmen to retail liquor, 303; quarrels of, with lieutenant governors, 306; governors rely on, for assistance, 307.

Magistrates, West Country, to enforce charter of 1634, 76.

Maine, Gulf of, 8; smuggling of foreign brandy in, 199-200.

Manufacturers, proposed by Kirke group, 89-90; sold to planters in exchange for fish, 153; Newfoundland as a market for, 189.

Marine life, 7, 13-16.

Mariners. See Seamen.

Market(s), France as, during religious wars, 32; more profitable in southern Europe, 34; direct voyages to, 37; English kept from usual, 60; free, in London, 65; absence of, in England for fish, 65-66, 333; decline of demand in foreign, 118; French competition in foreign, 141; loss of, in France, 141; Newfoundland fish sold to foreign, 144; poor, after 1680, 169; Portugal as sure, 228; France controls foreign and domestic fish, 230; complaints in European, over poor quality of fish, 317, 319; competition of European fish in Portuguese, 319.

Maryland, smuggling of, 200-201.

Mason, Capt. John, obtains Scottish financial assistance, 44.

Massachusetts Bay, suppression of smuggling in, 200.

Masters, Capt. John, estimates value of trade, 312.

Mayors, of West Country towns to execute fishery regulations, 72-73, 76.

Meat, imported by New England traders, 192.

Mediterranean, plant life of Newfoundland similar to, 5; Dutch inaugurate direct voyages to, 37; Barbary pirates in, 61-62, 169; naval protection needed in, 180.

Medley, Capt. Henry, R. N., governor of Newfoundland, lack of ability of, 291.

Melcombe Regis. See Weymouth.

Mercantile marine, employment of Newfoundland-trained seamen in, 171; smuggling by personnel of, 198; cases of officers and men, of, 299.

Mercantilism, foundations of, during Interregnum and Restoration, 92; Newfoundland's place in theories of, 144-145, 202-203; Londoners influenced by, 147; theories of, not accepted by crown, 188, 189; indifference of Bolingbroke ministry to, 238, 244; unfavorable reaction to treaty of advocates of, 240-243; wanes, 330-331; never finds application in Newfoundland, 334-335.

Merchants, English, interest of, in Newfoundland, 22, 35-36, 47; desire peace with Spain, 36; represented on shipping committee, 70; fear conflict between Kirke group and France, 84; help pay expenses of West Country agents, 85; demand wars with Holland and Spain, 92; influence of, upon Charles II, 92-93; complain of decline of fishery, 118; exchange fish for manufactures, 153; control export trade to Newfoundland, 189-190; reasons for opposing colonial trade, 194; propose improvements in fishery regulations, 216-217; fear New England

INDEX

trade, 222; hold Irish responsible for disorder, 301; lose by trans-Atlantic races, 318; point out possibilities in Labrador, 321.

Merchants, French, good financial condition of, 184; maintain contact with Frenchmen in Newfoundland, 255; undersell British in Italian market, 318.

Merchants, of London. *See* London, merchants of.

Merchants, New England, develop organization in Newfoundland, 221.

Merchants, Newfoundland, involved in quarrels of officers, 212; protest business activities of officers, 229; resent erection of naval office, 327; increasing influence of, at St. John's, 331; victimize poor, 331.

Merchants, West Country, agree with Kirke group for transportation of fish, 80-81; oppose impost on foreign buyers, 80; employ share system, 90; complaints of, 316.

Micmac Indians, migration of, 6.

Militia, creation of, 212, 217; refuses to coöperate in march on Placentia, 212; creation of, proposed, 217.

Mining, colonization as a means of developing, 24; Kirke group proposes to engage in iron, 90.

Molasses, importation of, 192, 194.

Monadnocks, of the interior, 4.

Money, used in New England trade, 192, 193.

Monopoly, contest over, in fisheries, 36-37, 38, 44, 69, 77-78, 91, 134, 151, 154, 334; in trade to southern Europe, 38, 233; legislation against, 49-53; of shipping, 51; on train oil imports, 64; as a

means of increasing national wealth, 87; Spanish commercial, in America, 92.

Montagnais Indians, migration of, 6.

Moody, Lieutenant, later Col. John, quarrels with Lloyd, 212; prohibits French ships from trading, 255; first lieutenant governor at Placentia, 305; dictator and magnate, 306.

Murder, trials for, at Newfoundland under jurisdiction of court of earl marshal, 74.

Muskerry, Lord. *See* McCarty, Gov. Robert.

Nantes, merchants of, 184.

Nationalism, commercial, supplants localism and medievalism, 53; economic, in relation to fisheries, 88-89, 90, 91.

Naval office, at Boston, 198-199; erected at Newfoundland, 327.

Navigation, obstructed by ice, 11; effect of fog upon, 17-18.

Navigation Act(s), applicability of, to Newfoundland, 199, 202; of 1650, 93-94; of 1651, 95; of 1660, 195-196; of 1663, 195-196, 257, 277; of 1696, 210.

Navy, Royal, Newfoundland fish as provision for, 34; inadequacy of, 62; supervision of fishery by officer, 127; jurisdiction of officers of, 132-133; planters offer quota of men to, 166; Newfoundland fishery a source of recruits for, 171; dependence of adventurers on, during wars, 171; crown's interest in building up, 172; objections of, to fishery controlled by planters or byboatkeepers, 174; adventurers require protection of, 180; shore activities of chaplains

of, 207-208; function of fishery in building up, 325-336.
Nets, suppled from New England, 193.
Newcastle, Duke of, secretary of state, 273, 288.
New England, similarity of Newfoundland to, 15; attempt to exclude interlopers from, 46; climate, 54; raids on shipping of, 62; Trade of, 109, 135, 142, 143, 166, 190-202, 220-224, 256-260, 325; French menace to, 116; fear that Newfoundland might develop along lines of, 117, 143, 333-334; influence of, feared, 143-144; emigration from Newfoundland to, 172, 178-179, 217, 224-227, 325-326; merchants trading to, 188; Congregationalists from, in Newfoundland, 208; profits of Newfoundlanders go to, 247; success of, attributed to share system, 248; responsible for conditions in Newfoundland, 266, 267.
New England Council, 46.
New England fishery. *See* Fishery, New England.
Newfoundland, location and description, 1-3; geology and topography, 3-5; flora and fauna, 5-6; Indians of, 6; undersea topography and hydrography of region, 7-13; marine life in region of, 13-17; climate, 17-18, 54; resources, 18; discovery of, 19-20; early fisheries at, 20-23; Parkhurst reports on, 22-25; Gilbert's activity in, 25-27; Guy's settlement in, 38 ff., lands disposed in, 46-47; Calvert's settlement in, 47-48; granted to Kirke group, 61, 78-82; threatened by Barbary pirates, 62; foreign buyers at, 67; fishing regulations and administration for, 71-77; king's right to legislate for, 72; Sir David Kirke, governor of, 82; plans of Kirke group for colonization of, 88-90; causes hampering settlement of, 87-88; commissioners for Commonwealth and Protectorate at, 99-100; permanent government for, proposed, 102, 114-116; legislation respecting, 113; restrictions upon inhabitants of, 129-130; report of Lords of Trade, concerning, 141-147; economic position of, 144-145; proposed government and military defense of, 170; floating population of, 177-178; as a market for English manufactures, 189; as a center for smuggling, 197-198; uncertain status of, 199, 202; religious influences in, 207-208; French raids on, 209-210; defense of, 217; negotiations over cession of, 235-240; confusion in, after 1713, 245 ff.; development of, retarded, 253; intemperance in, 254; increased influence of inhabitants of, 254-255; migration from, 259-260; crime and disorder in, 263; character of people of, 266; civil government established in, 269; potentialities of, 271; beginnings of colonial society in, 272; enters period of development, 273; royal governor appointed for, 274-275; government of, reorganized to include Labrador, Anticosti, and the Magdalen Islands, 304; end of an era in, 304; lack of demand for participation in government of, 309; condition of society in, 309; defense of, regarded important, 329; feudal control in, passes, 332; trade of, branch of Great Britain's foreign commerce, 333; limitations of, for settlement, 333;

vital function of, in training seamen for royal navy, 335; insular situation of, determines course of economic and political development, 336.

Newfoundland Act of 1699, terminates charter period regulation, 181, 204; passage of, 210, 213-214; description of, 213-214; general disregard of, 215-216, 224, 273; parliament asks crown to enforce, 226; necessary amendments to, refused, 226; commercial disagreements not covered by, 251; government unable to enforce, 261; amendments advocated, 262-263; reasons for failure of, 264; Board of Trade adheres to, 266; legislation proposed to replace, 268; general trading not covered by, 271; governor not to violate, 275; jail tax levied in violation of, 278; tax on fish contrary to, 279; favored position of western adventurers under, 280; administration of justice under, 285-286, 296-297; crown reluctant to ask alteration in, 286; claim of western adventurers in respect to, 288; relinquish privileges established by, 290; prevents use of royal prerogative, 292; reports required concerning, 296; powers conferred on commodores under, 307-308; failure of, contributes to change of policy, 334; purpose of certain regulations of, 335.

Newfoundland Banks. *See* Banks.

Newfoundland Company (London and Bristol Company), organized, 36, 38-39; objects of, 39; disputes of, with fishermen, 40-44; reorganized, 44; fails to obtain changes in charter, 44-45; land sales, 46, 48-49; decline and disappearance of, 49; monopolistic tendency of, 49-52; failure of, 53, 54; complaints of fishermen against Kirke similar to those made against the, 82-83; financial difficulties of, 88.

Newfoundlanders. *See* Planters.

New Jersey, plant life of Newfoundland similar to, 5.

Newman, George, register of prize court, 217-218.

New York, French menace to, 116; merchants trading to, 188; trade of, 199, 201, 314; expedition from, to relieve St. John's, 329 n.

Nicholson, Col. Francis, John Moody procures commission in Regiment of Foot of, 212.

Normans, early fishermen at Newfoundland, 21.

Norris, Capt. Charles, commodore, reports of, 194, 197.

Notre Dame Bay, 3.

Nova Scotia, similarity of Newfoundland to, 1; banks off, 8; course of Arctic Current along coast of, 10; colonization of, 46, 47, 61, 78; negotiations with France over, 235-237; proposed transfer of Newfoundlanders to, 247, 266, 274; ignored, 269; Placentia removed from jurisdiction of governor of, 275; influence of, on Newfoundland, 305; indignation in, at delay in calling assembly, 309.

Noye, William, attorney general, opinions of, 72, 73.

Oceanography, of Newfoundland region, 8-13.

Ochiltree, Lord, colony of, 61.

Officer, preventive, Cumming appointed, 218, 220-221, 225; Lewes appointed, 326-327.

Officer, prize, engages in business, 229.
Officers, Military, character and quarrels of, 211-212; at St. John's engage in trade, 212.
Oleron, laws of, charter of 1634 similar to, 77.
Osborn, Capt. Henry, R. N., first royal governor of Newfoundland, 275; instructions of, 275, 276-277; divides authority with Beauclerk, 276-277; commission to, 275; administration of, 277-283; continues as governor, 284; ability of, 288; instructions of, to lieutenant governor at Placentia, 305.
Outports, religious inactivity in, 208; Irish Catholics outnumber Protestants in southern, 302; surrogates appointed in remoter, 307; vessels owned at, 314; search at, for contraband, 323; emigration carried on from, 325.
Oxford, Thomas, agent for planters, 165-168; not influenced by mercantilist theories, 243.
Oyer and terminer, commissioners of, 297, 298; court of, held at St. John's, 299.

Packing, methods of, 56.
Palmer, Sir Henry, admiral at Plymouth, 29 n.
Parkhurst, Anthony, reports on Newfoundland, 22-23; scheme for settlement and operation of fishery, 23-25, 35; mentioned, 32.
Parliament, bills for free fishery in 49-52; West Country representation in, 49-50, 210; train-oil monopoly called to attention of, 64; inability of western adventurers to secure legislation in, 71; opposition and indifference of, checks proprietary groups, 87; legislates for Newfoundland fishery, 113; adventurers turn to, in 1699, 181, 209; passes Newfoundland Act of 1699, 212-213; considers changes in Act of 1699, 215-216; has limited interest in Newfoundland, 226; rejects commercial treaty, 238-239; gives no attention to fishery, 268; crown avoids clash with, 276, 292, 297; opposition to government not established by, 284; growing power of, 285; crown's attitude toward, 286.
See also Commons, House of; Lords, House of; *and* Legislation.
Parmenius, Stephen, adverse to colonization, 27.
Parrett, ———, London merchant, opinion of, concerning Newfoundland, 121; agent of adventurers, 166.
Passenger, William, commodore, report of, 266-267.
Passengers, carriage to Newfoundland restricted, 129-130, 132; encouraged, 177; carried away by New Englanders, 259-260.
Peas, imported from New England, 192.
Pease, Capt. ———, governor of Avalon, 106-108.
Peck, English pamphleteer, quoted, 230.
Peckham, Sir George, favors colonization, 27.
Pembroke, Earl of, grantee of Newfoundland, 61; applicant for charter, 78.
Penalties. *See* Fines.
Pennsylvania, smuggling from, 201.
Percy, Commodore, on presence of Irish Catholics, 300.
Pescaria, Terra de, early name of Newfoundland, 20.

INDEX 385

Philadelphia, participates in carrying trade, 314.

Philip II, policies of, injure Spanish fishery, 30.

Phillips, Richard, governor of Nova Scotia and Placentia, proposal of, 265-266, 269; responsibility of Gledhill to, 273; consulted, 274.

Physiography, of Newfoundland, 3-5.

Pickling, method of, fish in the bank fishery, 56.

Pilchards, overstock of, 65-66; bought by foreigners, 67.

Piracy, English resort to, 28, 31; Newfoundland Company complains of, 41, 42; Calvert aids fishermen against, 48; protection sought against, 137, 180; efforts at prevention of, 217.

Pitt, William, given estimate of Newfoundland trade, 312.

Placentia, included in Calvert's Avalon, 47; French settlement and garrison at, 115, 182-187; adventurers fear New England traders more than French at, 191; Franciscan mission at, 208; English expedition against, 210-211; Lloyd proposes attack on, 212; Moody, commander at, 212; French vessels engaged at, 231; governor eliminates friction at, 232; English covet, 233; proposal of draft treaty respecting, 235; Governor Phillips asks for increased power at, 266, 269; Gledhill interferes with fishery and trade at, 273, 305; titles to be cleared at, 274; placed under royal governor, 275; Osborn at, 277; fishing admirals assume judicial powers at, 280; garrison at, 290; justice of peace at, 295; Hamilton asks how to preserve order at, 295; lieutenant governors at, 305, 306; fishing curtailed at, during War of the Austrian Succession, 313; vessels owned at, 314; effect on British of removal of French from, 328; fortifications of, rehabilitated, 329.

Placentia Bay, important fishing resort, 3; temperature of seawater near, 11; good spawning place for cod, 16.

Plankton, described and classified, 13-15.

Plantations. *See* Colonies.

Plantation trade. *See* Trade, Colonial.

Planters, effect of arrival of, 39, 40; appropriate property of fishermen, 40-41; advantages of the, in fishery, 41; conflict with adventurers, 42; desire jurisdiction over fishermen, 44; Falkland advertises for, in Ireland, 47; left in Newfoundland after collapse of proprietary grants, 49, 55; begin fishing earlier than West Countrymen, 57; injure fishery and Indians, 74; taverns conducted by, prohibited, 75; excluded from participation in administration of fishery, 76-77; restrictions laid upon, by Kirke charter, 81, 82; rights conceded to, 83; restrictions on, hamper settlement, 88; Kirke accused of partiality to, 98; regulations of 1652 and 1653 respecting, 101; allies of London merchants, 108; seek reorganization of fishery, 110-111; seek civil government in Newfoundland, 114-115, 135; controversy over removal of, 117, 133, 145, 149-151, 157-158, 165-167, 226, 274; fishery managed by, feared, 117-118; blamed for decline of fishery, 118, 120-121; development of fishery

386 BRITISH FISHERY AT NEWFOUNDLAND

conducted by, threatens West Country, 120-121; London merchants favor, 121-125; contest of, over fishery regulations, 126 ff.; restrictions proposed upon, 129, 134; seduced by France, 137; defeated in investigations of 1675-1676, 140-141, 147; competition of, in Newfoundland fishery, 142; poverty of, 150; destructiveness of, 150-151; benefits from presence of, 152-153; Berry reports on, 153-154; Downing upholds, 154, 156; Poole defends, 160; condition of, in fishery, 160, 162-164, 169-170, 180-181; lack of pioneering spirit among, 165; demands of, 166; offer quota of men for navy, 166; crown more lenient toward, 166-168; well represented, 167-168; in relation to nursery for seamen, 171-172, 174; injured by emigration, 178; attitude toward French, 182-183, 185, 188; as consumers of imported goods, 189-192; interest of, in riparian land, 206-207; condition of, in 1699, 206-209; arms and ammunition distributed to, 212; involved in quarrels of military officers, 212; effect of Act of 1699 on, 214; losses of, during war, 228; increase share of catch, 245-246; unwilling to occupy south coast, 246; pay higher wages, 248; forced to buy unneeded goods, 252; ruined by debts, 252; more influential in fishery and commerce, 254; Irishmen seek employment from, 301; first to call attention to possibilities of Labrador, 321; dependent on plantations for provisions, 325; social condition of, 331-332; British fear encouragement of 333;

competition of, prevents monopoly, 334.

Plant life, origin and character of, 5.

Plymouth, interest of, in colonization, 47; fishing ships from, 55; Newfoundland trade of, under Charles I, 63; foreign shipping at, 66, 67; agents of, 67, 71-72; copy of western charter of 1634 at, 73; mayor of, empowered to enforce charter of 1634, 76; merchants of, complain against Kirke, 96-98; merchants of, complain of losses, 111; shipping losses of, 121; merchants of, complain of Irish and New England trade, 193.

Pollexfen, John, agent for adventurers, 156.

Poole, Newfoundland trade of, 63, 314; shipping losses of, 121; claim of merchants of, 288; complaints from, 316.

Poole, Capt. Sir William, commodore, reports of, 159-160.

Population, of Newfoundland scanty, 142; impermanence of, 177-178; increase in, 189; data on, 311-313.

Pork, imported from New England, 192.

Port aux Basques, distance of, from North Sydney, N. S., 2 n.; French reported settling near, 328.

Portugal, discovery of Newfoundland by, and early interest of, in island and fishery, 20-21; activity of, in fishery, 23, 31, 55, 56; loss of fishery to England, 30-31; as market for fish, 31, 163, 193, 228, 319; proposed monopoly on trade with, 38; protection of market fleet bound for, 180; France favorably situated to sell to, 185; salt illegal trade with, 196, 220; ships from, at Newfoundland,

197; ports of call in, for Newfoundland trade, 315; complains of quality of Newfoundland fish from, 317; Great Britain, maintains balance of trade with, 333.

Potash, manufacture of, proposed, 90.

Prerogative, Royal, adventurers fear extension of, 62; king's right to exercise, over Newfoundland, 72; attempt to extend, in America, 168; crown refuses to exercise, 292.

Presbyterians, Scottish, in Newfoundland, 208.

Prices, fall in, for fish in Spain, 119; need of better quality fish at lower, 122; adventurers accused of fixing, 154; for fish and oil, 161-164, 169; English competition with French, 186; rise in, of English commodities, 189-190; foreigners at Newfoundland offer better, 197; War of Spanish Succession raises, 221; remain the same, but quality of fish declines, 318; British trade suffers from high, 318, 319.

Prior, Matthew, negotiates with France, 236-239; not influenced by mercantilists, 243.

Privy Council, Newfoundland Company complains to, 41; reviews rivalry of fishermen and planters in 1618 and 1619, 42-43; considers grievances of fishing interests, 43-44; committee of, reports on Newfoundland Company's request, 44-45; adventurers complain to, 64; orders of, respecting carrying trade and fishery, 65, 66; investigation of 1633-1634, 67-71; directs circulation of charter of 1634, 73; hearings before, on Kirke patent, 77, 79; actions of,

in Kirke dispute, 80-84; considers affairs of West Country, 111; attempts to enforce charter of 1661, 114; studies civil-government proposals, 115, 116, 127; receives complaint against byboatkeepers, 127-128; approves additions to charter, 128; changes criminal procedure and powers of commodore, 131-132; enjoins obedience to amendments of 1671, 133; proposes removal of planters, 133; refers petitions to Lords of Trade, 135, 155; approves report of Lords of Trade, 146-147; suspend restrictive clauses, 158; authorizes investigation of fishery, 159; demands enforcement of charter, 167, 173; orders governor and fort for Newfoundland, 170; considers strengthening of administration, 215-216; Board of Trade submits draft bill to, 268; prestige of, declines, 285.

Prize Court, established at St. John's, 217-218; foreign ships condemned in, 314.

Prizes, agencies authorized to seize and condemn, 217-218; agents for, 118, 218.

Profits, maintenance of fishing fleet reduces, 89; considered inadequate, 89; share-system of, 90; reduced after 1680, 169.

Property, protection of, 75, 81, 113; confiscation of, prescribed for violation of charter of 1634, 76; of fishermen destroyed by planters, 83; disputes over, in fishing rooms, 283.

Proprietors, Newfoundland Company disposes of lands to other, 46-48, 49; and restrictions upon, 81-82, 88; complaints against, 82-84; differences between, and adventurers, 86-89; control by, barred, 125; Lord Baltimore makes last claim of, 332; reasons for crown's refusal to recognize claims of, 334.

Protectorate, *see* Interregnum.

Protestants, feeling of, toward Catholics, 301; Catholics outnumber, 302; intolerance of, 303.

Provisions, Gilbert's levy for, 26; stored at Newfoundland, 40; fish as, for troops, 59; appropriation of, forbidden, 75; Kirke group propose to furnish, 89; adventurers' capital tied up in, 89-90; furnished from West Country, 90; to be procured for entire voyage before leaving England, 130; Ireland and New England in trade in, 135, 143; cost of, 163; cheaper in France than in West Country, 184; planters dependent upon England or other sources for, 189; colonial trade in, 194, 199; annual importation of, 312; Newfoundlanders dependent on plantations for, 325.

Puritans, early influence of, in Guy's colony not continued, 207.

Pyle, Capt. William, commissioned to govern Newfoundland, 100.

Pyrenees, Treaty of, 185.

Quary, Robert, reports on illegal trade, 201.

Quintal, the measure of the catch and sale of fish, 57-58.

Races, trans-Atlantic, result in losses, 318.

Raleigh, Sir Walter, 29.

Randolph, Edward, surveyor general of customs, attempts to check smuggling, 198-202; remarks of, 219.

Rayner, Capt. John, deputy governor of Avalon, 106-108, 195.

Red Indian Lake, home of Beothuks, 6.

Regulations, fishery, by orders in council ineffective, 65; proposed, 68; of 1634, 72-77; of Interregnum, 101; enforced by commodore of convoy during, 103; struggle for control centers on, 110; Privy Council attempts to enforce, 114; amendments of 1671, 129-134; contest over, in 1675, 136-141; report of Lords of Trade on, 141-147; adventurers dissatisfied with new, 154-155; attempt to settle, in 1677, 160-161; planters willing to coöperate with adventurers in proper, 166; ineffectiveness of, by charter demonstrated, 180, 209; results of loose character of, 202-203; of Act of 1699, 214-215; proposed changes in 1703-1705, 217; proposals to strengthen, 263-265.

Religion, fear of New England influence upon, 143; development of, in Newfoundland, 207-208.

See also Catholics, Church of England, Congregationalists, Presbyterians, Protestants.

Restoration, significant epoch in English commerce, 92; reactionary tendencies of, 104.

Revolution, American, 309.

Rice, Rev. Jacob, clergyman at St. John's, 208.

Rich, Col. Robert, 104-105.

Richmond and Lenox, Duke of, 44.

Rivers, of Newfoundland, 3-4.

Robinson, Capt. Robert, cited, 111 nn.; candidate for governorship, 116 n., 126, 166; proposes civil government at Newfoundland 126-127; effect of petition of,

INDEX

133-134; report of, 164, 196; attempts to enforce Navigation Acts, 195-196.

Rochelle, La, merchants of, 184.

Rodney, Capt. George Bridges, R. N., governor of Newfoundland, admistration of, 296-297; alarmed over Irish Catholics, 302; reports illegal trade stopped, 327.

Rooms, fishing, method of selecting, 26, 57, 75-76, 81; first choice of, ignored by Guy, 40; fishermen forced from, by planters, 40; disputed occupancy of, 42; appropriated by planters, 83; disputes over property in, 283.

Roope, Col. ——, Royal Engineers, 211, 263.

Rouen, merchants of, 184.

Rum, imported by New England traders, 192; traffic in, 193, 194.

Rupert, Prince, 97.

Rushworth, Bryan, register of vice-admiralty court, 225.

Ryder, Sir Dudley, attorney general, advises crown, 297, 298.

Ryswick, Peace of, 211.

Sailors, see Seamen.

St. Christophers, proposed removal of planters to, 133.

St. George's River, 3-4.

St. Jean de Luz, fishing vessels from, 184.

St. John's, temperature of sea water near, 11; fog at, 18; Gilbert at, 26; Sir John Berry at, 149, 150, 151-152; English shipmasters at, 150-151, 160; shipmasters at, advise Poole as to conditions of fishery, 160; fortifying of, considered, 170; New Englanders engage in passenger traffic at, 178; fire at, 207; Church of England at, 208; capture of, by French, 208, 210; English fort at, 209, 211, 212; court for pirates and prize cases at, 217-218; public houses in, 254; as commercial center, 254; civil government erected at, 255 n.; opposition of "Trading Men" at, 293; defense of, 293; court of oyer and terminer at, 299; murder of Keen at, 300; Catholics outnumber Protestants at, 302; Irish forbidden to retail liquor at, 303; political influence shifts from West of England to, 273; jail tax levied at, 278; tax collectors defied at, 280; justices uncertain of powers at, 281; jail at, 286; convoy protection at, 313; vessels owned at, 314; vice-admiralty court at, 323, 326, 327; efforts of merchants at, to control illegal trade, 323-324; lieutenant governors at, 305, 306; customs house established at, 327; recaptured and regarrisoned, 329; increasing influence of merchants of, 331.

St. Lawrence, Gulf of, 1; rivers flowing into, 4; Arctic Current in, 10; characteristics of, 11; poor spawning place for cod, 16; French fishery in, 21, 56, 172, 182; islands in, conceded to France, 236.

St. Lawrence River, islands in, conceded to France, 236.

St. Malo, fishing vessels of, 184, 186, 231; merchants of, 184.

St. Mary's Bay, 3.

St. Paul Island, 11-12.

St. Pierre Bank, 8.

Salee Rovers. See Barbary Pirates.

Salmon, reported by Cabots, 19.

See also Fishery, salmon.

Salt, making of, proposed, 24, 89; stored in Newfoundland during winter, 58; appropriation of, for-

390 BRITISH FISHERY AT NEWFOUNDLAND

bidden, 75; dependence upon foreigners for, 89-90; decline in number of salt ships, 120; only supply not required to be purchased in England, 130; French government provides, 187; importation of, permits illegal trade, 195; French have better grade of, 232.

Salting, method of, 56; labor employed in, 58; right of English fishermen to engage in ashore, 129.

Sandys, Sir Edwin, 38.

Schools, 207.

Scilly Islands, fishing fleet intercepted at, 29.

Scotland, Newfoundland Company assisted by merchants of, 44; proposed settlement of Nova Scotia by, 46, 47; export of fish by ships of, 65; export trade from, 96, 190; illegal trade with, 198-201, 219; Presbyterians from, in Newfoundland, 208; merchants from, establish factory in Newfoundland, 219.

Scurvy, prevalence of, 58; planters cure fishermen of, 153.

Scutt, Benjamin, agent for adventurers, 156, 166.

Seal, as indicators of field ice, 13; reported by Cabots, 19; data on production of, oil, 311, 313-314, 320; skins become important item in export trade, 313-314.

Seamen, economic position of, would be improved by colonization, 24; escape impressment, 59; number of, at Newfoundland, 68, 69; recruited in West Country, 90; paid on share system, 90; provision for continuous supply of, 130; inadequately trained in fishery, 174; predicted decline in number of, 176; desertion of, at Newfoundland, 177, 197; emigration to New England, 178, 217, 245; attempt to curtail emigration of, 217; number of, employed in Newfoundland fishery, 231; impressment of, 316.

See also Seamen, nursery for; Fishermen; and Labor.

Seamen, Nursery for, the fisheries as, 33-34; projected by France, 61; decline of, prophesied by West Country, 68; threats to prosperity of, 83, 114; loss of, as blow to royal navy, 116; attempt to encourage growth of, 129-130; importance of, 137; as factor in Newfoundland policy, 170-180; claims of London merchants regarding, 188; affected by illegal trade, 223; British, increased by elimination of French, 234; effect of lawlessness on, 273; importance of fishery as, 329-330, 335; handicaps of, 335; vital to nation, 335.

Seasons, fishing, how determined, 15; length of, 56-57, 58.

Seines, used to take bait and cod fry, 40, 57; prohibited for taking bait, 113.

Servants, *see* Labor.

Settlement, recommended by Parkhurst, 23-25; Gilbert's, 25-26; fishermen oppose, 27; schemes for, 27-28; attempt of Newfoundland Company, 36, 39; advocated by Whitbourne, 45; traditional opposition to, 54; result of commercial competition, 61; of Canada proposed, 69; restricted by Kirke charter, 81-82; of Newfoundland hampered, 87-88; Kirke's plan for, 88; unfitness of Newfoundland for, 117; crown opposes, 134; proposals for, rejected, 141, 142.

Settlements, English, limited number of, expected, 52; zone of, 55;

INDEX

descriptions of, 57; unsuitable for civil government, 142.
Settlements, French, at Placentia subsidized by government, 182; condition and growth of, 186-187.
Seven Years' War, close of, marks end of an era in Newfoundland, 304; lack of reports of governors during, 309; improvement in fishery preceding, 313; condition of British trade at close of, 318; sealing becomes important industry after, 320; expansion northward during, 321; French occupation of St. John's during, 329; work on fortifications at Placentia and St. John's during, 329.
Shares, western adventurers pay labor on, 89, 90; of deserters to be forfeited, 130; abandoned, 248.
Shaw's *Practical Justice of the Peace*, text for magistrates, 277.
Sherwell, Thomas, 52.
Shetland, Newfoundland fishery classed with, 22.
Shipbuilding, French subsidies for, 185; in plantations, 314.
Shipmasters, individual initiative of, 35-36; object to abandoning old fishing customs, 43; fishing, annoyed by Londoners, 51; keep extra boats in Newfoundland, 58; protest against impressment of crews, 59; privileges of first arrival among, 75-76; accused of destroying stages and cookrooms, 84; ordered to collect impost from foreigners, 86; work on share system, 90; restrictions proposed on, 129-130; favor destruction of fishing conveniences, 150-151; leave men in Newfoundland, 151; oppose removal of planters, 152; Poole consults with, 160; attitude toward nursery for seamen, 177; interest of, in church at St. John's, 208; oppression of planters by, 208-209; admonished by Board of Trade, 216; commissioned to try pirates, 217; engage in liquor traffic, 254; West Country, called "kings," 261; disrespectful to justices, 284; arrogant attitude of, 285; oppose jail tax, 286; magistrates act arbitrarily toward, 288; deplore presence of Irish, 301; warned to cure fish properly, 317; rivalry of, 318; smuggling of, 323; colonial, required to give bond, 325; English, encourage emigration, 325-326; resent erection of naval office, 327.
Shipowners, restrictions upon, 129-130; dependent upon royal navy, 171; financial burdens of, 316; lose by trans-Atlantic races, 318.
Shipping, colonial, at Newfoundland, 164, 192, 314-315; registry of, 314; classes of, 315; routes and ports of call of, 315.
Shipping, English, frequency of, at Newfoundland, 22, 55; attacked by French and Barbary pirates, 59, 61-62; amount of, in Newfoundland traffic, 63-64, 69; trade confined to, 64-65; competition of foreign ships with, 66-67; of London, inadequate, 68; proposed restriction of carrying trade to, 70; employment of, advocated, 89; injury threatened to, by byboatkeepers, 114; decline in, 118-119, 120-121, 141, 174, 190; losses of, during wars, 121; restrictions upon, trading to Newfoundland, 129-130; distress of, relieved by planters, 153; participation of, in illegal trade, 195, 196, 219, 220; classes of, 314-315; routes and ports of call of, 315.

392 BRITISH FISHERY AT NEWFOUNDLAND

Shipping, foreign, engage in carrying trade, 35, 37, 66; attempted elimination from carrying trade, 67-71; reported by Robinson, 164; participation in illegal trade, 195; provision of Act of 1660 respecting, 195; commodores directed to seize, engaged in illegal trade, 196-197; at Newfoundland toward end of seventeenth century, 197.

Shipping, French, number and distribution of, at Newfoundland, 184-185; subsidized, 185; Berry reports on, at Newfoundland, 186; Parkhurst reports on, 23; no direct voyages before 1610, 35; at Newfoundland, 37; foreigners to be excluded from, 65-66; reasons for increase of, at Newfoundland, 68; engaged in illegal trade, 195, 196; data on, 311-315.

Shipping, Irish, in trade to Spain, 24-25; in Newfoundland trade, 314; character of, 315.

Shipping, Newfoundland, amount and character of, 314, 315.

Shipping routes, 315.

Ships, fishing, English, proposed conversion into carriers, 24; to pay for protection, 41; accommodations for, 52; dates of departure and arrival of, 56-57, 58; Dutch built, 68; regulations respecting, 75; proprietors employ own, 83; fitted out in West Country, 90; mortgaging of, 121-122; numbers of, 161-162, 164, 169, 251; decline in use of, as carriers, 164; planters not desirous of ousting, 166; cost of equipment of, 247; kind employed in bank fishery, 249-250; English operations of, after 1729, 310; data on, 312-315.

Ships, fishing, French, fitted out more cheaply than English, 184; size and numbers of (1676), 187; cargoes of, condemned at Newfoundland, 220.

Ships, sack, foreign, impost on fish bought by, proposed, 80; an incentive to improved product, 122; masters advise Poole, 160; number of, 162, 163; carry sixty per cent of catch by 1680, 164; large number of, 169; competitive-bidding advantages of, 169; engage in barter, 193; scarcity of, 229, 251; ports of call of, 315; competition of, responsible for decline in quality of fish, 318.

See also Carrying trade *and* Shipping.

Ships, salt, import brandy and wines into Newfoundland, 191; carry on illegal trade from Europe, 195, 219-220; acquire foreign goods after leaving England, 196; force sale of unneeded goods, 252.

Shoals, on Newfoundland banks, 9.

Sikes, Walter, commissioned to seize Kirke, 98 n.; commissioned to govern Newfoundland, 100.

Silver, Newfoundland Company hopes to find, 49.

Slaves, refuse fish as food for, 192.

Smith, James, judge of vice-admiralty court, 225.

Smith, Capt. Thomas, R. N., governor of Newfoundland, complains of lack of authority, 292; occupied with military duties, 293; appoints military officers as magistrates, 306; estimates value of Newfoundland trade, 312; appoints first naval officer and deputies in customs service, 327; lands marines to defend St. John's, 329.

Smuggling, extent of, and lack of facilities for checking at New-

foundland, 197-198; Edward Randolph attempts to check, 198-202; between New York and Newfoundland, 199; problem of preventing, important, 209; investigation of, 218-219; during War of Spanish Succession, 219-227; shipmasters increase, 245; highly organized, 256-257; lack of customs machinery for suppressing, 257, 322; necessity for vice-admiralty courts to curb, 323; efforts to control, 323-324, 326-328.
See also Trade, illegal.
Society for the Propagation of the Gospel, 208.
Soil, ill-suited to agriculture, 5-6, 333; probable effect of, on Calvert's abandonment of Avalon, 48; Lords of Trade emphasize infertility of, for settlement, 143.
Sole, reported by Cabots, 19.
Solicitor General, opinions of, 105-106; conducts negotiations between Kirke group and adventurers, 80; directed to consider charges against Kirke, 82-83; Sir Heneage Finch, considers and upholds Baltimore's claim to Avalon, 105-106.
Somersetshire, interest of, in fishery, 49-50; representation of, in parliament, 49-50.
Southampton, merchants of, enter agreement with Gilbert, 26; Gilbert's favoritism to fishermen from, 26; interest of, in Newfoundland fishery, 36; fishing ships from, 55; Newfoundland trade of, under Charles I, 63; mayor of, empowered to enforce charter of 1634, 76.
South Sea Company, attempts control of fishery, 230 n.

Sovereignty, English, claimed over Newfoundland, 25, 26, 74, 137.
Spain, Basques from, at Newfoundland, 21; size of fishing fleet of, in 1578, 23; war with, 28-30; rise and fall of fishery of, 30-31; as market for fish, 31, 34-35, 38; English seek to improve trade with, 36; Dutch trade with, 37, 60; proposed monopoly of trade with, 38; fishermen of, make wet fish on banks, 55; wet fish cured after arrival in, 56; effects of war with, 59, 95-96, 102-103, 121, 171; foreigners purchase fish for, 67; Dutch-built ships prohibited from entering, 68; reasons for war with, 92; curtailment of market in, 109, 119; amount of fish shipped to, 163; protection of fleet bound for, 180; France admitted to markets of, 185; exports of, 192; New Englanders trade with, 193; source of illegal imports, 196; ships from, buy fish at Newfoundland, 197; entices English seamen to desert, 197; ports of call in, for Newfoundland trade, 315; complaints from, of quality of Newfoundland fish, 317; competition of European fish in markets of, 319; British balance of trade with, 333.
Spanish Company, organized, 38.
Spanish Succession, War of, embargoes on Newfoundland shipping not general until, 179-180; Newfoundland during, 204-244; outbreak of, 211; military events in Newfoundland during, 211; effect of, 221; French rivalry during, 234; Newfoundlanders entrenched in management of fishery during, 229; preliminary peace conferences, 234-235; Prior-Desmarais draft agreements, 236-

237; Treaty of Utrecht concludes, 238; St. John's garrisoned during, 329.

Spawning, of cod, 13; location of, places, 16.

Spirits, see Liquor.

Stages, Gilbert confirms titles to, 26; maintained for conveniences of fishermen, 40; first choice of, ignored by Guy, 40; use of, for landing fish, 56; fishing admirals have first choice of, 75; destruction of, 84, 150-151; fishermen's right to construct, 129; to be occupied by members of same ship's company, 130.

Star Chamber, see Privy Council.

Stockfish, from Iceland, 19.

Stock-raising, colonization as a means of developing, 24.

Storage, of fishermen's goods permitted, 81; of fish assured, 129.

Storerooms, maintained by adventurers, 40.

Storms, 17.

Strait of Belle Isle, see Belle Isle, Strait of.

Street, Robert, one of commissioners to govern Newfoundland, 100.

Stuart Rebellion, 296.

Sturton, Rev. Erasmus, minister of Guy's colony, 207.

Subsidies, to French at Placentia, 182, 185, 187; effect of, discounted, 185.

Sugar, imported by New England traders, 192.

Surrogates, governors appoint in outports, 307; lack time to visit all communities, 308.

Survey, of fishing harbors proposed, 133.

Surveyor General. See Edward Randolph.

Swamps, of interior, 6.

Swanley, Capt. Robert, Baltimore's lieutenant at Ferryland, 106.

Tackle, fishing, supplied from New England, 193.

Tar, imported from New England, 192.

Tariff, France places, on dried codfish, 184-185.

Tavener, Capt. William, opinion of, 252; accused of illegal trading, 255; mentioned, 263; submits estimates on fishery, etc., to crown, 311-312; reports French settlement near Gaspey, 328.

Taverns, prohibited by charter of 1634, 75; licensed by Kirke, number and location of, in 1726, 254; license system for, 264.

Tax, export, paid by foreigners, 68; jail, levied at St. John's; 278, 279, 286-287.

Taxation, early Portuguese attempt, of cod, 20; bill to remove, on fish in Commons, 50; customs duties only form of, permitted on Newfoundland products, 82; on cod caught by Englishmen at Newfoundland prohibited, 113, 125; of fish and train oil proposed, 116-117; importance of, necessary to support civil government, 139.

Teignmouth, Newfoundland trade at, 314; complaints from, 316.

Terra de Cortereal, early name for Newfoundland, 20.

Terra de Pescaria, early name for Newfoundland, 20.

Terra Nova River, 4.

Thames, idle shipping in, 67.

Theft, jurisdiction for cases of, 74.

Thoroughgood, Capt. Thomas, commissioned to seize Kirke, 98.

Timber, resources of Newfoundland,

INDEX

5; procured by fishermen, 40; destruction of, 58, 150-151, 165; conservation of, 75; regulations on cutting, 81, 129.

Tithes, bill to prevent exaction of, on fish, 50.

Tobacco, sold at taverns licensed by Kirke, 83; imported in New England vessels, 192; becomes leading article of import after 1675, 194; illegal export of, 198-201; direct and indirect traffic in, from Virginia, 258.

Topography of Newfoundland, major features of, 3-5; submarine, 7, 9.

Topsham, Newfoundland trade of, 63, 314; complaints from, 316.

Trade, Board of. *See* Board of Trade.

Trade, from England to Newfoundland frequent by 1548, 22; Parkhurst's scheme to encourage, 24-25; course of, before 1610, 34-35; London and Bristol Company to engage in, 36; course of, after 1610, 37; proposed monopoly in, 38; conditions in, under Charles I, 60; attempt to exclude foreigners from, 64-69; medieval conception of, in West Country, 91; conditions in, during early years of Restoration, 108-109, 111-112, 118-121; decline in, predicted under planter-controlled fishery, 139; impediments to, 140-141; in Newfoundland fish lies outside colonial commercial system, 144; planters accused of harming, 166; conditions in, after 1680, 168-170; growing competition to English, in Newfoundland, 189-191; between New England and Newfoundland, 164, 189-194; economic function of Newfoundland, 213-214; benefited by elimination of France, 233-234; West Countrymen turn from fishery to, 245, 248, 249; conditions in, after War of Spanish Succession, 256-259; New Englanders cause decline in British, 267; has future in Newfoundland, 272; general prosperity of, 310; statistics on, 311-315; routes of, 315; character and value of, 324-325; similarity of rice to codfish in, 332-333; contributions of, 336.

Trade, illegal, predicted, 139; character and extent of, 194-202; during War of Spanish Succession, 217, 218-222; continues, 255; need of vice-admiralty court to check, 291; suppressed by Byng, 292; lieutenant governor to stop, at Placentia, 305; attempts to check, under civil government, 321-324, 326-328; crown unwilling to act in spite of, 335.

Traders, colonial, oppression of planters by, 208-209.

Tradesmen, in West Country interested in prosperity of fishery, 90.

Training School for Seamen, *see* Seamen, Nursery for.

Train oil (codliver oil), made ashore, 40; stolen by pirates, 41; methods of making, 56; measures of, 58; casks and barrels for, stored in Newfoundland, 58; Greenland Company's monopoly on imports of, 64; most profitable branch of Newfoundland trade, 69; protection of equipment for, 75; appropriation of, forbidden, 75; taxation of, proposed, 116; returns from, 120; English fishermen assured freedom to make, 129; data on production of, 311-313.

Trapping, restricted, 6; colonization as means of developing, 24.

Trespassey, Vaughan's settlement at, 46-47; French prizes taken at, 48; limit of English and French fishing zones, 55; French rendezvous at, 186; public houses in, 254; vessels owned at, 314.

Trespassey Bay, 3.

Treworgie, John, commissioned to seize Kirke, 98 n.; commissioned to govern Newfoundland, 100.

Trinity Bay, important fishing resort, 3; good spawning place for cod, 16.

Trinity House, objects to foreign carrying trade, 65-67; represented on shipping committee, 70.

Truro, fishing ships from, 55.

Turks, *see* Barbary pirates.

Ushant, fishing fleet intercepted off, 29.

Utrecht, Treaty of, 238-243; advantages of, to French, 242-243; French settlement on south coast in violation of, 328.
See also Spanish Succession, War of.

Vanbrugh, Capt. Philip, R. N., governor of Newfoundland, lack of ability of, 291; draft commission of, 297; reports on illegal trade, 323.

Vaughan, Sir William, Newfoundland Company disposes of land to, 46; establishes colony at Trespassey, 46-47; settlements founded by, remain, 55; financial difficulties of, 88.

Venice, ambassador of, prophesies conflict between English proprietors and France, 84.

Vice Admirals, of certain counties, jurisdiction of, 74, 76.

Vice-admiralty Court, erected in Newfoundland, 225; officers of, 225; failure of, 225; need of, 291; steps taken to erect, 291-292; lack of, to try offenders, 322; in Boston, 323; established at St. John's, 323, 326; reduces amount of illegal trade, 327.

Virginia, free fishing at, provided in Commons bill, 50; French menace to, 116; smuggling of tobacco from, 201.

Virgin Rocks, a shoal, described, 9; icebergs held by eddies at, 13.

Wabana, iron ore field on Bell Island, 49.

Wade, Capt. Caleb, plan of, for seizure of foreign ships and goods, 256; mentioned, 263.

Wages, of seamen, development of Newfoundland trade would improve, 24; advocated to replace share-system, 89; would result in unemployment in West Country, 90-91; of deserters to be forfeit, 130; real, at Newfoundland compared with England, 150; payment of, avoided, 151; adventurers attempt to enforce standards of, 154; unpaid, force fishermen to remain in Newfoundland, 177; higher, in New England attract seamen and fishermen, 178; objections to system of, 248.

War, of religion, attacks on French commerce during, 32; effect of, during reign of Charles I on fishery, 59, 63-64; on West Country, 63-64.
See also Austrian Succession, War of the; Spanish Succession, War of the; Seven Years' War.

Ware, Sir James, 106.

Watson, Admiral Charles, details investigator, 294; commands fleet at Louisbourg, 295; commissioned

INDEX

governor, 295; reports on Irish Catholics, 295, 302; reconstructs magisterial system, 296; revokes commissions of military officers as magistrates, 306; reports illegal trade stopped, 327.

Webb, Capt. James, R. N., governor of Newfoundland, death of, 304; instructions of, 304 n.; orders search in outports for contraband, 323.

West Country, benefit of colonization to, 24-25; ports of, capture French vessels, 32; uninterested in carrying trade, 35; attitude toward foreign carrying trade, 35, 37, 38, 60, 66-71; localization of fishing industry in, 36; mayors of, ports warned not to interfere with plantation, 45; representation of, in parliament, 49-50, 210; opposes monopoly in fisheries, 49-52; Newfoundland fishery organized by, 53; ships from, ports, 55; shipping, dates of sailing and arrival Newfoundland, 56-57; coasts of, raided by Barbary pirates, 62-64; foreign buyers in, 67; represented at investigation by Privy Council, 67; charter of 1634 to be circulated in, 73; mayors of certain ports in, to enforce charter regulations of 1634, 72-73, 74; customs of managing fishery perpetuated by charter of 1634, 74; jurisdiction of vice admirals of certain counties in, 74, 76; gains complete control over fishery, 76-77; effect of Kirke's application upon, 78; relations between Kirke group and fishermen from, 82-83; commission to take evidence in, 84; narrow vision in, 89; economic dependence of, on fishery, 90-91; prosperity of, affected by Spanish War, 96; economic depression in, 119-121; more power granted to mayors and magistrates of, 129-131; economically dependent on fishery, 138-139; trade of, dependent upon planters, 166; mayors of ports of, called on to enforce charter, 173; reasons for decline in export trade of, 189-190; ships of, engaged in illegal trade, 196; represented on Commons committee on Newfoundland bill, 213; fishery removed from hands of mayors and magistrates of, 214; effect of war with France on fishermen of, 227-229; favors French concessions, 238; Board of Trade continues favorable to, 267, 268, 269, 271, 272; Osborn arouses ire of, with jail tax, 278; Board of Trade calls for opinion of, 288; indifference of, 288-289; statements from, questioned, 289; complaints of merchants of, unsubstantiated, 290; merchants of, relinquish struggle for privileges, 290; Newfoundland trade in ports of, 314; hard hit in war with France, 315-316; total catch of, remains stationary, 317.

See also Adventurers, western.

West Countrymen, *See* West Country; Adventurers, western; *and* Fishermen.

Western adventurers. *See* Adventurers, western.

Western Charter. *See* Charter, western.

West Indies, proposed removal of Newfoundland planters to, 133; economic position of Newfoundland different from that of, 144; ships bound for, relieved at Newfoundland, 153; extent of trade with (1691), 193; rum from, 194;

illegal trade of, 194; refuse fish carried to, 325.
West of England. *See* West Country.
Weston, ——, Lord Treasurer, 65-66.
Weston, Nicholas, interest of, in Newfoundland, 34-35.
Weymouth, fishing ships from, 55; mayor of, to enforce charter of 1634, 76.
Whale Hole, 9.
Whale oil, data on production of, 311; important item of export trade, 313-314.
Whaling, Portuguese, at Labrador and in Hudson Strait, 20; Spanish and French Basques, at Newfoundland, 21; Basque, 23; Basques continue, 31. *See also* Whale oil.
Whitbourne, Capt. Richard, to investigate conditions at Newfoundland, 41; favorable to the Newfoundland Company, 42; his pamphlet on Newfoundland, 45.
Wigmore, William, naval customs officer, 327.
William and Mary, proposal for governor to, and forts at, Newfoundland under, 170; lack of interest in French competition before, 189; threatened reversal of royal policy under, 209; extension of royal authority over plantations by, 213.
Wind, effect of, on ocean currents, 11; on icebergs, 13; direction and velocity of, 17.
Wines, sold at taverns licensed by Kirke, 83; New Englanders accused of importing, 191; French and Spanish exported, 192; New Englanders procure, from sack ships, 193; imported into Newfoundland by English ships, 196; navy officers ordered to prevent smuggling of, 322-323; forms large part of illegal imports, 323.
Wrecks, caused by indraft of currents, 10.
Wyborne, Capt. ——, R. N., on French settlement at Placentia, 187.
Wynn, Capt. Edward, governor of Avalon, 48.

Zone, English fishing, described, 55; foreigners prohibited from fishing in, 129; supervision over, by royal navy, 133; congestion in, 164; expansion beyond, blocked by France, 164.
Zone, French fishing, described, 55.

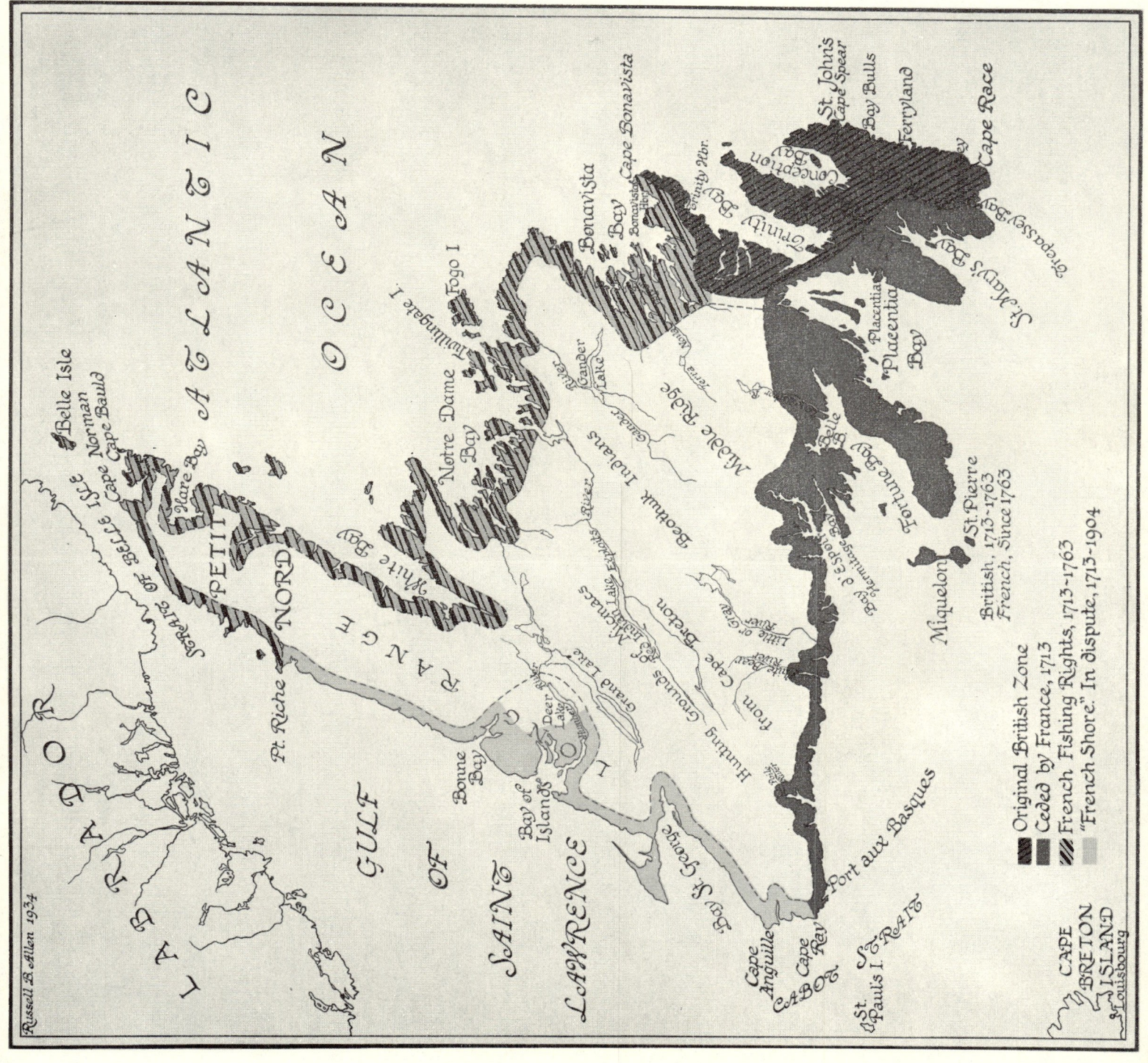